The Shaggy Steed of Physics

David Oliver

The Shaggy Steed of Physics

Mathematical Beauty in the Physical World

With 71 Illustrations

Springer-Verlag
New York Berlin Heidelberg London Paris
Tokyo Hong Kong Barcelona Budapest

David Oliver
The Fields of Michael and Gabriel
Meadows of Dan, VA 24120 USA

Cover. Such disparate natural objects as a planet orbiting the sun and the hydrogen atom manifest the same hidden symmetry: the symmetry of the three-sphere. The three-sphere has the remarkable ability to turn itself into a family of nested tori as shown here in this illustration by Huseyn Koçak and David Laidlaw.

Library of Congress Cataloging-in-Publication Data
Oliver, David.
The shaggy steed of physics : mathematical beauty in the physical world / David Oliver.
p. cm.
Includes bibliographical references and index.
ISBN 0-387-94163-0. — ISBN 3-540-94163-0
1. Physics. I. Title.
QC21.2.O45 1993
530—dc20 93-33015

Printed on acid-free paper.

Production managed by Ellen Seham; manufacturing supervised by Vincent Scelta.
Photocomposed copy prepared from the author's TeX files.
Printed and bound by Edwards Brothers, Inc., Ann Arbor, MI.
Printed in the United States of America.

9 8 7 6 5 4 3 2 1

ISBN 0-387-94163-0 Springer-Verlag New York Berlin Heidelberg
ISBN 3-540-94163-0 Springer-Verlag Berlin Heidelberg New York

To

Duncan Foley

who is himself a prince

keen as midsummer's keen beyond
conceiving mind of sun will stand,
so strictly (over utmost him
so hugely) stood my father's dream

—e e cummings

Preface

The universe embraces with sensual presence. No thought, no word is needed to behold sky, sun, moon, stars. It is not in reflection but in experience that one inhales the earthy smells of damp woods, becomes dizzy under the sweep of the starry night sky, basks in the sun's warmth, rolls in ocean breakers, tastes salt. The sensual universe is open, available, immediate.

But beneath its sensuality, the universe has an interior landscape that lies hidden. Though veiled, the inner nature of the cosmos draws us. It invites us to find the underlying source of the motions of sun, moon, and stars, search out the primal form of all matter, and discover the unity beneath the diversity of the earthy substances that excite the senses. The inner nature of the cosmos is a landscape of exquisite beauty which appropriately underlies the exterior splendor of our experience. This interior landscape of matter and motion which lies both in and beyond the senses—the physical face of the universe—is mathematical.

That the world we see, smell, and touch is deeply mathematical is a fact of unceasing amazement to me. The natural world presents us mathematical symmetries of compelling beauty; and a rich vein of mathematics, inspired by these raw materials from the natural world, is potently physical. The physical and mathematical resonate, creatively nourishing one another. Although the sensual face of the world presents itself to us quite openly, its physical face, hidden as it is in mathematics, must be gained by following a path of reflection and study.

The physical world opens to us through *mechanics*, the science of motion. Mechanics is the wellspring from which physics flows, the source out of which a multitude of applications and elaborations from astrophysics to elementary particles emerges. Insight into the interior nature of the physical world begins with mechanics, with a vision of the motion that underlies all the objects of the world and the subject of this book.

For whom is this book intended? It is intended for lovers. As a young man, I was drawn by love of the physical world and the things one could make from it through knowledge of its interior, mechanical behavior. I followed my fascination into the study of physics and engineering. I found myself not only conceiving and developing nuclear and electric rockets and sources for power in space with this knowledge but also sharing that vision by teaching other young men and women like myself.

Then, slowly, as flickering symptoms scarcely cross across one's consciousness but steadily rise up in the body to reveal a raging disease, the things I was doing with the physical world rose up to confront me. I had begun with delight in scientific discovery and engineering craft. I ended up developing sources of massive energy and power for a new generation of space weaponry—well before the movie and the reality of Star Wars. I had been drawn by the beauty of the starry night sky. I ended up in the first ambitious attempt of the human race to fill that sky with devastation. My innocence gave way to the realization that mechanics is a science both wonderful and terrible. Mechanics is the vehicle of all physical theory. Mechanics is the vehicle of war. The two have been inseparable.

Mechanics originated in the earliest myths of human origins. Cain was a man born into matter, a worker of the soil given to wonder about force and motion. He took a piece of the natural world and mechanically fashioned it into a plowshare. Then he took that plowshare, split Abel's head, and simultaneously became the inventor of the sword. Physics, a reveler in dualities, has never escaped its primal duality in the sword and the plowshare.

In the time of the Greeks our science received its name: *mechanics* (*μηχᾰνή*)—*a device for delusion in warfare.* The practitioners of Cain's art had moved on in mechanical prowess. They beat their plowshares into the armed bowels of mechanical horses that could simultaneously split a multitude of heads. Mechanics, originating as delight in force and motion, suffered to become the handmaid of humanly wrought power and delusion in warfare.

Seventeenth century Venice, an early superpower, showed the world how to organize mechanical wondering and tinkering under the minions of the arsenal; and the arsenal proliferated beyond renaissance imagination to become the superarsenals of the modern world. The arsenal claimed the most distinguished physicists and engineers of the mid-twentieth century in our own recapitulation of the plowshare and the sword. These men and women hammered Albert Einstein's beautiful plowshare of nature's fields of motion, $\mathcal{E} = mc^2$, into the most hideous transformation of the sword yet to appear.

Is it possible to have a celebration of mechanics with that dreadful heritage? Let me answer with a story from the mountains in which I now live. A notorious local man was known as "the awfulest man." Not only was he scheming and exploitative; he could be dangerous. When he died, folks came to the funeral more to hear what the preacher would have to say about the awfulest man than to mourn. The preacher began with silence. Then, looking squarely at the congregation, he broke into a dreamy smile and said, "Didn't you just *love* to hear that man whistle!"

As surely as a wondrous side of the awfulest man could be recognized by the mellifluous trilling that announced his presence, mechanics too can be recognized as a wondrous, if blighted science. Mechanics has a voice in praise of the creation. Mechanics can dance—other than to the tune of the piper. After fleeing the work of the arsenal, I remembered and longed for the ethereal whistle of mechanics; and I began to write.

Looking to the redemption of the dreaded science, I turn from humanly fabricated delusion to naturally created wonder. Look to mechanics as the trance of nature rather than the delusion of invincible safety. Think of mechanics as a gift of nature rather than the purveyor of devices to threaten and dominate. So set aside the monstrosities into which the plowshare has been beaten and gather round. Let me tell the story of the inner nature of the world from the scale of the solar system down to the scale of the atom. My story speaks of both the physical world and human imagination, of the heavens and the elements and the individual human beings past and living entranced with the nature of the universe.

Telling about physics is a part of story-telling appropriate to the modern world. It is in the story that the collected knowledge of a culture is gathered, celebrated, and passed on—and physics is no exception. It is my hope that this story of the physical world may touch and delight others, particularly those just coming of age in physics and mathematics and caught up in its fascination. It is to you that I especially offer this book as a spirited companion; for I wrote the book I would have loved to have placed in my hands as a young student.

My tale of mechanics is not an alternative to rigorous study and the exercise of problem-solving required to master physical principles. Nor is it intended as a diversion from masterful texts such as Arnold Sommerfeld's *Mechanics*, the ebullient works of V. I. Arnol'd in mechanics, or the *Mechanics* and the *Quantum Mechanics* of L. D. Landau and E. M. Lifshitz. I offer my telling of this traditional story as a soulful companion to the textbooks.

Experienced old-timers know the stories intimately. I would be delighted if they find a thing or two to catch their fancy in my rendering of the tale. (Stories, especially the most traditional, are never retold in precisely the same way.)

It is my hope that those whose main interests lie outside physics may also be drawn into this story of the mathematical splendor of the world; and I welcome you. Mathematics is a central mode of expression in this kind of story-telling. Mathematics is part of our language, an important means of perception and discernment. I suggest that mathematical physics is not only a language but a visual art whose structure, symmetries, and iconography reflect the symmetry and beauty of the world that it describes.

Basic undergraduate mathematics for science, engineering, and mathematics students provides a sufficient level with which one may enter into the story. For those readers with deeper interest and greater mathematical preparation I offer a set of notes with more extensive (and in some cases more advanced) mathematical development. These notes constitute the second half of the book so they do not interrupt the flow of the story. The notes are distinguished from the main text by boldface titles and equation numbers such as **Note 1**, Eq. (**1.1**).

This story would not be worth telling if it were not a story of mystery. The sword of mechanics proclaims the profane. But the plowshare of mechanics parts the earth revealing the sacred in matter. Each time the story is retold, glimpses of that mystery become possible. I pray that will happen here.

David Oliver
The Fields of Michael and Gabriel
Autumn, 1993

Acknowledgments

I want to especially acknowledge George D. Purvis III who generously responded to my request for graphics of hydrogen atom eigenstates. Surprisingly, detailed graphical renderings of the hydrogen atom, even in this computer age, are not readily available. George Purvis's gallery of states which appears in Chapter 5 and the Notes fill a real need; and I thank him for it. I also wish to acknowledge Christopher Oliver who supplied the computer program for the probability densities from which the hydrogen states were constructed.

I am grateful to Guido Sandri of Boston University for his effusive encouragement. He shared drafts of this book with his students, assured me they loved it, and thereby gave me the courage to continue. Louis L. Bucciarelli of the Massachusetts Institute of Technology was a faithful, if skeptical, reader of an early version of the manuscript. He also kindly supplied me with images of Mendeleev's Periodic Table of 1872. Endless thanks go to Duncan Foley of Barnard College. He patiently read through each draft of the book, steadfastly challenged me whenever the exposition got murky, and encouraged me to hold fast to the vision of physics that got me writing and bring it faithfully into print. Deficiencies that remain are despite his best endeavors.

Contents

Chapter 1

The Shaggy Steed of Physics

In the story of Prince Conn* a peaceful and prosperous kingdom has fallen into chaos. A stepmother's intrigue has cast a deadly spell over the youthful monarch and his realm. Helpless to find the way to save his life and restore harmony to the kingdom, the prince is directed by a druid to a small shaggy horse. The shaggy steed offers to conduct the prince upon a quest through harrowing trials of fire and flood and faithfully bring him to a magical realm where he will procure powers with which he can break the spell and restore the harmony of his land.

The shaggy steed proves to be an extraordinary mentor to the prince. They embark upon a fabulous journey in which he brings the young man to an understanding of the forces hidden in the universe and in himself. Under the tutelage of this creature from the natural world the prince is prepared to enter the magical kingdom and then return to rule his own realm.

But at the threshold of success, the horse, whom the prince now loves dearly, will go no further with him. Instead, he bids the prince take a knife and a vial of ointment from a hiding place within his ear. Then he directs the youth to slay him, flay his hide, and wrap himself in it. It is through his sacrifice that the prince will be able to make his way unscathed into the magical kingdom.

"Never!" proclaims the prince in horror.

But the prince's mentor is implacable. "If you fail in this final task," he implores, "you will perish and I shall endure a fate worse than death. I only ask that after you achieve your goal you return to my body, put away the vultures, and cover it with the ointment."

Reluctantly, the prince takes up the knife and uncertainly points it at his teacher. Instantly, as if impelled by powers beyond the young man's hand, the knife plunges into the throat of the horse and the dreaded act is done.

With broken heart, the prince makes his way through the magical king-

* "The Story of Conn-eda" in W. B. Yeats (ed.), *Irish Fairy and Folk Tales*, Dorset Press (1986), New York.

dom shrouded in the hide of his teacher, procures the magical artifacts, and prepares to return to his own kingdom. But first he searches out the carcass of his slain guide. Driving off the vultures, he annoints the remains. Thereupon the dead flesh undergoes a magical transformation, achieving the form of a noble prince like himself. In a joyous reunion the prince recognizes both his former teacher and a new, unexpected companion.

I am reminded of this tale as we set out to learn of the motion, symmetry, and beauty of nature; for our own experience of the physical world is similar to that of Prince Conn. Like the world of the prince, the world of physics can be plunged into chaos. Familiar understanding can crumble in contradiction as happened when deepening inconsistencies and new observations with the telescope undermined the long tradition of the Ptolemaic theory of the universe.*

In the midst of the confusion and uncertainty of such a breakdown, a new idea appears. At first it seems unpromising, even false as did Copernicus's vision of a heliocentric universe to the affirmed practitioners of Ptolemaic astronomy. Yet once we grasp its power and beauty, we embrace the new theory with conviction. We grow to affirm it as a cornerstone. We even fall in love with it. Then, in the face of relentless probing and deepening observation of the world, our beloved theory leads us to a crisis of contradiction. We must give it up.

The calculations of Kepler and the telescope of Galileo opened the way to the classical mechanics of Isaac Newton. Newton's laws of motion reigned supreme for over two hundred years achieving a brilliant synthesis in the principle of least action through the work of Leonhard Euler, Joseph-Louis Lagrange, C. G. J. Jacobi, and William Rowan Hamilton. Then the discovery of the atom thrust Newtonian mechanics into crisis. In the midst of this crisis Max Planck and Neils Bohr proposed the bewildering idea of the *quantum*. And quantum mechanics came to supersede Newtonian mechanics.

In our time we acclaim the quantum mechanics of P. A. M. Dirac and Richard Feynman. But it is a remarkable fact that the classical mechanical heritage of Newton, Kepler, and Galileo and particularly that of Euler, Lagrange, Jacobi, and Hamilton is both potently present and wondrously transformed in quantum mechanics. The new theory rises from the old. Like Prince Conn, we affirm the new knowing we embrace a metamorphosis of the old.

There is a further connection to the shaggy steed appropriate to my own telling of the story of the physical world. It is in the smallness and homeliness of the mentor-horse. The teacher I have selected is a small and homely bit of physics. Our guide will be the *two-body problem*, a simple but mathematically exact description of the motion of two bodies that interact

---

* Giorgio de Santillana has painted a sweeping and poignant portrait of this breakdown in *The Crime of Galileo*, U. of Chicago Press (Chicago and London), Midway Reprints, (1976).

through the inverse-square force of gravity and electricity (the force acting between the two bodies has a strength that is proportional to the inverse-square of the distance of separation between them). The two-body problem is the Shaggy Steed of Physics; and it shall carry us upon a mathematical journey in which the unity and beauty of the heavens and the elements will be revealed.

The two-body problem first appears as a little problem. It flows by as one among many textbooks shower down upon students. But those willing to look more deeply will find a jewel casting its glow over the celestial realm of Kepler and Newton to the quantum realm of Bohr and Dirac. In its classical form the two-body problem gives us the Kepler ellipse, the exemplary motion of the heavens of our solar system. In its quantum-mechanical form it yields the hydrogen atom, the prototypical element. In these two achievements alone the two-body problem is a moving paradigm of the unity and beauty of the physical world. But there is more.

In its origin in a pristine variational principle and in its beautifully knit symmetries the two-body problem heralds the affirmed attributes and cherished aesthetics of unified theories of all forces and particles. Symmetry describes the way objects, when transformed, nonetheless present the same identical shape. For example, the symmetries of a sphere are rotations. Rotate a sphere from one angular position to another and one has the same thing with which one began. Symmetries reflect themselves in invariants: quantities that do not change under the action of the symmetry. The angles specifying the orientation of the sphere change as it is rotated, but the radius of the sphere does not. The radius is an invariant of rotational symmetry.

The symmetries of two-body motion with inverse-square forces are most remarkable. They are both real and imaginary *four-dimensional* rotations. Just as the rotational symmetry of the familiar sphere embedded in three-dimensional space is a generalization of the rotational symmetry of the sphere in two dimensions (the circle), we shall see that a sphere in four-dimensional space is a generalization of the sphere embedded in three-dimensional space. All are the same object with an invariant radius, but in different dimensions. A sphere with a real radius gives rise to ordinary trigonometric functions. When the radius is imaginary, the sphere has imaginary rotation angles that give rise to hyperbolic rather than trigonometric functions.

The four-dimensional rotational symmetry of the Kepler problem is intimately connected to the two majestic achievements of twentieth century physics: quantum theory and relativity theory. When the rotations are real, the radius of the sphere is the Planck constant, the emblem of all quantum motions. When the rotations are imaginary, they peal with divine laughter. The four-dimensional rotational symmetry is then similar to that of the relativistic space–time discovered by Albert Einstein.

The motion of two bodies with inverse-square forces and Einsteinian relativity have nothing directly to do with one another. But because of the surprising coincidence of their symmetries we shall find that unique manifestation of relativistic symmetry—the spinor—to also be one of the natural expressions of two-body celestial motions. (Spinors are quantities similar to vectors discovered by P. A. M. Dirac in working out the relativistic description of the electron.) Moreover, the rotational symmetry of the two-body problem is a paradigm for the unitary symmetries of quarks, the most elementary of particles, and pointed the way to them. All these diverse aspects of physics are reflected in the motion of two bodies bound by the inverse-square force.

The symmetries of the two-body problem enfold a beautiful set of topological relationships that open our eyes to the way in which the topology of orbits powerfully summarizes motion—vision that cannot be had by simply looking at solutions of differential equations. The four-dimensional rotational symmetry heralded by the two-body problem with inverse-square force is extraordinary in its topology: a four-dimensional rotation (and its symmetry sphere in four dimensions) may be decomposed into a pair of three-dimensional rotations (and their symmetry spheres in three dimensions). This decomposition is exceptional; rotations in all other dimensions cannot be decomposed into rotations in lower dimensional spaces. Only a space of four dimensions has this property.

Though the world of physics it illuminates is vast, the two-body problem has limits. Like Prince Conn's steed, a faithful and informative guide is respectful of its limitations. At the boundary of its realm the motion of two-bodies, like the prince's steed, must be given up.

One of Isaac Newton's laws of motion proposed in the *Principia Mathematica* of 1666 was that the force experienced by one body is equal and opposite to that experienced by the other. This is true for gravitational forces in the solar system and electrical forces in the atom for which motion takes place at a small fraction of the velocity of light. But in the realm of velocities near the speed of light electrical and gravitational forces are no longer instantaneously equal and opposite. Instead, they propagate from body to body at a finite speed, the speed of light. In that revolutionary discovery the space and time established by Galileo and Newton is revealed as only a limiting form of Einstein's even more graceful vision of relativistic space and time.

In the regime of motion in which forces propagate from body to body at a finite speed, two-body interactions cannot exist. New particles enter the picture which carry the propagating force. Motion near the speed of light inherently couples infinitely many bodies, a situation radically different from the motion of two bodies.

In the ordinary world of electricity and gravity with velocities small compared to the velocity of light, the two-body problem reaches another limit at precisely *two* bodies. Ordinary motion beyond two bodies—even that with just three bodies such as two planets and the sun—is not just quantitatively different from that of two bodies; it is qualitatively different. There is a singular rift between two-body motion and the motion of many bodies. Only two-body motion may be described by simple mathematical objects such as the ellipse of Kepler. On the other hand, the motion of more than two bodies—the so-called *many-body problem*—cannot be expressed in the most general functions of classical mathematical analysis, infinite series of algebraic polynomials.

The motion of two bodies bound by the inverse-square force possesses a crucial property: it preserves the integrity of the initial conditions. Orderly transformation of the initial conditions by the subsequent motion of the bodies occurs because the motion of two bodies is insensitive to small fluctuations. Neighboring trajectories emanating from closely neighboring initial points always remain close to one another. Orderly behavior like this is describable by conventional mathematical functions.

Whereas two-body motion preserves the integrity of the initial conditions in the sense described above, many-body motions generally do not. A small change in the initial conditions of many-body motion may lead to a dramatic divergence in the resulting trajectories. This sensitivity destroys the information contained in the initial condition in the subsequent course of the motion. A new feature which is impossible with two-body electrical and gravitational forces appears which challenges predictability: complex turbulent motion better known as chaos.

Suddenly a surprising and paradoxical freedom springs from the deterministic laws of mechanics. Motion frees itself from the pattern and structure imprinted in the initial conditions by escaping into chaotic behavior. Many-body motion is governed by a deterministic set of mathematical equations; but these equations yield random rather than deterministic behavior. One encounters one of the most remarkable and still puzzling aspects of physical law—freedom and determinism coexisting together.

The vision of the world held by our forebears in antiquity and in the medieval era was one in which the heavens and the elements formed a cosmic whole. This vision was inspired by order and symmetry; but the order and symmetry that the ancients and medievalists envisioned were not rooted in penetrating observation of the natural world. Like Prince Conn's quest, our journey begins with disintegration, the disintegration of a medieval kingdom of astronomy, philosophy, and politics, a disintegration spurred by new observations of the natural world obtained through instruments like the telescope.

We then describe the new unity that began to replace the old, the vision of a universal law sustaining all the motions of the universe. The law of motion flows from a principle of extraordinary simplicity—the principle of least action. We direct this principle to the classical dynamics of two bodies. The classical regime is crowned by celestial mechanics.

Next we consider the extension of mechanics into the quantum world of microscopic particles. The quantum regime is crowned by the hydrogen atom. We then come to the limits of the two-body problem whereupon we give up our guide before the relativistic nature of gravitational and electrical forces on the one hand and the rich structure of many-body motion with its freedom within constraint on the other. And in this transformation of the two-body problem into the many-body problem we encounter a surprising reunion with our familiar guide.

Chapter 2

The Heavens and the Elements

From the earliest times humankind has been drawn to the unity of the creation. The ancients understood the large-scale bodies of the heavens and the smallest bits of elemental matter as intimately related parts of a cosmic whole. They sought to explain that unity with the geometrical and religious concepts of their era.

The medieval world view derived from the Greeks was, for its time, a thing of great beauty. The universe was seen as a series of seven concentric spheres surrounding the earth. Each was the habitation of one of the moving heavenly bodies. Proceeding outward from the earth they were the moon, Mercury, Venus, the sun, Mars, Jupiter, and Saturn. Each sphere was the seat of one of the seven chemical elements: silver, mercury, copper, gold, iron, tin, lead. Each heavenly sphere resonated with a note of the western musical scale: *re, mi, fa, sol, la, ti, do.* Learning in the Middle Ages was itself congruent with this cosmic order. The seven disciplines were grammar, logic, rhetoric, arithmetic, music, geometry, and astronomy.

This vision of the universe also provided the medievalists with an explanation of the unity between spirit and matter. The region beyond the outer spheres of the universe was the Empyrean—a realm of pure spirit thought to be inhabited by angelic creatures much as men and women inhabit the purely material earth at its center. For the popular mind God too was localized in space. Beyond the sphere of Saturn God dwelt upon His throne in the Empyrean just as the king dwelt above the nobles, peasants, and fields of his realm.

In the hierarchy from pure spirit in God's enveloping realm to the material earth at its center the heavenly spheres embodied progressively greater proportions of spirit to matter. Human beings came into existence with the movement of their souls from God through the spheres down to earth. Upon death the soul migrated back out through the cosmos losing its material elements appropriately to each sphere until it arrived in God's realm.

The medieval insight that the cosmos was unified in beauty and simplicity was correct. But the medieval understanding of the interplay between spirit and matter and the mechanism proposed for the cosmic unity of the heavens and the elements were in error. Beautiful theories can sometimes end in error.

Beginning with the renaissance, the ambition of a unified vision of the spiritual and material aspects of the cosmos and human existence gave way to the more narrow quest for an understanding of the physical world, a narrowing of focus that continues to this day.

Through a Glass Darkly

The vision of the universe as a system of spheres reveals the human capacity to see geometry in nature. These geometrical structures, even for the ancients, were more than stationary artifacts. The heavens were in motion. The planets were wandering stars. The geometrical reckoning of the heavens went beyond the ordering of the heavenly bodies in a coherent geometry. That same geometry could be used to predict their motions.

The planets were understood to be bodies moving on circular orbits that lay upon the celestial spheres. A radius and a speed of rotation (or angular velocity) could be deduced for each planet. With these two properties the motions were not only describable; they were predictable.

But the perfect circle did not perfectly predict the motions of heavenly bodies. There were serious discrepancies in the predictions afforded by a model of planetary orbits moving on concentric circles. The modifications that were made to the description of the heavens of spheres and circles came to be known as the Ptolemaic system. This system was named after Claudius Ptolemaeus, the second century compiler of the *Almagest*, a monograph of the best astronomical formulas of the time.

The Ptolemaic system embraced the circle as the geometrical form of the heavens and accepted the earth as the center of the universe. But it introduced many additional circles to bring the predictions in accord with the observations. Instead of a planet moving on single circle rotating at a single angular velocity about the center of the universe (called the deferent), the planet was proposed to move on a second circle (called an epicycle) which itself rotated on the deferent. The resulting motion of the planet was no longer a perfect circle but a swirling rosette as shown in Fig. 2-1.

Each planet was burdened with two circles and two angular velocities. This cumbersome arrangement proved useful because the epicycle had the subversive effect of displacing the true center of rotation of a planet from the earth (the assumed center of the universe). The displacement of the center of rotation hidden in the mathematics ruined the notion of a "center of the universe;" but it vastly improved the predictions of early astronomy.

The prediction of the orbit shape was a challenge for Ptolemaic astronomy; but the prediction of the *speed* of a heavenly body over the course of its orbit was a more trying challenge. Heavenly bodies were to move

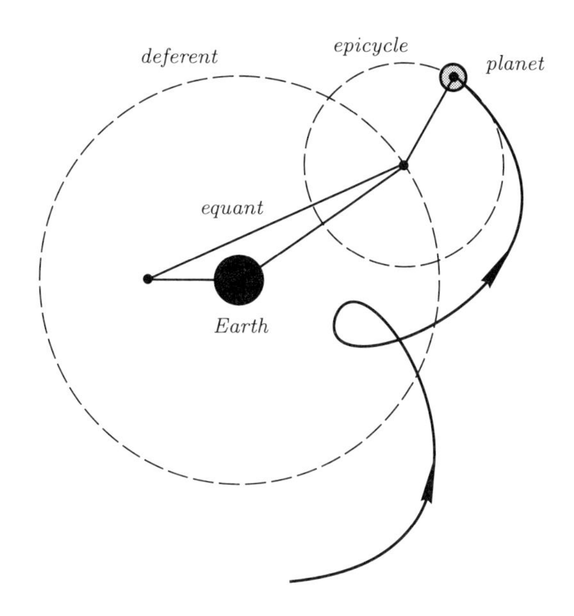

Figure 2-1. The Elements of Ptolemaic Astronomy. The earth is the center of the universe and the sun is one of the planets. Planets move on epicycles which themselves move on deferents centered on the earth. The epicycle moves with the equant which rotates uniformly about a center displaced from the center of the earth.

at a uniform speed of rotation on their circles in correspondence with the classical idea of perfection. But uniform speed, like the perfect circle, failed to match the observations.

Ptolemaic astronomers sought to correct this discrepancy by introducing the equant. The equant was a ray centered on a point slightly displaced from the center of the earth and drawn to the center of the epicycle as shown in Fig. 2-1. The equant was to rotate at the uniform speed of the heavenly spheres and the epicycle tracked it while moving on its deferent as shown in Fig. 2-1. As a result, the planet acquired a nonuniform speed that better matched the predictions. The Ptolemaic description of the heavens was one in which circles rode upon circles that did not lie at the center of the universe.

Galileo brought measurement and the sharp eye of visual observation, enhanced by instruments like the telescope, to this ancient conception of the world. So too did Tycho Brahe. Even with the tinkering of epicycles and equants, Ptolemaic astronomy could not describe the planetary motions to the degree of accuracy with which they could be measured. If the medieval and Ptolemaic view of the world's behavior did not conform to careful and rigorous observation, it would be denied its beauty. By Greek and medieval standards themselves, that which is not true can be neither good nor beautiful. The heavenly spheres, displaced from their center on the earth and cast in Ptolemaic epicycles and equants, had already suffered the loss of much of their beauty.

While Galileo and Brahe devoted themselves to rigorous observation of heavenly motions, Nicholai Copernicus and Johannes Kepler were inspired by the belief that the cosmos *really was beautiful.* A cumbersome Ptolemaic model was a flawed image of the cosmos even if its orbit predictions could be brought into close correspondence with observations. Yielding the geometric centrality of the earth as the center of the universe, Copernicus temporarily regained the beauty of the ancient world view by establishing the sun at the center of the solar system. The Ptolemaic system seemed saved by the simple device of placing the sun at the center. But epicycles, equants, and deferents, though simplified, could not be fully dispensed with (particularly in the motion of a nearby planet like Mars with an eccentric orbit); and even with them Ptolemaic predictions still departed from the accurate observations of Tycho Brahe.

Kepler himself had been a superb practitioner of Ptolemaic astronomy. He painstakingly constructed a combination of deferents, epicycles, and equants that provided the most accurate prediction of the motion of Mars yet achieved. Kepler's Ptolemaic model culminated in a representation of the motion of Mars that matched Tycho's observations with less than 8 min of error in arc. This difference is scarcely noticeable by the naked eye. Nonetheless, Kepler took this small discrepancy as certain evidence that the Ptolemaic basis of astronomy was fundamentally in error.*

Kepler then found that a single geometric shape, the *ellipse* with the sun at one of the foci, fit the actual orbits of all the planets almost perfectly. (The foci are the generating points of the ellipse displaced from the center by a distance that is a measure of its eccentricity.) An ellipse with the sun at a focus could account more accurately for the heavenly motions than the ponderous Ptolemaic universe with its deferents, epicycles, and equants.

Kepler whent on to show that this new mathematical description had even greater powers. In his search for a method of predicting the nonuniform speeds of the planets correctly he found that a planet moved at a speed over its orbit for which the time taken to sweep out a fixed amount of area is always the same. Visualize the planet moving on its ellipse as shown in Fig. 2-2. Imagine wedges of equal amounts of area cut out of the elliptical surface contained by the orbit. The time it takes the planet to sweep over the elliptical path along the edge of each wedge is always the same. As a result, the planet moves at a nonuniform speed, sweeping rapidly over the portion of its orbit near the focus at the center and more slowly about the distant focus.

---

* Kepler wrote, "Divine Providence granted us such a diligent observer in Tycho Brahe that his observations convicted this Ptolemaic calculation of an error of 8′; it is only right that we should accept God's gift with a grateful mind.... Because these 8′ could not be ignored, they alone have led to a total reformation of astronomy." [*Johannes Kepler Gesammelte III*, 178, Munich (1937–)]

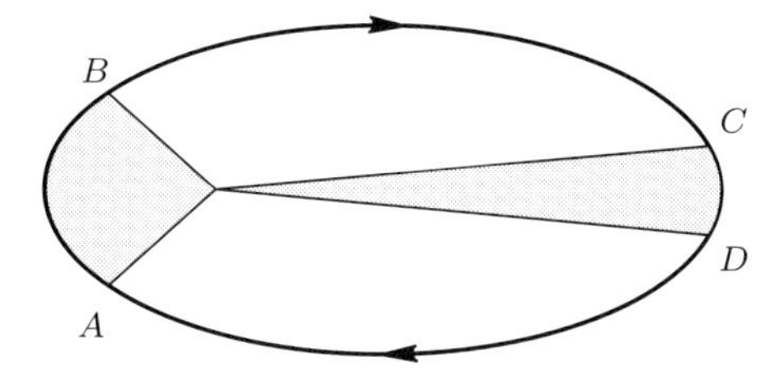

Figure 2-2. Kepler's Second Law. A planet sweeps out equal areas of its ellipse in equal times. The two wedge-shaped shaded areas are equal. The planet takes the same time to cover the distance from A to B as that from C to D. As a result, the planet moves at a nonuniform speed over its orbit.

It is a tribute to Kepler's respect for the primacy of well-verified observation against his quest for the underlying geometry of these observations that he stuck with the second law even as he felt it lacking. He was drawn to the second law by its stunning prediction of the periods of the planets. But he found it wanting because he could not see its connection to a deeper, all-embracing geometry. We now know there is such a deeper basis (first clarified by Newton); and it will be revealed in the course of this book.

The explanatory power of the Kepler ellipse seemed inexhaustible. Kepler also found that the time for a complete orbit of any planet was inversely proportional to the $\frac{3}{2}$ power of the mean distance of the planet from the sun. He wrote, "I first believed I was dreaming; but it is absolutely certain and exact that the ratio which exists between the periodic times of any two planets is precisely the ratio of the $\frac{3}{2}$ powers of the mean distances."

Kepler pointed the way to a deep-seated law that governed *all* motion. The insight of the ancient astronomy that nature was built upon beauty and simplicity was correct; but that beauty and simplicity were not rooted in spheres and uniform motion. The inner beauty and simplicity were resident in an underlying physical law. Here was a new kind of perfection that permeated the entire heavens and did not depend upon the specific properties of any planet, a beauty more mystical than that of the heavenly spheres. Isaac Newton would explicitly exhibit that law: *the gravitational force binding the bodies of the solar system is inversely proportional to the square of the distance of separation between the bodies.*

The Pattern of the Elements

Copernicus, Kepler, Galileo, and Newton shed great light upon the large-scale cosmos. But what of elemental matter itself? What were the ultimate microscopic constituents out of which the stuff of the cosmos was made? With the rending of the celestial spheres the musical resonance between the heavenly spheres and the elements was silenced. The vision of celestial bodies and elemental matter as a cosmic unity fell into oblivion as the investigation of elemental matter followed an earth-bound course far removed from the motions of heavenly bodies.

A small number of elements with distinctly varied strength, maleability, lustre, melting and boiling points, and combustible properties were found to build up all matter. The classical set of elements—silver, mercury, copper, gold, iron, tin, lead—were all metals. New elements such as carbon, bismuth, and sulfur expanded those known by the medievalists. The gases hydrogen, oxygen, and nitrogen were well identified by the late eighteenth century. In the middle of the nineteenth century well over fifty elements were known.

By this time it had become clear that elements were associated with multiples of an indivisible unit weight. The weight of a particular kind of element provided a useful way of labeling the elements since chemical reactions consumed reactants and produced products with weights in well established ratios. The distinguishing feature of an element came to be that it was a macroscopic aggregation of a single kind of unit. Matter was not continuous and infinitely divisible. It was built up of irreducible elements with invariant properties.

The various elements fell into groups with similar chemical and physical properties. Repeating patterns of properties followed increasing multiples of atomic weight. An arrangement of atomic weights in octaves (*Newland's Octaves*, 1850) neatly grouped elements with similar properties but with different atomic weights. By the end of the century, the Russian chemist Dimitri Ivanovich Mendeleev had brilliantly constructed a *Periodic Table of the Elements* in which the repeating patterns of the elements were revealed.

Mendeleev's table, published in 1872, is shown in Fig. 2-3. It gave birth to the modern Periodic Table shown in Fig. 2-4. Mendeleev recognized the natural grouping of the elements in eight fundamental groups of the first two periods; but the third and fourth period—as well as the higher periods—were less clear. This was because after two periods, elements group themselves in eighteen groups rather than eight as shown in Fig. 2-4. Moreover, the eighth group, which was ultimately recognized as the abode of the noble gases, could not be properly filled in 1872 because the nobles were inaccessible to the chemical methods of the time.

Using the pattern in the table, Mendeleev predicted the existence of elements which had not yet been discovered. For example, he proposed the element he called *ekaboron* with atomic weight 44 to fill the vacant slot in row IV of the third column. The element, a silvery white metal with pinkish cast in air, was discovered only a few years later (1879) in Scandanavia and has come to be known as Scandium, element number 21 of Fig. 2-4. Mendeleev also corrected errors in the published weights of elements because they did not properly match the patterns in the table.

The Periodic Table provokes obvious questions. For example, some of the most reactive elements are those in the first column—the alkali metals

Tabelle II.

Reihen	Gruppe I. — R^2O	Gruppe II. — RO	Gruppe III. — R^2O^3	Gruppe IV. RH^4 RO^2	Gruppe V. RH^3 R^2O^5	Gruppe VI. RH^2 RO^3	Gruppe VII. RH R^2O^7	Gruppe VIII. — RO^4
1	H=1							
2	Li=7	Be=9,4	B=11	C=12	N=14	O=16	F=19	
3	Na=23	Mg=24	Al=27,3	Si=28	P=31	S=32	Cl=35,5	
4	K=39	Ca=40	—=44	Ti=48	V=51	Cr=52	Mn=55	Fe=56, Co=59, Ni=59, Cu=63.
5	(Cu=63)	Zn=65	—=68	—=72	As=75	Se=78	Br=80	
6	Rb=85	Sr=87	?Yt=88	Zr=90	Nb=94	Mo=96	—=100	Ru=104, Rh=104, Pd=106, Ag=108.
7	(Ag=108)	Cd=112	In=113	Sn=118	Sb=122	Te=125	J=127	
8	Cs=133	Ba=137	?Di=138	?Ce=140	—	—	—	— — — —
9	(—)	—	—	—	—	—	—	
10	—	—	?Er=178	?La=180	Ta=182	W=184	—	Os=195, Ir=197, Pt=198, Au=199.
11	(Au=199)	Hg=200	Tl=204	Pb=207	Bi=208	—	—	
12	—	—	—	Th=231	—	U=240	—	— — — —

Figure 2-3. The Mendeleev Periodic Table of 1872. The chemical elements are not without order. They may be arranged in groups with similar chemical and physical properties forming the eight columns of the table. Similar elements have a repeating pattern in their atomic weights giving rise to repeating rows [*Liebigs Ann. Chem.* Suppl. viii, 133–229 (1872)].

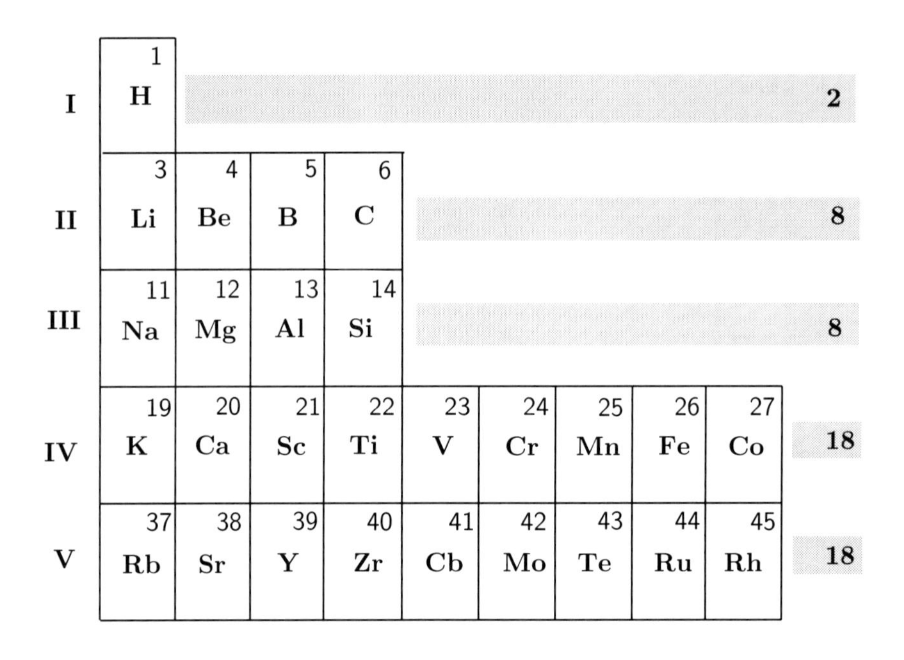

Figure 2-4. Periodic Table of the Elements. The diversity of the chemical elements is imprinted with a fundamental pattern. Each column of the table groups elements with similar chemical properties from the extreme chemical reactivity of the alkali metals on the far left to the inertness of the noble gases on the far right. When

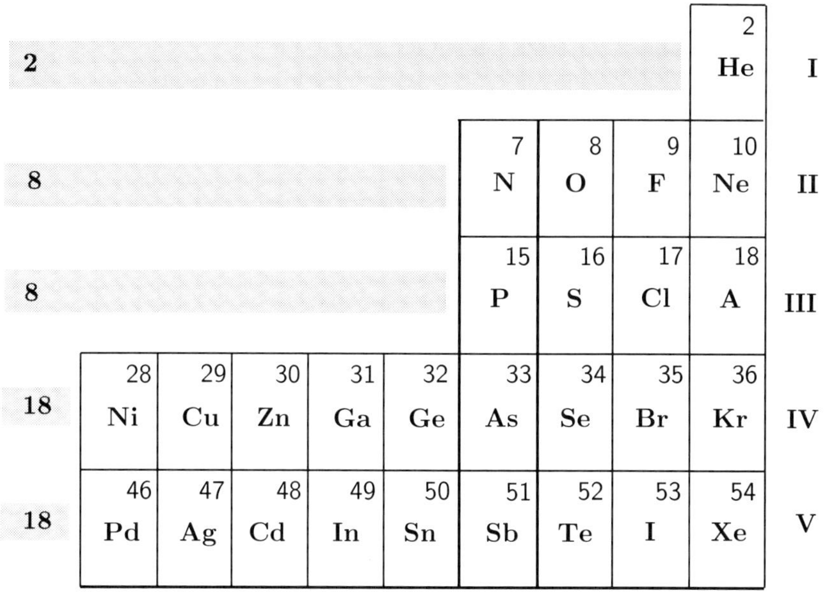

2									2 **He**	I
8						7 **N**	8 **O**	9 **F**	10 **Ne**	II
8						15 **P**	16 **S**	17 **Cl**	18 **A**	III
18	28 **Ni**	29 **Cu**	30 **Zn**	31 **Ga**	32 **Ge**	33 **As**	34 **Se**	35 **Br**	36 **Kr**	IV
18	46 **Pd**	47 **Ag**	48 **Cd**	49 **In**	50 **Sn**	51 **Sb**	52 **Te**	53 **I**	54 **Xe**	V

ordered by their atomic number (which is roughly one-half the weight), progressively higher numbered elements are found to be similar in chemical and physical properties to lower numbered elements and may be grouped with them forming rows. The rows periodically repeat in the pattern $\mathbf{2} + \mathbf{8} + \mathbf{8} + \mathbf{18} + \mathbf{18} + \cdots$.

such as lithium, sodium, potassium, and rubidium. These elements are chemically similar. They are soft metals at the pressures and temperatures at the earth's surface. Put any of them in the presence of oxygen and fiery results ensue. Put them in water and hydrogen gas is released. Yet their atomic weights differ dramatically. Rubidium with an atomic weight of 85 is an order of magnitude more massive than lithium. How can elements of such disparate atomic weight possess nearly identical chemical properties?

If the elements are ordered by number which is directly correlated with atomic weight (the number is roughly one-half the weight), these same elements reveal a persistent pattern: **H** (hydrogen) $1 = 1$, **Li** (lithium) $3 = \mathbf{2} + 1$, **Na** (sodium) $11 = \mathbf{2} + \mathbf{8} + 1$, **K** (potassium) $19 = \mathbf{2} + \mathbf{8} + \mathbf{8} + 1$, **Rb** (rubidium) $37 = \mathbf{2} + \mathbf{8} + \mathbf{8} + \mathbf{18} + 1$. The period is indeed eight in the pattern of Newland's octaves; but only for two repeats: it then jumps to repeats of eighteen.

All these highly active elements have one thing in common: their atomic numbers exceed the periods in the pattern $\mathbf{2} + \mathbf{8} + \mathbf{8} + \mathbf{18} + \cdots$ by 1. Remarkably, the elements that possess this very pattern are those with properties most radically different from the alkali metals: the noble gases on the opposite side of the table. The nobles such as **He** (helium), **Ne** (neon), **A** (argon), **Kr** (krypton), and **Xe** (xenon) are inert—they do not chemically combine with other elements; but they are almost identical to the alkalis in atomic number. **He 2** precedes **Li** $3 = \mathbf{2} + 1$. **Ne 10** $= \mathbf{2} + \mathbf{8}$ precedes **Na** $11 = \mathbf{2} + \mathbf{8} + 1$. **A 18** $= \mathbf{2} + \mathbf{8} + \mathbf{8}$ precedes **K** $19 = \mathbf{2} + \mathbf{8} + \mathbf{8} + 1$, and so on. The small difference of one unit of the alkali metals over the completed pattern blocks of the noble gases signals a radical chemical difference from inertness to extreme reactivity.

A New and Deeper Unity

The Periodic Table was a scheme of heavenly spheres writ small. It described patterns in the microscopic world just as the heavenly spheres described the patterns of the wandering stars. Periods of **2**, **8**, and **18** were deeply etched in the Periodic Table. Could there be be an underlying unification of these numbers in a law of mechanics just as the laws of mechanics provided a simple unifying explanation for the orbits and transit times of the heavenly bodies?

The sun, the moon, and the planets exhibit themselves as bodies in motion bound by forces. But when one examines a fragment of gold or a lump of charcoal, one does not see bodies in motion bound by forces. One looks upon a contiguous piece of material. The character of gold and carbon at the scale of the human hand does not appear related to motions like those of the planets. The character of materials on this level are manifested in chemical and physical properties. Gold is lustrous, readily conducts heat, and is heavy. Carbon burns easily with oxygen.

The description of the chemical and physical properties of gold, carbon, or any element in terms of bodies in motion bound by forces—or, as we shall now say, *particles* in motion—does not arise until one considers the ultimate divisibility of matter. Take hand-sized lumps of gold or carbon and imagine them subdivided into smaller and smaller pieces. What happens in the limit of finer and finer subdivision? A long history of experiment has revealed that an irreducible particle that possesses the properties of gold, carbon, or any other element is finally encountered. Break that particle down to a finer scale and one no longer has gold, carbon, or the element with which one began. This irreducible particle of element is the atom. Every element is a macroscopic aggregation of a single kind of atom.

In progressing from the macroscopic level of the human hand down to the microscopic level of the atom, one leaves the contiguous world of lustre, heat conduction, and combustion and encounters once again a discrete world of particles, motion, and force that images the motions of the heavenly bodies. The atom is mostly space just as the solar system is mostly space. It consists of a tiny positively charged nucleus and a constellation of distantly orbiting, negatively charged electrons bound to the nucleus by the electrical force much like planets bound to the sun.

The rhythmic properties and patterns of the Periodic Table are a manifestation of the law of motion with an electrical inverse-square force mathematically identical to that of the gravitational inverse-square force. The unity of the heavens and the elements is reborn from the womb of motion itself. It is in their law of motion that one glimpses the cosmic unity of the heavens and the elements.

The unity of the cosmos inheres in a law of motion, a law hinted at by the pattern and order in matter and in the orbits of heavenly bodies, a law first quantitatively rendered by Kepler and Newton. But Newton's discovery is more than law. It is beautiful law. The subsequent work of Euler, Lagrange, Hamilton, Jacobi, and others in the eighteenth and nineteenth centuries revealed that Newton's law of motion derives from an even more fundamental principle: the principle of least action. The action principle weaves motion in a magnificent mathematical tapestry, a work of art of sheer simplicity in its central design, yet richly elaborated in detail so fine, so astonishingly complex, it cannot be exhausted by even the most powerful functions of mathematical analysis. It is a beautiful thing to behold; and it is our goal to do so as this story of the physical world unfolds.

The law of motion embraces both the heavens and the elements; for a universal law of motion is law unto the small as well as large. But it would not be until this century that the law of motion governing the heavens would be extended into the realm of the elements revealing the profound connections—and dramatic differences—between the quantum mechanics of the elements and the celestial mechanics of the solar system.

One says "law" but one does so with a daunting sense of its inadequacy. "Law" is both the right word and the wrong word to describe the principle underlying the dynamics of nature. The rich texture, structure, and complex pattern of the physical world are ultimately traceable to strikingly simple mathematical statements about the nature of all motion. It is the potent rendering of all motion under a simple mathematical principle that resonates with our concept of law.

But there are contradictions in "law" as the defining category of the patterns and order one sees in nature. Law invokes the sense of an inevitable routine. Law dictates determinism. Yet the richness of the cosmos belies—dare one say—"mechanical" subservience to a mathematical law book. Nature manifests spontaneity and surprise as much as pattern and law.

The final task of the Shaggy Steed of Physics will be to reveal that the rigid husk of the law of motion contains unexpected seeds of freedom within. The law of motion inherently gives rise to behavior that is, from a mechanical point of view, unpredictably open. This discovery in which the determinism and constraint of the law of motion birth creative diversity is a profound development in mechanics. In it, physical law is at one with music and poetry: a rigid rule of rhyme or rhythm provides a ground for creative inspiration at precisely those points where it is broken.

The law of motion is not a static statute. The law of motion is a sublime mathematical fabric reflecting the creation. Its unfolding implications of cosmos and chaos to this day surprise and illuminate us. It has been subject to the epochal evolution from classical mechanics to quantum mechanics while at the same time it possesses a seamless continuity flowing back through Feynman and Dirac to Hamilton and Lagrange, to Newton, to Galileo and Kepler, and ultimately to the Pythagoreans.

The structure of the world revealed by the mechanics birthed by Kepler, Galileo, and Newton is physical, material. The physical world is illuminated by mathematics. Mathematical physics grasps the motions of the heavens and the elements through bodies with properties such as mass and charge that interact with forces that depend in a geometrical manner on the motions of the bodies. Yet it partakes of the same divine attributes which prompted the celebration of the heavenly spheres: beauty and truth.

Physics has come to reveal the unity of the richness, diversity, and complexity of the natural world by piercing through these things for the kernel of explanation. Physics grasps the entire natural world with what seems precious little: the motions of a small number of elementary particles that interact by means of an equally small number of forces. In the case of the elements and the large-scale motions of the solar system, two kinds of particles (electrons and nuclei) bearing two properties (mass and charge) corresponding to two forces (gravity and electricity) suffice.

This is one side of the mystery. The other side is that everything material—all the elements and molecules, the animal, vegetable, and mineral world, and the motions of solar systems—spring from the interaction of these particles and forces. And in that the physical world is revealed in a mystical light: an abundant, interleaved richness of space and matter—in stars, planets, waters, earth, and living creatures—all made of the same elementary particles blooming riotously out of a simple law which itself dissolves in freedom. It is a unity more wondrous than that of the celestial spheres. And one fully spiritual. For when one renders the fullness of respect to the truth in the material world, matter becomes luminous with spirituality, revealing the mystery of the Creator in a way both more immanent and more abundant than that which held the imagination of the Greek and medieval mind.

Chapter 3

The Law of Motion

The motion of the universe is not assigned a fixed score. Nature orchestrates motion in a far more graceful manner with the *action principle.* According to the action principle, the motion of the universe is free save for this requirement: in their movement between any two points in space and time, particles of matter take, from the infinite number of possible paths, just those for which their *action* has the least possible value. From this simple principle come the planetary orbits of heavenly bodies and the structure of chemical elements.

As we progress on this journey through mechanics we shall encounter the many faces of this mechanical quantity in nature called *action.* And we shall draw out the implications of the overarching principle, *action shall be a minimum,* for the heavens and the elements. Ultimately we shall find the action and the law of motion things of astonishing simplicity: the many faces of the action are measures of the lengths of paths the objects of the world trace out in their motion; the law of motion the simple requirement that these paths are always the shortest routes possible between any two points.

The Warp and Weft of Motion

The primitive strands of experience out of which motion is woven are space and time. It is with these elementary perceptions that one begins the ascent to the principle of least action.

At the human scale it takes three independent pieces of information to pin down a location in space. The location of a point within a multi-story building requires three numbers: specification of the floor of interest and two distances from any two sides of the building to a point on that floor. The location of an aircraft in flight requires three numbers: a longitude, a latitude, and an altitude. Our spatial world is a three-dimensional world.

Why three? No one knows why the human-scale world is a three-world. A world of only one or two dimensions would be too constricted for any rich texture in creation; a world of many dimensions too free, too densely filiated. Perhaps a world with more than one or two dimensions (but not vastly many more) provides for a maximum diversity of form and structure without degenerating into a tangle of monotonous chaos.

To reckon motion one needs a basis for the description of space—a reference frame in which one may locate and describe the positions of the objects of the world. Three independent lengths x_1, x_2, x_3 may be used to single out a point in space. The triplets (x_1, x_2, x_3) are said to constitute a manifold. A manifold is the set of all points where particles in motion may be located.

A manifold of points is still a formless space. To give form to the manifold, one may specify that the lengths be laid out along mutually perpendicular directions. To be perpendicular, the line from the origin of the coordinates to the point singled out (denoted x in Fig. 3-1) must depend in a particular way upon the three lengths x_1, x_2, x_3 that specify that point. The form of that dependence and the meaning of perpendicular go hand in hand.

The specification of a point by mutually perpendicular coordinate axes and an ancient geometrical object—the triangle—are intimately connected in the reckoning of space. Any of the three lines x_1, x_2, x_3, taken in pairs, form two sides of a triangle. These lines are perpendicular when the triangle they form has a third side given by the rule of Pythagoras, *the square of the third side is the sum of the squares of the two adjacent sides.* If the lines of length x_1 and x_2 are the two adjacent sides, the length l of the third side of this triangle is given by the Pythagorean rule as $l^2 = {x_1}^2 + {x_2}^2$ as shown in Fig. 3-1.

Now consider the line of length l and the coordinate line of length x_3 which are also adjacent sides of a triangle. The third side of this triangle is the line of length x. By invoking the Pythagorean principle again, one finds $x^2 = l^2 + {x_3}^2$ and the fundamental rule of Euclidean space,

$$x^2 = {x_1}^2 + {x_2}^2 + {x_3}^2. \tag{3.1}$$

The specification of the manner in which the distance between points in space is related to the component lengths that define those points is called the *metric* of the space. The metric endows three independent coordinates with an internal structure. In the case of the metric (3.1) the manifold has Euclidean structure.

The position of a point in space is specified by three lengths structured by the Euclidean metric (3.1). Such a threefold quantity obeying the Euclidean metric is a position vector specifying points in space and is symbolized as

$$\mathbf{x} = (x_1, x_2, x_3),$$

where x_1, x_2, and x_3 are the three orthogonal lengths that specify the direction and length of the vector. The length of the vector $\mathbf{x}$ is its magnitude

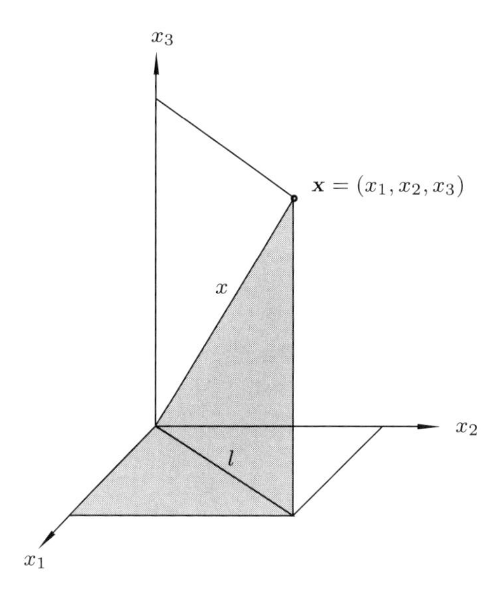

Figure 3-1. The Euclidean Description of Space. A point in space is specified by a vector $\mathbf{x}$ whose components are equal in number to the three dimensions of the space. The length of the vector $\mathbf{x}$ is x and may be computed in terms of its component lengths by first computing the side length l of the triangle formed by x_1 and x_2. The triangle formed by the sides of length l and x_3 then has a side x given by the Pythagorean law as $x^2 = {x_1}^2 + {x_2}^2 + {x_3}^2$.

and is symbolized as x or sometimes as $|x|$. In Euclidean space the square of the length of a vector is equal to the sum of the squares of its component lengths.

We have established a reckoning of space; but ours is a world in motion. The objects one identifies at particular locations in space do not remain there. They change their positions. Another coordinate beyond those that mark off the spatial location of objects is called into being. Time is this coordinate; but it is important to note that it is our experience of *change* which prompts our sense of time.

We shall join Newton and Galileo in the seemingly unquestionable assumption that the flow of time is absolute. Time flows in a clock or hourglass independently of the objects one locates in space. Likewise, time does not depend upon the spatial location of the clock in which it is recorded. As a result of that assumption, space and time do not intertwine in the measurement of distance. The three spatial coordinates plus the time coordinate constitute four coordinates. The three spatial coordinates are united by the Euclidean metric; but the time coordinate is aloof from this union. Time goes its own way. In the mechanics bequeathed to us by Galileo and Newton the time t at which a measurement of position is made does not directly enter into the calculation of length of the position vector according to Eq. (3.1). Only the three spatial components x_1, x_2, x_3 enter into the computation. The four coordinates consisting of space and time are regarded as three spatial coordinates which are a function of time.

There is no guarantee that the space of the physical world is Euclidean or that the flow of time is absolute. The conception of space and time is a product of human imagination. Our concrete experience with triangles shapes the imagery of Euclidean space; and the world conforms to Euclidean space quite well under ordinary, earth-bound conditions. But the space of the world need not be Euclidean under conditions beyond the human-scale experiences in which it is grounded.

The assumption of the Euclidean structure of the physical world was never explicitly acknowledged until Henri Poincaré, Albert Einstein, and others questioned it in the early part of this century. Einstein showed it to be arbitrary, a limiting form of a more fundamental geometry and metric valid at speeds small compared to the speed of light. Since the motion in both the solar system and the atom takes place at a small fraction of the speed of light, the ancient rule shall guide our journey through the heavens and the elements. Space is Euclidean. Its structure is that of a metric inspired by the Pythagorean triangle rule.

Mechanical Genesis

Outside of time there is only one vector—position in space. With time, the position vector evolves and changes. It becomes a function of time $\mathbf{x} = \mathbf{x}(t)$. In a world of time as well as space a second elementary vector exists. This is the velocity, $\dot{\mathbf{x}} \equiv d\mathbf{x}/dt$. The primal objects of motion are the two elementary vectors, position and velocity.

With the existence of a second elementary vector, interaction with the first becomes possible. The two elementary vectors may be combined to generate a family of mechanical quantities. The simplest is the projection of one vector onto the other. This projection yields a single quantity from two three-component quantities. This quantity is the *scalar product*. A vector is a three-component object but a scalar is a single quantity. For any two vectors, for example $\mathbf{x}$ and $\dot{\mathbf{x}}$, the scalar product is symbolized as $\mathbf{x} \cdot \dot{\mathbf{x}}$ and formed as a sum of products of the components which produce it as

$$\mathbf{x} \cdot \dot{\mathbf{x}} = \sum_i x_i \dot{x}_i = x_1 \dot{x}_1 + x_2 \dot{x}_2 + x_3 \dot{x}_3.$$

If the two vectors are perpendicular, the scalar product is zero. If both vectors are identical, the scalar product produces the square magnitude of the vector,

$$\mathbf{x} \cdot \mathbf{x} = {x_1}^2 + {x_2}^2 + {x_3}^2.$$

Albert Einstein introduced a useful notation for indicating the summations that occur in operations with products of vectors. In Einstein's convention a *repeated* subscript implies summation over the range of the subscript. Thus, the scalar product $\mathbf{x} \cdot \dot{\mathbf{x}} = \sum_i x_i \dot{x}_i$ can be compactly written in subscript notation without the summation symbol as

$$x_i \dot{x}_i = x_1 \dot{x}_1 + x_2 \dot{x}_2 + x_3 \dot{x}_3.$$

When the subscripts are distinct, as in $x_i \dot{x}_j$, no summation is to be performed. When the subscripts are the same, the products are summed. Note that when the subscripts are distinct, their identity must be preserved. But once a scalar product is indicated by making the subscripts identical, the subscript pair indicating the sum can be changed at will, $x_i \dot{x}_i = x_j \dot{x}_j = x_k \dot{x}_k$.

A plane can be defined by a single vector. It is that plane in which all lines perpendicular to that vector and intersecting it at some point must lie. But in three-dimensional space a plane can also be defined by two vectors as the common plane in which both lie. Thus, the existence of two vectors in three-dimensional space naturally implies the existence of a third vector perpendicular to each of them. This vector is their *vector product.*

The vector product of two vectors such as $\mathbf{x}$ and $\dot{\mathbf{x}}$ (sometimes called the cross-product) is itself a vector with three components and is symbolized as $\mathbf{x} \times \dot{\mathbf{x}}$. The three components of the vector product are antisymmetric pairs of the components of the vectors that produce it:

$$\begin{aligned}
(\mathbf{x} \times \dot{\mathbf{x}})_1 &= x_2 \dot{x}_3 - x_3 \dot{x}_2, \\
(\mathbf{x} \times \dot{\mathbf{x}})_2 &= x_3 \dot{x}_1 - x_1 \dot{x}_3, \\
(\mathbf{x} \times \dot{\mathbf{x}})_3 &= x_1 \dot{x}_2 - x_2 \dot{x}_1.
\end{aligned}$$

The components of the vector product are structurally identical. Any component may be obtained from a previous component by advancing all subscripts by 1 in the cyclic order 123123.... A component of the vector product in a given direction does not contain any of the components of its constituents in that direction. This is the manifestation of the orthogonality of two vectors with respect to their vector product. The antisymmetric pairing of component products requires that the vector product of two vectors which are parallel vanish; for if two vectors such as $\dot{\mathbf{x}}$ and $\mathbf{x}$ are parallel, their components are fixed ratios of one another, $\dot{x}_1/x_1 = \dot{x}_2/x_2 = \dot{x}_3/x_3$. For the same reason, the vector product of a vector with itself is identically zero. If the order of the terms is reversed in a vector product, the vector product changes sign, $\mathbf{x} \times \dot{\mathbf{x}} = -\dot{\mathbf{x}} \times \mathbf{x}$.

The vector product may be compactly expressed in subscript notation utilizing the *alternating tensor* ϵ_{ijk} as

$$(\mathbf{x} \times \dot{\mathbf{x}})_i = \epsilon_{ijk} x_j \dot{x}_k, \qquad i, j, k = 1, 2, 3.$$

The alternating tensor is a simple object.* Its components take only the values 0, 1, or -1. If two or more of the ijk are identical, then $\epsilon_{ijk} = 0$. If the ijk "wind forward" in cyclic order 123, 231, 312, then $\epsilon_{ijk} = 1$. If the ijk "wind backward" in anti-cyclic order 321, 213, 132, then $\epsilon_{ijk} = -1$. Einstein's convention for repeated indices implies summation over the indices j and k in the vector product $\epsilon_{ijk}x_j\dot{x}_k$ so that, for example, one has $(\mathbf{x} \times \dot{\mathbf{x}})_1 = \epsilon_{1jk}x_j\dot{x}_k = x_2\dot{x}_3 - x_3\dot{x}_2$.

What kind of dynamical structure naturally arises when one combines the two primal vectors $\mathbf{x}$ and $\dot{\mathbf{x}}$? The products $\mathbf{x} \times \mathbf{x}$ and $\dot{\mathbf{x}} \times \dot{\mathbf{x}}$ both vanish while the product $\mathbf{x} \cdot \dot{\mathbf{x}}$ is just one-half the time derivative of $\mathbf{x} \cdot \mathbf{x}$. This leaves only three unique products which may be formed from position and velocity:

$$\mathbf{x} \cdot \mathbf{x}, \qquad \mathbf{x} \times \dot{\mathbf{x}}, \qquad \dot{\mathbf{x}} \cdot \dot{\mathbf{x}}.$$

The first is the square magnitude of the position vector itself which has already been described. The second is a fundamental quantity in the description of motion. Multiplied by the particle mass m, it is the angular momentum $\boldsymbol{J} = \mathbf{x} \times m\dot{\mathbf{x}}$. Because the angular momentum is a vector product, it is perpendicular to both the position and velocity vectors. The angular momentum vector defines a plane in which the position and velocity vectors lie.

The last product is the square magnitude of the velocity vector. Multiplied by one-half the mass m of the particle, it has come to be known as the kinetic energy, $T = m\dot{x}^2/2$. These dynamical quantities may also be expressed directly in terms of the particle linear momentum $\boldsymbol{p} = m\dot{\mathbf{x}}$ instead of the velocity as $\boldsymbol{J} = \mathbf{x} \times \boldsymbol{p}$ and $T = p^2/2m$.

Three components of position $\mathbf{x}$, three components of linear momentum $\boldsymbol{p}$, three components of angular momentum $\boldsymbol{J}$, and the kinetic energy T are the fundamental objects created by the motion of a particle. Motion in three-dimensional space naturally gives rise to these ten (each vector contains three) dynamical quantities. We shall find this tenfold set of dynamical quantities appearing again and again in the unfolding story of motion.

We now have the elementary vocabulary of mechanics. We have yet to speak of the law of motion. Yet much of the dynamical structure of motion—linear momentum, angular momentum, and energy—is naturally implied by the position and velocity vectors and their composition.

---

* The alternating tensor components are explicitly

$$\epsilon_{1jk} = \begin{pmatrix} 0 & 0 & 0 \\ 0 & 0 & 1 \\ 0 & -1 & 0 \end{pmatrix}, \quad \epsilon_{2jk} = \begin{pmatrix} 0 & 0 & -1 \\ 0 & 0 & 0 \\ 1 & 0 & 0 \end{pmatrix}, \quad \epsilon_{3jk} = \begin{pmatrix} 0 & 1 & 0 \\ -1 & 0 & 0 \\ 0 & 0 & 0 \end{pmatrix}.$$

The Blank Canvas of Mechanics

A reference frame from which one may observe and describe the motion of the world consists of space and time coordinates $(\mathbf{x}, t)$. But the world exists quite independently of any reference frame. The reference frame is our creation—and there are arbitrarily many of them from which we may observe any motion. The reference frame is essential for observation of the world. Without it one sees nothing with certainty. Yet what one sees should not be an artifact of the reference frame.

Motion presents a formidable challenge to this principle; for the dynamics which one experiences as well as the mathematical form of the law of motion depend upon the frame of reference in which one observes the motion. One may plant one's reference frame on the moon, on a ship moving on the ocean, on a spinning merry-go-round, or on a soaring bird. Moreover, one may keep time on different clocks in each of these reference frames. The motion which is observed in all these reference frames is governed by a universal law. But that law will not exhibit a universal form when expressed in the coordinates $(\mathbf{x}, t)_1$, $(\mathbf{x}, t)_2$, ... of each of these reference frames. Although the law of motion is universal, its form in various reverence frames is not.

Is there a unique reference frame that is truly universal, one from which the motions in all other reference frames may be reckoned? The nineteenth century physicist and philosopher Ernst Mach suggested that such a reference frame was given by nature. He proposed that the universal reference frame was one in which the "fixed stars" were at rest. We now know the stars are in violent motion. So Mach's proposal does not provide us a naturally given reference frame in which the law of motion will reveal its universal form.

Since nature does not provide a universal reference frame attached to natural objects, one must proceed indirectly. One must conceive such a reference frame by incorporating just those properties of motion that are universal in all reference frames while avoiding those that are unique to particular frames. This was the approach followed by Isaac Newton who conceived of such a universal reference frame as an "absolute space,"—a blank canvas stretched across space and time devoid of all local references to motion. It is upon this blank mathematical canvas that one invites nature to paint the law of motion.

Peculiarly local properties of reference frames are their origins and their directions. One therefore blanks out the canvas of space–time with a wash of symmetry that purges it of all unique points of origin and all unique lines of direction. The symmetry of origins is that there is no unique point in either space or time from which motion must be reckoned. No point in the universe or in time has a special status over any other. No point on the canvas of motion can be painted in as a unique origin.

Direction is a more subtle notion; for there appear to be two ways in

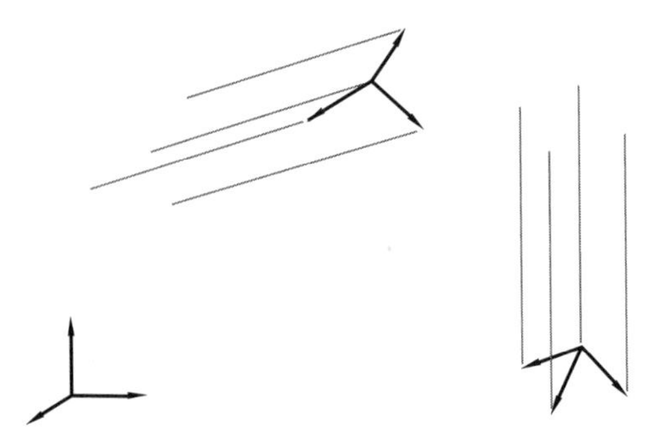

Figure 3-2. Galilean Relativity. The mathematical form of the law of motion is identical in all reference frames displaced from one another, uniformly translating with respect to one another, and rotated with respect to one another.

which it may be defined. Both arise out of experience. Move adamantly forward and one establishes a translationally based direction fixed by one's velocity. Point at a star and one establishes a rotationally based direction fixed by the angle one's arm makes with the horizon. In one case a velocity specifies a direction; in the other an angle specifies direction. In the mechanics that came into being through Galileo and Newton both kinds of directional symmetry, one translational and the other rotational, are included. Symmetry with respect to directions means that there is neither a unique velocity nor unique rotational orientation from which motion must be reckoned. No unique line of direction can be drawn on the canvas of motion.

The symmetry of the law of motion with respect to origins, translational directions, and rotational orientations means that it has the same mathematical form in all reference frames that have been translated and rotated with respect to one another. Similarly, the law of motion is the same when expressed in the times of all clocks whose time origins have been translated with respect to one another.

The space and time in which the law of motion is independent of translations of the space or time coordinates is said to be *homogeneous*. The space in which the law of motion is independent of rotations of the coordinates is said to be *isotropic*, after the Greek prefix ἰσο meaning "equal" and τροπός meaning "rotation." Reference frames in which the law of motion reveals its universal form are homogeneous (there is no preferred origin) and isotropic (there is no preferred direction) as indicated in Fig. 3-2.

The reference frames in which the law of motion takes its universal form are not attached to any particular object in nature (such as the earth or a star) but are rather created from the insight that homogeneity and isotropy are the universal symmetries underlying all motion. Reference frames in which the law of motion has its universal form are homogeneous and isotropic. Nature affirms this insight by presenting us with a pro-

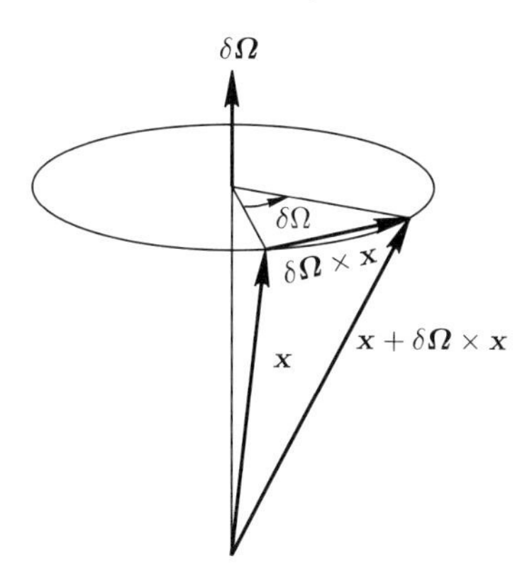

Figure 3-3. Rotation and the Vector Product. The vector $\delta\boldsymbol{\Omega}\times\mathbf{x}$ is perpendicular to both $\delta\boldsymbol{\Omega}$ and $\mathbf{x}$. For infinitesimal rotations $\delta\boldsymbol{\Omega}$, the sum $\mathbf{x}+\delta\boldsymbol{\Omega}\times\mathbf{x}$ has the same magnitude as $\mathbf{x}$ but its direction has been changed by a rotation through an angle $\delta\Omega$.

foundly simple law of motion in all reference frames that partake of these elementary symmetries.

The requirements of homogeneity and isotropy may be expressed in the statement that the law of motion has the same form in all reference frames $(\mathbf{x}, t)$ and $(\mathbf{x}', t')$ that differ in translational displacement and rotational orientation. It is both simple and sufficient to state this condition for reference frames that differ infinitesimally: those displaced and rotated by infinitesimal distances $\delta\mathbf{a}$, times $\delta\tau$, velocities $\delta\boldsymbol{u}$, and angles $\delta\boldsymbol{\Omega}$. Homogeneous and isotropic reference frames are related by the transformations

$$\begin{aligned} \mathbf{x}' &= \mathbf{x} + \delta\mathbf{a} + \delta\mathbf{u}t + \delta\boldsymbol{\Omega}\times\mathbf{x}, \\ t' &= t + \delta\tau. \end{aligned} \tag{3.2}$$

The law of motion has precisely the same form whether expressed in the coordinates $(\mathbf{x}, t)$ of one reference frame or the coordinates $(\mathbf{x}', t')$ of another related by the transformations (3.2).

In the transformations (3.2) that endow reference frames with homogeneity and isotropy $\delta\mathbf{a}$ and $\delta\tau$ represent shifts in the origins of space and time. The velocity $\delta\boldsymbol{u}$ represents a uniform translation of the space coordinates. The rotational transformation vector $\delta\boldsymbol{\Omega}$ is a constant vector about which the coordinates are rotated. Since $\delta\boldsymbol{\Omega}\times\mathbf{x}$ is perpendicular to both $\mathbf{x}$ and $\delta\boldsymbol{\Omega}$, the sum $\mathbf{x}+\delta\boldsymbol{\Omega}\times\mathbf{x}$ results in a rotation of $\mathbf{x}$ about the axis of $\delta\boldsymbol{\Omega}$ as shown in Fig. 3-3. The infinitesimal vector product therefore represents a rotation. The direction of $\delta\boldsymbol{\Omega}$ is along the axis about which the rotation takes place in the pointing sense of a right-hand screw. For infinitesimal rotations the magnitude $\delta\boldsymbol{\Omega}$ represents the angle of rotation about its axis and the sum $\mathbf{x}+\delta\boldsymbol{\Omega}\times\mathbf{x}$ has the same magnitude as $\mathbf{x}$. Only the direction changes. For finite $\boldsymbol{\Omega}$, the transformation $\mathbf{x}+\boldsymbol{\Omega}\times\mathbf{x}$ is not a pure rotation because the vector product is not only a rotation of the original vector; its magnitude is also changed.*

The law of motion is the same in all reference frames differing by uniform shifts of their origins, uniformly translating at a constant velocity with respect to one another, rotated through a constant angle with respect to one another, and whose clocks are shifted in time by a constant amount from one another. The group of all transformations obeying Eqs. (3.2) is called the Galilean relativity group.†

Ten symmetry parameters (each vector again contains three) describe the Galilean group. They are the space origin shift $\mathbf{a}$, the space translation $\boldsymbol{u}$, the space rotation $\boldsymbol{\Omega}$, and the time origin shift τ. The blank canvas of mechanics can be thought of as a ground of motion formed of the most primitive, wide-ranging symmetries—an openness to all origins, all translational directions, and all rotational orientations.

Interaction

The Western vision of the physical world gives precedence to the whole as made of elemental parts. In this vision of mechanics, *interaction* is a fundamental feature of the natural world; for if the whole is made of elemental parts, the character of the whole arises out of the way in which the elemental parts interact. The interaction of elemental matter is the glorious drama in which the world is rendered.

We call the interactions between particles of matter "forces." The two known large-scale interactions of matter are those of gravity and electricity. How can this interaction be represented? At this point one stands at a great juncture in human imagination. On the one hand one can attempt to describe gravity and electricity as entities in themselves generated by properties of particles that are their *raison d'être*: *mass* for gravity, *charge* for electricity. The interaction between material particles is then described in terms of forces which the particles exert upon one another by virtue of their mass and charge. The force fields may be computed for an ensemble of particles and inserted into a law of motion to find their effect upon the movement of the particles. This is the approach of classical mechanics—the mechanics first grasped by Isaac Newton and elaborated by Joseph-Louis Lagrange, C. G. J. Jacobi, and William Rowan Hamilton.

On the other hand one can adopt a truly geometrical point of view and insist that force is an artifact of Euclidean geometry. In this view mass

---

* Though it is not required here, a rotation through finite $\boldsymbol{\Omega}$ can be obtained in an almost magical manner. One exponentiates the infinitesimal transformation $x'_i = x_i + \epsilon_{ijk}\delta\Omega_j x_k$ to obtain the finite rotation $x'_i = e^{\epsilon_{ijk}\Omega_j} x_k$.

† This term recognizes Galileo's insight that bodies in motion with uniform velocity are free of forces (later codified in Newton's first law of motion) and seems to have originated in P. Frank, *Ber. Akad. Wiss. Wien.* IIa, **118**, 382 (1909).

and charge do not create forces. They create geometry. Mass and charge induce a curvature in space which invalidates the Euclidean law (3.1). A new metric of space shaped by the presence of mass and charge rather than the Euclidean metric becomes the basis of the law of motion. This was Einstein's way; and it is essential for motions with velocities of the order of the speed of light. Since the heavenly motions of our solar system and the microscopic motions of the electron in the atom take place at velocities only a small fraction of the speed of light, we shall follow the Newtonian approach.

In the Newtonian picture mass and charge fill space with potential energy. The gravitational potential energy V of two masses m_1 and m_2 located at the points $\mathbf{x}_1$ and $\mathbf{x}_2$ is directly proportional to the masses of the bodies and inversely proportional to the distance of separation $|\mathbf{x}_2 - \mathbf{x}_1|$ between them,

$$V = -\frac{m_1 m_2}{|\mathbf{x}_2 - \mathbf{x}_1|}.$$

The force $\boldsymbol{f}$ acting on a particle whose position vector is $\mathbf{x}$ and described by a potential energy $V(\mathbf{x})$ is

$$\boldsymbol{f} = -\boldsymbol{\nabla} V.$$

The potential V is a scalar; but its derivative with respect to the vector $\mathbf{x}$ is a vector $\boldsymbol{\nabla} V = \partial V/\partial \mathbf{x}$ called the gradient,

$$\boldsymbol{\nabla} V = (\partial V/\partial x_1, \partial V/\partial x_2, \partial V/\partial x_3).$$

The force field that two gravitating masses create through the potential $V = -m_1 m_2/|\mathbf{x}_2 - \mathbf{x}_1|$ is an attractive inverse-square field first proposed by Isaac Newton. The force acting on the particle at $\mathbf{x}_1$ is $\boldsymbol{f} = -\partial V/\partial \mathbf{x}_1$ or

$$\boldsymbol{f} = -\frac{m_1 m_2}{|\mathbf{x}_2 - \mathbf{x}_1|^2}\left(\frac{\mathbf{x}_2 - \mathbf{x}_1}{|\mathbf{x}_2 - \mathbf{x}_1|}\right). \tag{3.3}$$

Each of the parts of the force law may be read off to reveal a significant property of the gravitational force. The first factor on the right of Eq. (3.3) shows that the gravitational force is an inverse-square force. The force is proportional to the inverse square of the distance of separation between the particles. The negative sign shows that this force is attractive. The second factor is a vector of unit length directed from the center of m_1 to the center of m_2. It shows that the gravitational force is a *central* force directed along the line of centers between the two particles. This follows from the fact that the potential energy field, consistent with the isotropy of space, depends only upon the magnitude of the distance of separation between the particles $|\mathbf{x}_1 - \mathbf{x}_2|$.

The ranges of the gravitational and electrical force are vastly different (the gravitational force between two electrons is forty two orders of magnitude smaller than the electrical force between them). Nonetheless the potential energy and force fields of gravity and electricity (for which charges move slowly compared to the speed of light) are structurally identical. The potential energy V and force $\boldsymbol{f}$ for a system of charges are precisely the same as those of gravity with mass m replaced by charge e. Force outside the scale of the nucleus is inverse-square. The potential energy and force for either field may be expressed in the form

$$V = -\frac{k}{|\mathbf{x}_1 - \mathbf{x}_2|}, \qquad \boldsymbol{f} = -k\frac{(\mathbf{x}_1 - \mathbf{x}_2)}{|\mathbf{x}_1 - \mathbf{x}_2|^3}, \tag{3.4}$$

where for gravity $k = m_1 m_2$ and for electricity $k = -e_1 e_2$.

For an ensemble of particles of masses m_α, $\alpha = 1, 2, \ldots$ interacting with gravitational forces, the potential energy function is

$$V = -\sum_\alpha \sum_{\beta<\alpha} \frac{m_\alpha m_\beta}{|\mathbf{x}_\alpha - \mathbf{x}_\beta|}.$$

The condition $\beta < \alpha$ is required so that interactions are not counted twice. The potential energy of a system of slowly moving charges interacting with electrical forces is the same as that for gravity with charges e_α replacing the masses m_α.

The interaction between particles described by their force and energy fields must be consistent with the underlying symmetries of space and time given by Eqs. (3.2). This means the potential energy field must be the same in any two reference frames where the particle coordinates $(\mathbf{x}'_\alpha, t')$ and $(\mathbf{x}_\alpha, t)$ are related by the homogeneous and isotropic transformations (3.2),

$$V(\mathbf{x}'_\alpha, t') = V(\mathbf{x}_\alpha, t). \tag{3.5}$$

It is again sufficient to verify this symmetry condition for infinitesimal differences between reference frames for which one expands $V(\mathbf{x}'_\alpha, t')$ about $V(\mathbf{x}_\alpha, t)$ so that the condition (3.5) becomes

$$\sum_\alpha \frac{\partial V}{\partial \mathbf{x}_\alpha} \cdot (\mathbf{x}'_\alpha - \mathbf{x}_\alpha) + \frac{\partial V}{\partial t}(t' - t) = 0.$$

One now makes this transformation homogeneous and isotropic by expressing $(\mathbf{x}'_\alpha, t')$ in terms of $(\mathbf{x}_\alpha, t)$ from Eqs. (3.2) for infinitesimal transformations generated by $\delta\mathbf{a}$, $\delta\mathbf{u}$, $\delta\boldsymbol{\Omega}$, and $\delta\tau$ with the result

$$\sum_\alpha \frac{\partial V}{\partial \mathbf{x}_\alpha} \cdot (\delta\mathbf{a} + \delta\mathbf{u}t) + \sum_\alpha (\mathbf{x}_\alpha \times \frac{\partial V}{\partial \mathbf{x}_\alpha}) \cdot \delta\boldsymbol{\Omega} + \frac{\partial V}{\partial t}\delta\tau = 0,$$

where use has been made of the identity $\partial V/\partial \mathbf{x}_\alpha \cdot (\mathbf{x}_\alpha \times \delta\boldsymbol{\Omega}) = (\mathbf{x}_\alpha \times \partial V/\partial \mathbf{x}_\alpha) \cdot \delta\boldsymbol{\Omega}$. Since the symmetry parameter variations are independent and need not vanish, the potential energy fields of Galilean relativity satisfy the symmetry conditions

$$\sum_\alpha \frac{\partial V}{\partial \mathbf{x}_\alpha} = 0, \qquad \sum_\alpha \mathbf{x}_\alpha \times \frac{\partial V}{\partial \mathbf{x}_\alpha} = 0, \qquad \frac{\partial V}{\partial t} = 0. \tag{3.6}$$

The first of the symmetry conditions (3.6) indicates that the sum of the forces acting on an ensemble of particles vanishes while the second shows that the total moment of the entire ensemble also vanishes. For the isolated interaction between any two particles, these conditions are contained in Newton's first law of motion: the force exerted by the first particle on the second is equal and opposite to the force exerted by the second on the first and the force between the two particles is directed along the line of centers that connects them.

It is a fundamental hypothesis of mechanics, verified down to the subnuclear level, that the symmetry between any pair of particles is preserved in the ensemble. Hence, all forces in nature are central forces and cancel in pairs in the sums in Eq. (3.6). Any potential $V(\mathbf{x}_\alpha)$ which depends only upon the magnitudes of the distances between particles (for which gravity and electricity are paradigms) will satisfy this condition.

The third of the symmetry conditions (3.6) shows that force and potential energy fields in Galilean relativity are *independent of time.* The interaction between particles is instantaneous. It can be seen from the forms of the force and potential energy (3.4) that any change in the position of one of the particles is instantly reflected in a change in the force acting on the other. The propagation of information about the change of a particle's position is infinitely fast. The infinite speed of propagation of forces is a hallmark of Galilean space–time.

The Gene of Motion

The array of particle trajectories and their momenta, the shape, pattern, and structure these trajectories create in space and time, in sum, the whole nature of a given motion issues from a single mechanical quantity. We shall work up to this quantity—the gene of motion—by first gathering together the properties of an ensemble of particles as a whole. The entire ensemble has a total mass m and a center of mass $\boldsymbol{X}$ defined as

$$m = \sum_\alpha m_\alpha, \qquad \boldsymbol{X} = m^{-1} \sum_\alpha m_\alpha \mathbf{x}_\alpha,$$

where the sum is over all the particles of the ensemble. The total linear momentum $\boldsymbol{P}$ and the total angular momentum $\boldsymbol{J}$ are also a sum of individual particle contributions:

$$\boldsymbol{P} = \sum_\alpha \boldsymbol{p}_\alpha, \qquad \boldsymbol{J} = \sum_\alpha \mathbf{x}_\alpha \times \boldsymbol{p}_\alpha.$$

Since $\boldsymbol{p}_\alpha = m_\alpha \dot{\mathbf{x}}_\alpha$, it follows that the total momentum can be expressed directly in terms of the center of mass velocity by $\boldsymbol{P} = m\dot{\mathbf{X}}$. In the event that the total momentum $\boldsymbol{P}$ is a constant, this equation may be integrated to yield

$$\boldsymbol{N} = \boldsymbol{P}t - m\boldsymbol{X},$$

where $\boldsymbol{N}$ is the constant of integration. The vector $\boldsymbol{N}$ fixes the motion of the center of mass of an ensemble whose total momentum $\boldsymbol{P}$ is uniform. It can be expressed in terms of the center of mass at some fixed time, for example, as $\boldsymbol{N} = -m\boldsymbol{X_0}$ where $\boldsymbol{X_0}$ is the center of mass at time $t = 0$. We shall call $\boldsymbol{N}$ the *mass-center* (which is to be distinguished from the center of mass $\boldsymbol{X}$).

The total kinetic energy is similarly a sum over particle contributions:

$$T = \sum_\alpha m_\alpha \dot{x}_\alpha^2/2 = \sum_\alpha p_\alpha^2/2m_\alpha.$$

The sum of the kinetic and potential energy is the total energy H of a system of particles,

$$H = T + V.$$

The difference between the kinetic and potential energy is known as the Lagrangian

$$L = T - V.$$

The Lagrangian is the gene of motion. All the dynamical quantities describing motion may be shown to be descended from the Lagrangian. The momentum of a particle $\mathbf{p}_\alpha$ is the derivative of the Lagrangian with respect to velocity, $\mathbf{p}_\alpha = \partial L/\partial \dot{\mathbf{x}}_\alpha$. The angular momentum of a particle $\mathbf{J}_\alpha$ may also be unfolded from the Lagrangian as $\mathbf{J}_\alpha = \mathbf{x}_\alpha \times \partial L/\partial \dot{\mathbf{x}}_\alpha$. The force acting on a particle likewise derives from the Lagrangian as $\partial L/\partial \mathbf{x}_\alpha$ and the moment as $\mathbf{x}_\alpha \times \partial L/\partial \mathbf{x}_\alpha$. Consistent with the potential energy condition given by the second and third of Eqs. (3.6), the Lagrangian of an isolated ensemble of particles satisfies the conditions

$$\sum_\alpha \frac{\partial L}{\partial \mathbf{x}_\alpha} = 0, \qquad \sum_\alpha \mathbf{x}_\alpha \times \frac{\partial L}{\partial \mathbf{x}_\alpha} = 0, \tag{3.7}$$

where the sums are over all the particles.

All mechanical quantities turn out to be expressible in terms of the Lagrangian and its derivatives with respect to two quantities, position and velocity. This is not by accident. It is a deep seated principle of nature that a state of motion is completely specified by a twofold set of conjugate coordinates such as the positions and velocities of the particles. The coordinates of motion are said to be conjugate because they come in pairs and there exists a conjugal relationship between the two coordinates that constitute the pair. In the case of position and velocity, one coordinate is the time derivative of the other.

Everything one could want to know about a system of classical point particles is obtainable from the pair of conjugate coordinates. This need not have been the case; nature might have in addition to $\mathbf{x}_\alpha$ and $\dot{\mathbf{x}}_\alpha$ utilized the accelerations of the particles $\ddot{\mathbf{x}}_\alpha$ as well as even higher derivatives to fix a state of motion. But the world is so constituted that the acceleration and all the higher derivatives of a particle's position are expressible in terms of a conjugate pair of coordinates such as $\mathbf{x}_\alpha$ and $\dot{\mathbf{x}}_\alpha$—and no more. The manifold of the motion of n bodies is a $6n$-dimensional collection of positions and velocities. It is with these coordinate pairs that one must build the Lagrangian $L = L(\mathbf{x}_\alpha, \dot{\mathbf{x}}_\alpha)$ from which all mechanical quantities flow.

The total linear momentum $\boldsymbol{P}$ and angular momentum $\boldsymbol{J}$ may be expressed in terms of the Lagrangian as

$$\boldsymbol{P} = \sum_\alpha \frac{\partial L}{\partial \dot{\mathbf{x}}_\alpha}, \qquad \boldsymbol{J} = \sum_\alpha \mathbf{x}_\alpha \times \frac{\partial L}{\partial \dot{\mathbf{x}}_\alpha}.$$

The mass-center $\boldsymbol{N}$ follows from the Lagrangian as

$$\boldsymbol{N} = \sum_\alpha \frac{\partial L}{\partial \dot{\mathbf{x}}_\alpha} t - m\mathbf{X}.$$

The kinetic energy is a homogeneous function of the velocities. It can therefore be expressed in terms of derivatives of itself with respect to velocity as

$$T = \tfrac{1}{2} \sum_\alpha m_\alpha \dot{x}_\alpha^2 = \tfrac{1}{2} \sum_\alpha \frac{\partial T}{\partial \dot{\mathbf{x}}_\alpha} \cdot \dot{\mathbf{x}}_\alpha.$$

Since $L = T - V$ and V is independent of $\dot{\mathbf{x}}_\alpha$, it follows that $\partial T / \partial \dot{\mathbf{x}}_\alpha = \partial L / \partial \dot{\mathbf{x}}_\alpha$ and the kinetic energy is given by the Lagrangian as

$$T = \tfrac{1}{2} \sum_\alpha \frac{\partial L}{\partial \dot{\mathbf{x}}_\alpha} \cdot \dot{\mathbf{x}}_\alpha.$$

The total energy may be written in T and L rather than T and V as $H = 2T - L$. With the kinetic energy also expressed in terms of the Lagrangian according to the above forms, the total energy is obtained from the Lagrangian as

$$H = \sum_\alpha (\partial L / \partial \dot{\mathbf{x}}_\alpha) \cdot \dot{\mathbf{x}}_\alpha - L$$

or

$$H = \sum_\alpha \boldsymbol{p}_\alpha \cdot \dot{\mathbf{x}}_\alpha - L.$$

The mechanical quantities $\boldsymbol{P}$, $\boldsymbol{N}$, $\boldsymbol{J}$, and H all issue from the Lagrangian. These quantities describe the ensemble as a whole and are ten in number. These ten global quantities are united with the ten symmetry parameters of Eqs. (3.2) by the action principle.

The Action Principle

The action is the grand quantity in the orchestration of motion; and it too is built from the Lagrangian. The action embraces all the particles of the motion in an integral of the Lagrangian over the time path of the ensemble,

$$S = \int_{t_1}^{t_2} L\,dt. \tag{3.8}$$

The Lagrangian—the gene of motion—plays the central role in the action and the resulting law of motion. The action is its integral and the principle of least action is simply this: of all the possible paths the particles may take between any two points in space and time, they take those paths for which the action S has the least possible value. All other equations of motion including Newton's celebrated law follow from this simple statement.

One might well ask why, of the various mechanical quantities, nature chooses the difference between the kinetic and potential energies as the Lagrangian? The best answer one can give to this question in Galilean space–time is that in the continual interchange of kinetic and potential energy, nature seeks to minimize any deviation between the two. However, in the relativistic space–time of Einstein, least action and the trajectories to which it leads have a deep and beautiful geometric significance. We shall see the geometric significance of the action principle in Chap. 6.

The action involves all the particles of the ensemble. It is useful therefore to make a natural extension in notation that describes the ensemble as a whole. The configuration of all the particles may be given by specifying the rectangular vector components $\mathbf{x}_\alpha$. But one may also use other systems of coordinates such as spherical-polar and other coordinate systems involving angles and lengths. The essential idea is that one must provide $s = 3n$ configuration coordinates to fix the positions of n particles. The ensemble is said to have s degrees of freedom. The generalized coordinates which fix the configuration of the particles therefore form an s-dimensional vector,*

$$q = (q_1, q_2, \ldots, q_s).$$

In the case in which the generalized coordinates are indeed rectangular, the components of q are just the components of the $\mathbf{x}_\alpha$ with the particles taken in a fixed order.

The velocity vector corresponding to the position vector is

$$\dot{q} = (\dot{q}_1, \dot{q}_2, \ldots, \dot{q}_s).$$

The Lagrangian $L = L(\mathbf{x}_\alpha, \dot{\mathbf{x}}_\alpha)$ becomes $L = L(q, \dot{q})$ and the generalized momenta are

$$p = \partial L/\partial \dot{q} = (p_1, p_2, \ldots, p_s).$$

---

* Generalized coordinate vectors are written in latin script with boldface reserved for three-vectors in Euclidean space.

One now regards all mechanical quantities as having s components arranged in this fashion. These vectors also have a subscript notation such as q_i, p_i, and follow the Einstein summation convention in the formation of scalar products where the sums now run over $i = 1, 2, \ldots, s$.

The advantage of generalized coordinates is that most expressions describing the ensemble have the same form as those for a single particle. A typical quantity like the total energy $H = \sum_\alpha \boldsymbol{p}_\alpha \cdot \dot{\mathbf{x}}_\alpha - L(\mathbf{x}_\alpha, \dot{\mathbf{x}}_\alpha)$ may be written simply

$$H(q, p) = p_i \dot{q}_i - L(q, \dot{q}).$$

(Note, however, the exception of the vector product which is only defined for three-vectors, $\boldsymbol{J} = \sum_\alpha \mathbf{x}_\alpha \times \boldsymbol{p}_\alpha$.)

Least action can be illustrated with the familiar image of a valley. At the minimum point of any function "at the bottom of a valley" the slope—the first derivative—vanishes. For a function of several variables, all the first partial derivatives vanish at a minimum point.

Another way of stating the condition of vanishing slope is that the *variation* of a function must vanish at a minimum point. The variation of a function $f(q)$ is given by

$$\delta f = \frac{\partial f}{\partial q_i} \delta q_i.$$

The variables q_i upon which f depends are said to be its *arguments*. The δq_i are the variations of the arguments. The variation of a function consists of a sum of the variations of its arguments with the first partial derivatives appearing as the coefficients of the argument variations. The variation of a function shows how small changes in the arguments feed into changes of the function. If a partial derivative with respect to a given argument q_i vanishes, the function suffers no variation from that argument no matter what the argument variation. Since the variations of the arguments are arbitrary, a vanishing variation δq necessarily requires that every partial derivative $\partial f / \partial q_i$ vanish at a minimum point. Most points of a function are not maximum or minimum points. The maximum or minimum points of a function can only occur at those special points where the first partial derivatives vanish.

Least action requires a vanishing variation. But in the case of the action, one is not looking for a single point where the action variation vanishes; one is looking for the law governing whole functions $q(t)$ for which the action variation vanishes. It is the embrace of whole functions in least action that gives the principle its power and sweep. A law of motion that simply required a function to have a maximum or minimum at some point in space and time would not be a universal law of motion. The action is therefore said to be a *functional*, a quantity that has a single value corresponding to an entire function. It is customary to symbolize the variation of a functional such as S as δS. Whereas simple maximization or minimization produces points, action minimization produces a rule governing entire sets of functions $q(t)$, the trajectories of the motion. This rule is the law of motion.

The determination of the general variation of the action S with Lagrangian L and the law that the functions $q(t)$ must satisfy for a vanishing action variation are the central achievements of the variational calculus whose development in mathematics was spurred by mechanics. As the example of a minimum point at the bottom of a valley illustrates, the variation of the action will be given by a sum of the variations of its arguments with the first partial derivatives of the action appearing as the coefficients. We have seen that nature fixes a state of motion by the positions and velocities of the particles in the Lagrangian $L = L(q, \dot{q})$. There are therefore two sets of arguments, the q and the $\dot{q}$, and two sets of derivative coefficients, the Lagrangian derivatives $\partial L/\partial q$ and $\partial L/\partial \dot{q}$. The action variation that results when the paths are given variations δq while still passing through fixed end points at $(q,t)_1$ and $(q,t)_2$ is

$$\delta S = \int_{t_1}^{t_2} \delta L \, dt = \int_{t_1}^{t_2} \left(\frac{\partial L}{\partial q_i} \delta q_i + \frac{\partial L}{\partial \dot{q}_i} \delta \dot{q}_i \right) dt. \tag{3.9}$$

A problem now arises. There are two different variations, $\delta\dot{q}$ and δq, so the two Lagrangian derivative terms are not on the same footing. To proceed further one must have them on the same footing. This may be done by transforming the term in $\delta\dot{q}$ into one in δq through integration by parts,

$$\begin{aligned} \frac{\partial L}{\partial \dot{q}_i} \delta \dot{q}_i &= \frac{\partial L}{\partial \dot{q}_i} \frac{d}{dt} \delta q_i \\ &= \frac{d}{dt} \left(\frac{\partial L}{\partial \dot{q}_i} \delta q_i \right) - \frac{d}{dt} \left(\frac{\partial L}{\partial \dot{q}_i} \right) \delta q_i. \end{aligned}$$

The first term on the second line, $d/dt[(\partial L/\partial \dot{q}_i)\delta q_i]$, may be immediately integrated and expressed in terms of its values at the end points $(\delta q, t)_1$ and $(\delta q, t)_2$. Since the path variations δq vanish at the end points, this term vanishes. One therefore finds an important equivalence for path variations that vanish at the endpoints:

$$\int_{t_1}^{t_2} \frac{\partial L}{\partial \dot{q}_i} \delta \dot{q}_i \, dt = - \int_{t_1}^{t_2} \frac{d}{dt} \frac{\partial L}{\partial \dot{q}_i} \delta q_i \, dt.$$

The action variation due to the path variations thus reduces to

$$\delta S = - \int_{t_1}^{t_2} \left(\frac{d}{dt} \frac{\partial L}{\partial \dot{q}_i} - \frac{\partial L}{\partial q_i} \right) \delta q_i \, dt. \tag{3.10}$$

Equation (3.10) contains the contribution to the action variation resulting from path variations. But particle paths are not the only arguments that generate variations of the action. The action and the Lagrangian also embody the underlying symmetries of motion. They must be invariant to origin shifts and translational and rotational orientation shifts of space and time. These variations are described by the symmetry parameters in the transformations (3.2).

The argument variations of the action thus spring from two sources, the path of motion described by the trajectories $q(t)$ in a given reference frame and the variations of the origin and the translational and rotational orientation of the reference frame itself. According to the transformation law (3.2), the variations of the origin and the translational and rotational orientation of the reference frame are described by the variations of the symmetry parameters $\mathbf{a}$, $\boldsymbol{u}$, $\boldsymbol{\Omega}$, and τ.

The total action variation δS therefore results from the path variations δq [that portion which Eq. (3.10) describes] and the variations of the ten parameters of the symmetry group, $\delta\mathbf{a}$, $\delta\boldsymbol{u}$, $\delta\boldsymbol{\Omega}$, and $\delta\tau$. Their coefficients then each vanish for a condition of least action. These coefficients, which are also known as the functional derivatives of the action with respect to the symmetry parameters, are developed in **Note 1** where the variational principle is discussed in detail. They turn out to be the ten fundamental mechanical quantities $\boldsymbol{P}$, $\boldsymbol{N}$, $\boldsymbol{J}$, and H.

The resulting action variation contains that given by path variations in Eq. (3.10) which we shall label δS_{path} as well as the contributions from the variations of the symmetry group parameters which we label $\delta S_{\text{symmetries}}$:

$$\delta S = \delta S_{\text{path}} + \delta S_{\text{symmetries}}.$$

These two components of the action variation are

$$\begin{aligned} \delta S_{\text{path}} &= -\int_{t_1}^{t_2} \left(\frac{d}{dt}\frac{\partial L}{\partial \dot{q}_i} - \frac{\partial L}{\partial q_i} \right) \delta q_i \, dt, \\ \delta S_{\text{symmetries}} &= -\Delta\boldsymbol{P}\cdot\delta\mathbf{a} - \Delta\boldsymbol{N}\cdot\delta\boldsymbol{u} - \Delta\boldsymbol{J}\cdot\delta\boldsymbol{\Omega} - \Delta H\delta\tau. \end{aligned} \tag{3.11}$$

In the second of Eqs. (3.11) Δ indicates the difference of values between end points of the path of any mechanical quantity before which it stands. For example, $\Delta\boldsymbol{P} \equiv \boldsymbol{P}_2 - \boldsymbol{P}_1$.

Least action leads to two kinds of conditions: *equations of motion* corresponding to the variations of the path proportional to δq in the first of Eqs. (3.11) and *conservation laws* corresponding to the symmetry parameter variations $\delta\mathbf{a}$, $\delta\boldsymbol{u}$, $\delta\boldsymbol{\Omega}$, $\delta\tau$ which appear in the second.

The principle of least action requires that δS vanish. A vanishing action variation with respect to the variation of the trajectories along the path δq in the first of Eqs. (3.11) leads to the *equations of motion* of the particles,

$$\frac{d}{dt}\frac{\partial L}{\partial \dot{q}_i} - \frac{\partial L}{\partial q_i} = 0. \tag{3.12}$$

These are the *Euler-Lagrange equations* of the variational calculus, partial differential equations which functions that minimize a functional integral must satisfy. For the Lagrangian of motion they are none other than Newton's second law of motion; for since the momentum is $p_i = \partial L/\partial \dot{q}_i$ and the force is $f_i = \partial L/\partial q_i$, Eq. (3.12) is actually

$$\frac{dp_i}{dt} = f_i. \tag{3.13}$$

The action principle for the paths of motion can be thought of as a principle that establishes a particular relationship between the Lagrangian partial derivatives along this path. The derivatives appearing in the total differential of the Lagrangian $L = L(q, \dot{q})$,

$$dL = \frac{\partial L}{\partial q_i} dq_i + \frac{\partial L}{\partial \dot{q}_i} d\dot{q}_i,$$

in the action principle are not independent. Their relationship to one another is fixed by the equations of motion

$$p_i = \partial L/\partial \dot{q}_i, \qquad \dot{p}_i = \partial L/\partial q_i.$$

The differential of the Lagrangian which satisfies the action principle is therefore completely expressible in terms of p and $\dot{p}$,

$$dL = \dot{p}_i dq_i + p_i d\dot{q}_i. \tag{3.14}$$

A vanishing action variation with respect to the Galilean symmetry parameter variations in the second of Eqs. (3.11) generates conservation laws, one for each symmetry parameter variation:

$$\Delta \boldsymbol{P} = 0, \qquad \Delta \boldsymbol{N} = 0, \qquad \Delta \boldsymbol{J} = 0, \qquad \Delta H = 0.$$

These laws are called conservation laws because the quantity governed by each conservation law is the same at both the beginning and end of the path. Since the path end points are arbitrary, the quantity is conserved during the course of motion between any two points. Such a quantity is an *invariant.* It never changes as the particles sweep along their paths.

It is interesting to observe that the variation of the action wrought by the symmetries consists of products of the symmetry parameter variations and their corresponding invariants. The product of a symmetry parameter and its invariant is an "action." Corresponding to the variations of the origin $\delta \mathbf{a}$ is the total linear momentum, the invariant $\boldsymbol{P} = \sum_\alpha \mathbf{p}_\alpha$. Similarly, the translational variations of velocity space $\delta \mathbf{u}$ produce the mass-center invariant $\boldsymbol{N} = \boldsymbol{P}t - m\boldsymbol{X}$. Space rotations $\delta \boldsymbol{\Omega}$ yield the total angular momentum, the invariant $\boldsymbol{J} = \sum_\alpha \mathbf{x}_\alpha \times \mathbf{p}_\alpha$. Time translations $\delta \tau$ lead to the invariant $H = p_i \dot{q}_i - L$, the total energy.

Pouring out of the condition of least action come Newton's law of motion and the ten invariants of total energy, linear momentum, mass-center, and angular momentum. Once established by initial conditions the total linear momentum, mass-center, angular momentum, and energy never change. As a flurry of other mechanical quantities dance in space and time, the invariants remain forever fixed. These ten invariants are always known for any mechanical system no matter how many particles compose it or how complex the force laws.

Conservation laws are a manifestation of the symmetries of space and time which the action and the Lagrangian possess. When these symmetries are subjected to the principle of least action, a corresponding set of invariants emerge, each invariant linked to an underlying symmetry. Corresponding to each parameter carrying a given symmetry is a conservation law yielding the invariant appropriate to that symmetry. This profound connection between symmetries and invariants which is forged in the principle of least action was first clearly portrayed by Amalie Emmy Noether* in the first part of this century (**Note 1**).

The emergence of invariants out of symmetry is one of the generous revelations of the physical world, a high moment in physics, and a mathematical discovery of rare beauty in its own right. The discovery of Emmy Noether shows us that nature manifests the symmetries of the world through invariants. In so doing it provides us the most powerful insight ever to come into physics. The story of the physical world has largely been woven about this many-splendored theme.

The Hamiltonian Heritage

The Lagrangian $L = T - V$ is the centerpiece of the action principle; but it is not an invariant. However, it has a near image, the total energy $H = T + V$, which is the leading invariant of mechanics and to which it is related by

$$H = p_i \dot{q}_i - L.$$

It is possible to formulate the action principle with the energy rather than the Lagrangian. When one does so, the law of motion takes a striking form first exhibited by William Rowan Hamilton. The total energy as the centerpiece of the action principle has come to be known as the *Hamiltonian.*

The action principle may be formulated with the Hamiltonian by expressing $S = \int L\,dt$ in terms of H rather than L:

$$\delta S = \delta \int_{t_1}^{t_2} (p_i \dot{q}_i - H)\, dt = 0.$$

* See "Emily Noether" by C. H. Kimberling in *Am. Math. Monthly* **79**, 136 (1972).

The equations of motion in the Hamiltonian formulation of the action principle may be found by eliminating the Lagrangian in favor of the Hamiltonian. This is most simply done by exploiting the fact that the equations of motion are directly linked to the partial derivatives of the Lagrangian as in Eq. (3.14). The same is true for the Hamiltonian. The equations of motion in Hamiltonian format are directly linked to the partial derivatives of the Hamiltonian. To find them, take the differential of the Hamiltonian–Lagrangian relationship,

$$dH = p_i d\dot{q}_i + \dot{q}_i dp_i - dL.$$

The differential of the Lagrangian may be eliminated using Eq. (3.14) showing that the differential of the Hamiltonian has the austere, antisymmetric form

$$dH = \dot{q}_i dp_i - \dot{p}_i dq_i.$$

The equations of motion may therefore be cast in the elegant form of the Hamilton equations:

$$\dot{q}_i = \partial H/\partial p_i, \qquad \dot{p}_i = -\partial H/\partial q_i. \tag{3.15}$$

If the Lagrangian and the Hamiltonian are the genes of motion, the Hamilton equations reflect their double-stranded structure. The coordinates are linked in pairs; and this pairing is an intertwined, antisymmetric, double strand of derivatives of the Hamiltonian with respect to positions and momenta.

Antisymmetric pairing of coordinates is a reflection of deep structure in mechanics induced by the action principle called *symplectic structure* after πλεκτός, the Greek word for *twined* or *braided.* Symplectic structure describes the unique way in which the action principle weaves the motion of the world as a many-plaited twining of positions and momenta.

A state of motion is fixed by a twofold set of coordinates. In the Lagrangian formulation of the action principle these coordinates are positions and velocities. One sees that in the Hamiltonian formulation these coordinates are positions and momenta. The Lagrangian and the Hamiltonian are equally valid carriers of the genes of motion, the Lagrangian $L = L(q, \dot{q})$ coded in a position-velocity alphabet and the Hamiltonian $H = H(q,p)$ in a position-momentum alphabet, with the translation from one to the other effected by $p_i = \partial L/\partial \dot{q}_i$ and $\dot{q}_i = \partial H/\partial p_i$. The paired set of coordinates satisfying Hamilton's equations are called *canonical coordinates* and Hamilton's equations are called the *canonical equations of motion.*

It is important to observe that the Hamilton equations are not themselves the trajectories of the particles. At the center of the law of motion one does not find trajectories but rather the Hamiltonian genes and the developmental machinery of the Hamilton equations. The Hamiltonian contains not trajectories but a mathematical code of particle positions and momenta from which trajectories may be constructed. The code must be read into the appropriate instructions from which the trajectories may be built. This is the function of the Hamilton equations. They read the Hamiltonian into a set of differential equations which conform to the law of motion. But one will not have the trajectories themselves until these differential equations are integrated. Integration is the developmental process of motion. The integrals of the Hamilton equations are the final expression of the trajectories encoded in the Hamiltonian—the fully formed motion in space and time.

The only general method available for constructing the trajectories from the Hamilton equations is that of numerical integration, a method which gives up expression in mathematical functions for expression in streams of data. This method is of practical importance; and it also shows explicitly how states of motion evolve into new states under the action of the Hamilton equations.

One numerically constructs the integrals of the Hamilton equations over a sequence of small steps in time Δt by approximating the time derivatives as finite differences, for example as

$$\dot{q} \approx \frac{q(t+\Delta t) - q(t)}{\Delta t}$$

with a similar expression for $\dot{p}(t)$. The Hamiltonian $H(q,p)$ and its derivatives may be evaluated with the state $q(t), p(t)$. Hamilton's equations may then be used to build the trajectories of q and p in incremental fashion,

$$q_i(t+\Delta t) = q_i(t) + \frac{\partial H}{\partial p_i}\Delta t, \qquad p_i(t+\Delta t) = p_i(t) - \frac{\partial H}{\partial q_i}\Delta t.$$

An initial state sets the ensemble into motion. Each state $q(t), p(t)$ when acted upon by the Hamilton equations creates a succeeding state $q(t+\Delta t)$, $p(t+\Delta t)$. Each succeeding state serves as the initial state for the next state. States follow states as Hamilton's equations develop, step by step, the entire trajectories enfolded in the genetic code of the Hamiltonian.

Notice that the action itself makes no appearance in the Hamilton equations; the trajectories may be found without ever dealing directly with it. The coordinates (q,p) and the Hamiltonian $H(q,p)$, not the action, are the players in the Hamilton equations. The action, however, is very much present behind the scene. And as we further probe the nature of motion we shall find the action making a dramatic reappearance on center stage.

Although any mechanical quantity $f(q,p,t)$ may be obtained from the canonical coordinates, its time rate of change along the path of motion is of interest in its own right. The time derivative along the path of motion or, as it is sometimes called, the *total time derivative* of any mechanical quantity is given by

$$\frac{df}{dt} = \frac{\partial f}{\partial t} + \left(\frac{\partial f}{\partial q_i}\dot{q}_i + \frac{\partial f}{\partial p_i}\dot{p}_i\right).$$

The rate of change along the path of motion first recognizes that a quantity may be an explicit function of time t and hence it possesses a time-rate of change described by $\partial/\partial t$. But more pointedly, the total rate of change incorporates the change in a quantity brought about by its being carried along the path of motion as a function of q and p. All mechanical quantities—even those not explicit functions of time—possess this rate of change.

Using Hamilton's equations (3.15) to eliminate $\dot{q}$ and $\dot{p}$, the total rate of change may be compactly expressed as

$$\frac{df}{dt} = \frac{\partial f}{\partial t} + [f, H], \tag{3.16}$$

where $[f, g]$ is the Poisson bracket of any two functions of the motion:

$$[f, g] \equiv \left(\frac{\partial f}{\partial q_i} \frac{\partial g}{\partial p_i} - \frac{\partial g}{\partial q_i} \frac{\partial f}{\partial p_i} \right). \tag{3.17}$$

The Poisson bracket is built from a pair of derivatives with respect to the position-momentum pair; and this pairing is again an antisymmetric manifestation of symplectic structure. The Poisson bracket of two functions f and g takes the character of the product fg of these functions. If f and g are both scalars, their Poisson bracket is a scalar. If only one of them is a vector, their bracket is a vector. If both are vectors, their bracket is the product of two vectors known as a second order tensor (a vector being a first order tensor).

The Poisson bracket is a mathematical form of prodigious power; and its presence is ubiquitous throughout mechanics. This is because the complex dynamics of the world are ultimately woven upon two sets of coordinates. The Poisson bracket is a composition of two functions. It builds up complex dynamical quantities from a more primitive pair in accord with the law of motion. We shall find that virtually the full content of the law of motion is contained within the Poisson bracket. As the first of many illustrations, the Hamilton equations of motion themselves acquire a simple and elegant expression in Poisson brackets as

$$\dot{q}_i = [q_i, H], \qquad \dot{p}_i = [p_i, H].$$

The most elementary Poisson brackets are those of the position and momentum vectors themselves. They are found to vanish for each vector separately. But the Poisson bracket of the fundamental pair of mechanical variables *with one another* is unity:

$$[q_i, q_j] = 0, \qquad [q_i, p_j] = \delta_{ij}, \qquad [p_i, p_j] = 0,$$

where $\delta_{ij} = 0$ for $i \neq j$ and $\delta_{ij} = 1$ for $i = j$.

While the Poisson bracket of different components of linear momentum always vanishes, the Poisson brackets of different components of angular momentum do not. Rather, angular momentum cyclically reproduces itself under the action of the Poisson bracket. The Poisson brackets of the angular momentum of a single particle are readily calculated to be

$$[J_1, J_2] = J_3, \qquad [J_2, J_3] = J_1, \qquad [J_3, J_1] = J_2,$$

and the Poisson bracket of any two identical components vanishes. The total angular momentum of the ensemble also has the same Poisson bracket behavior. This result may be naturally expressed through use of the alternating tensor ϵ_{ijk} as

$$[J_i, J_j] = \epsilon_{ijk} J_k. \tag{3.18}$$

Equation (3.18) also shows that the angular momentum is perpendicular to its own Poisson bracket, $\boldsymbol{J} \cdot [\boldsymbol{J}, J_j] = J_i[J_i, J_j] = \epsilon_{ijk} J_i J_k \equiv 0$. As a result, the magnitude of the angular momentum and any one of its components has a vanishing Poisson bracket,

$$[J^2, J_j] = 0. \tag{3.19}$$

(Since the Poisson bracket is based on derivatives, it obeys the product rule: $[J^2, J_j] = [J_i J_i, J_j] = 2J_i[J_i, J_j]$.)

The Poisson bracket of functions of the motion with the angular momentum is the most intriguing of the operations induced by Poisson brackets. A direct calculation shows the rectangular position and momentum coordinates, like the angular momentum itself, are cyclically reproduced by Poisson brackets with the angular momentum,

$$[x_i, J_j] = \epsilon_{ijk} x_k, \qquad [p_i, J_j] = \epsilon_{ijk} p_k. \tag{3.20}$$

As a result, the scalar magnitudes x^2, p^2, and $\mathbf{x} \cdot \boldsymbol{p}$ all have vanishing Poisson brackets with the angular momentum according to the same argument as that leading from Eq. (3.18) to Eq. (3.19). An important property of the angular momentum Poisson bracket is therefore apparent: it vanishes for all scalar functions. This follows from the fact that a scalar function of two vectors such as $\mathbf{x}$ and $\boldsymbol{p}$ can only depend upon scalars formed from the vectors in the combinations x^2, p^2, and $\mathbf{x} \cdot \boldsymbol{p}$.

The results (3.19) and (3.20) may be used to reveal an even more remarkable property of the angular momentum. Angular momentum cyclically reproduces not only itself but any vector function $\boldsymbol{F} = \boldsymbol{F}(\mathbf{x}, \boldsymbol{p})$ of the positions and momenta of the motion:

$$[F_i, J_j] = \epsilon_{ijk} F_k.$$

The Poisson bracket of any scalar with the angular momentum is instantly known: it vanishes. The Poisson bracket of any vector function of motion with the angular momentum is also instantly known, no matter how complex the vector: it is the cyclic-reproduction of itself.

This magical property arises because a vector function $\boldsymbol{F}(\mathbf{x},\mathbf{p})$ can only be formed from the vectors $\mathbf{x}$, $\mathbf{p}$, and $\boldsymbol{J} = \mathbf{x} \times \mathbf{p}$ in the form

$$\boldsymbol{F}(\mathbf{x},\mathbf{p}) = a\mathbf{x} + b\mathbf{p} + c\mathbf{x} \times \mathbf{p},$$

where a, b, and c are scalar functions of $\mathbf{x}$ and $\mathbf{p}$. The Poisson brackets of the scalar functions with the angular momentum vanish while $\mathbf{x}$, $\mathbf{p}$, and $\boldsymbol{J} = \mathbf{x} \times \mathbf{p}$ cyclically reproduce themselves—and the vector function $\boldsymbol{F}(\mathbf{x},\mathbf{p})$.

The cyclic Poisson bracket, illustrated here by the angular momentum, is a deep and recurring presence in mechanics. Cyclic structure is a manifestation of rotational symmetry. Since angular momentum is the invariant arising from the rotational symmetry of space, it is not surprising that the angular momentum possesses a cyclic Poisson bracket. But other mechanical quantities will be found that also possess cyclic Poisson brackets with the structure of Eqs. (3.18) and (3.19) revealing that rotations are far-reaching symmetries of the physical world embracing more than the angular momentum. Indeed, we shall find that all the symmetries of two-body motion with inverse-square force are rotational.

The Mathematical World of Motion

One naturally thinks of motion in the configuration space of the particles. But nature tells us that a state of motion is described by not one but a twofold set of coordinates. The manifold of momentum coordinates is the other half of the complete space of motion. For a single particle, the space of motion is a six-dimensional manifold consisting of its three position coordinates and its three momentum coordinates. For a constellation of n particles, the state of motion is specified by a point in a $2s$-dimensional manifold whose coordinates are the s positions and the s momenta of the particles where $s = 3n$. The space of motion is therefore always an even-dimensional manifold and has come to be known as the *phase space.* Phase space is the mathematical world in which the motion of the world is imaged.

Motion in configuration space is a projection of the more comprehensive tableau acted out in phase space. Motion fills phase space with phase trajectories $q(t)$, $p(t)$. The phase trajectories when projected onto configuration space yield particle trajectories. (One thinks of shining a light on a string representing a phase space trajectory looping through space and projecting its shadow on a plane.) This leads to an interesting way in which the structure of a trajectory in configuration space is created. For example, a phase trajectory which proceeds in a single direction may project onto configuration space with reversals as shown in Fig. 3-4 *(a).* Phase trajectories free of intersections may project into configuration space with intersections *(b).* Other more complex folds of the phase trajectory project

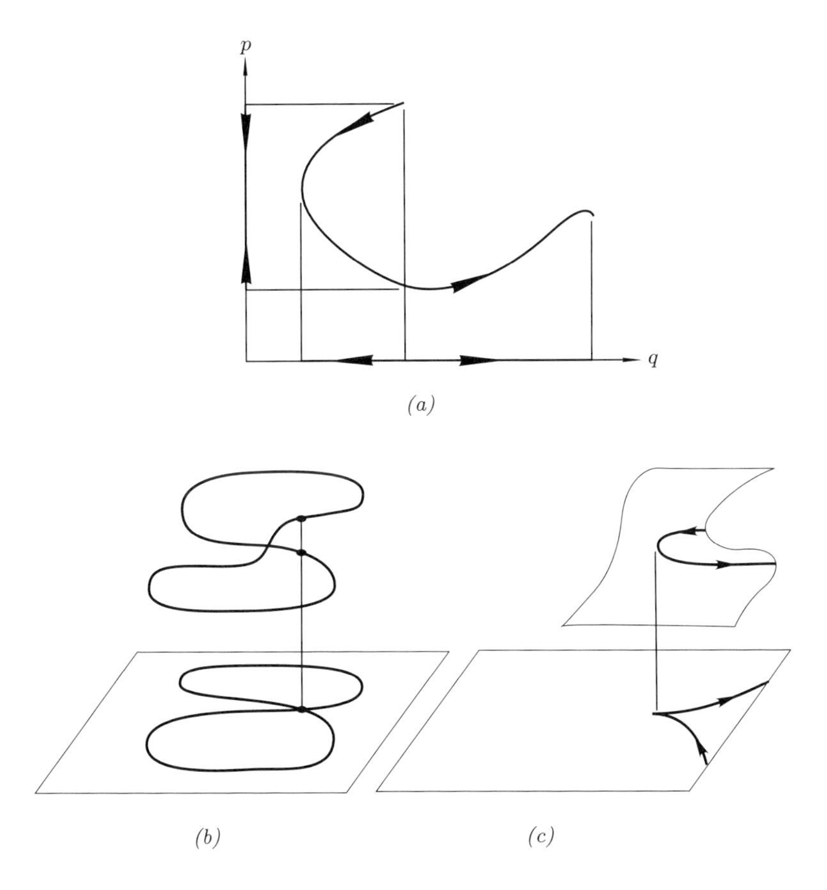

Figure 3-4. Trajectories in Configuration Space as Projections from Phase Space. A phase trajectory without reversals may possess reversals when projected into configuration space *(a)*. A smooth, non-intersecting trajectory in phase space may project with intersections *(b)* and singularities such as cusps *(c)*.

onto configuration space as trajectories with singular points such as cusps as shown in Fig. 3-4 *(c)*.

Geometrical figures in phase space are not just transformed images of the positions of particles in configuration space. Since half the coordinates of phase space are momenta, the phase trajectories are composed of both positions and momenta (or any appropriate conjugate pair of canonical coordinates). Curves, angles, surfaces, and volumes are built from the point sets of both position and momentum. The geometrical objects in phase space are truly dynamical rather than simply configurational. They embody momentum in their structure as well as the shapes created by the particle positions.

The essence of configuration space is its Euclidean structure. The essence of phase space is its symplectic structure. (The Euclidean and symplectic properties of a space are independent.) Phase space is inherently symplectic in a manner we shall define more precisely, but it is rarely Euclidean. Although each position and momentum vector creates a Euclidean subspace with its three components, the metric relationships between the position and momentum components of a single particle as well as those between the positions or momenta of different particles are not in general Euclidean. The only generic structure of phase space is its symplectic structure.

It is useful to distinguish the two sets of canonical coordinates q and p in separate vectors for many purposes. But the unique properties of phase space become apparent when one makes the final leap to a single $2s$-vector, the state vector of phase space,

$$\xi = (q, p).$$

The first s components of this vector are the position coordinates of all the particles and the following s components are their momentum components taken in the same order. A state of motion of the complete constellation of particles is specified by the phase vector ξ. Corresponding to the phase vector is the phase velocity,

$$\dot{\xi} = (\dot{q}, \dot{p}).$$

When one describes motion with the phase vector ξ, the symplectic structure is carried by a $2s \times 2s$ antisymmetric matrix known as the *symplectic*,

$$J = \begin{pmatrix} 0 & I \\ -I & 0 \end{pmatrix},$$

with I the identity or unit matrix (of dimension s in the above expression). The matrix J is the signature of symplectic space with the notable properties

$$J = -J^{\dagger} = -J^{-1},$$

where $J^{\dagger}$ is the transpose of J obtained by interchanging its rows and columns. The symplectic matrix also has a square which is the negative $2s \times 2s$ unit matrix,

$$J^2 = -I = -\begin{pmatrix} I & 0 \\ 0 & I \end{pmatrix}.$$

The symplectic J has the defining property of inducing a vanishing scalar product on any phase space vector*:

$$\xi J \xi = 0. \tag{3.21}$$

* It is customary to omit the subscripts in simple products of vectors and matrices. With Einstein's notation explicitly written out, Eq. (3.21) reads $\xi_i J_{ij}\, \xi_j = 0$.

A symplectic space is a space of vectors that satisfy Eq. (3.21). This condition parallels the condition for Euclidean space in which the identity I plays the role of the symplectic J:

$$\mathbf{x} I \mathbf{x} = x^2.$$

In both cases, a quadratic form (a form involving sums of second order products of the coordinate vector) is an invariant of the space.

The Poisson bracket acquires a natural expression in symplectic space. It is the symplectic product of the derivatives of functions $f(\xi)$ and $g(\xi)$,

$$[f, g] = \frac{\partial f}{\partial \xi} J \frac{\partial g}{\partial \xi}.$$

Because the fundamental property of symplectic space is $\eta J \eta = 0$ for any vector η, one immediately sees that the Poisson bracket of any function with itself vanishes since such a form is just the defining form of symplectic space.

This general form shows that the Poisson bracket of the phase vector ξ itself with any function $f(\xi)$ is

$$[\xi, f] = J \frac{\partial f}{\partial \xi}.$$

The Hamilton equations involve precisely this form. The time derivatives of the coordinates that constitute the phase vector $\dot{\xi} = (\dot{q}, \dot{p})$ are the Poisson bracket of the phase vector with the Hamiltonian:

$$\dot{\xi} = [\xi, H] = J \frac{\partial H}{\partial \xi}. \tag{3.22}$$

The Poisson Bracket: Motion as Flow

The classical motion of the world is mathematically captured in particle-points. It is interesting that the motion of these points creates a mathematical image in phase space that is the flow of a fluid. The motion of particles sets the manifold of points in phase space flowing. This is illustrated by the Hamilton equations for which all points $q(t)$, $p(t)$ are swept to new locations $q(t + \Delta t)$, $p(t + \Delta t)$ in the time interval Δt. The cumulative tracks of these motions are the streamlines of the particle trajectories.

The Poisson bracket plays the central role in the flow of mechanical quantities. If one traces a region through phase space in which a function $F(\xi)$ such as the energy, angular momentum, or any other is constant, one traces out a hypersurface, a submanifold one dimension less than the full

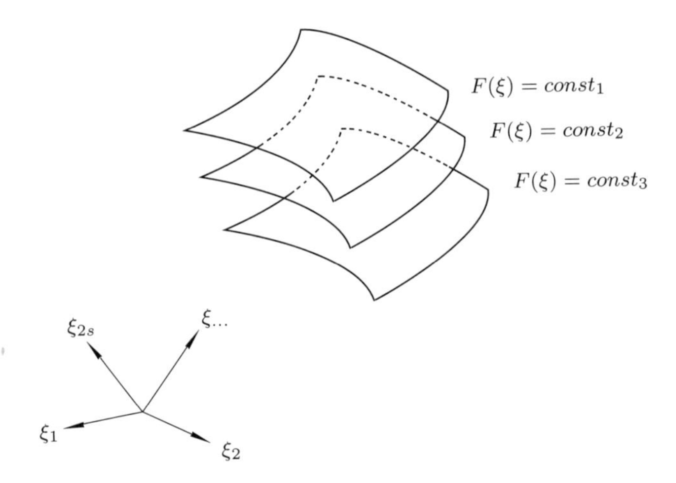

Figure 3-5. Mechanical Quantities as Level Surfaces in Phase Space. A mechanical quantity $F(\xi)$ exists in phase space as interleaved level surfaces one less dimension than the dimension of phase space.

manifold of phase space as shown in Fig. 3-5. Submanifolds upon which a function is constant are called the level surfaces of the function.

A level surface has at each point a vector normal to the surface given by $\partial F/\partial \xi$ as shown in Fig. 3-6. Such a surface also has vectors orthogonal to the normal vector that lie wholly within it. These are its tangent vectors. How are these tangent vectors determined? Remarkably, they are determined by the Poisson bracket. The Poisson bracket $[\xi, F]$ of the phase vector ξ itself with any function $F(\xi)$ turns out to be a vector tangent to the level surface of F as indicated in Fig. 3-6.

The Poisson bracket $[\xi, F]$ can be seen to lie in the level surface of F by projecting the normal vector $\partial F/\partial \xi$ on the vector $[\xi, F]$ (i.e., by taking the scalar product of the two vectors) and finding that it vanishes,

$$\frac{\partial F}{\partial \xi} \cdot [\xi, F] = \frac{\partial F}{\partial \xi} J \frac{\partial F}{\partial \xi} \equiv 0.$$

The Poisson bracket can be thought of as a machine that takes as inputs the phase vector ξ and the function $F(\xi)$ and produces as output the vector $[\xi, F]$ which is tangent to the level surface of F.

The fluid nature of the tangent vectors on the level surface can now be portrayed. At each point one has a local tangent vector $[\xi, F]$. One may think of this tangent vector as the "velocity" $d\xi/d\tau$ of a point of fluid lying on the surface,

$$\frac{d\xi}{d\tau} = [\xi, F], \tag{3.23}$$

where τ is a parameter analogous to time. Beginning at a point, one may trace out a curve $\xi(\tau)$ as the parameter τ varies which at every point has

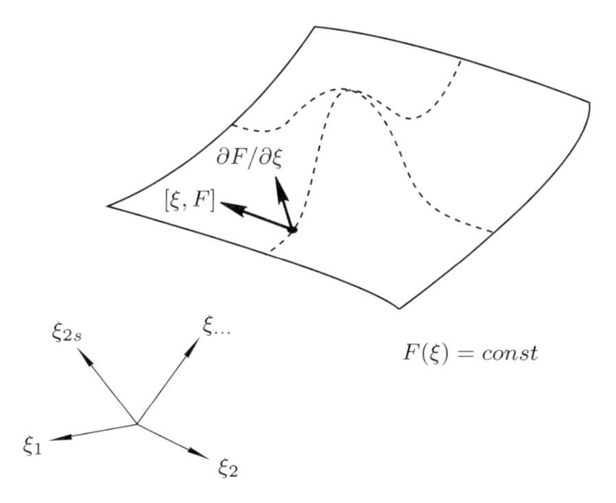

Figure 3-6. Normal, Poisson Bracket, and Tangent Vector. The normal vector to the level surface $F(\xi) = const$ is $\partial F/\partial \xi$. The Poisson Bracket of the phase vector ξ with the function $F(\xi)$ is a tangent vector $[\xi, F]$ of the level surface $F(\xi) = const$.

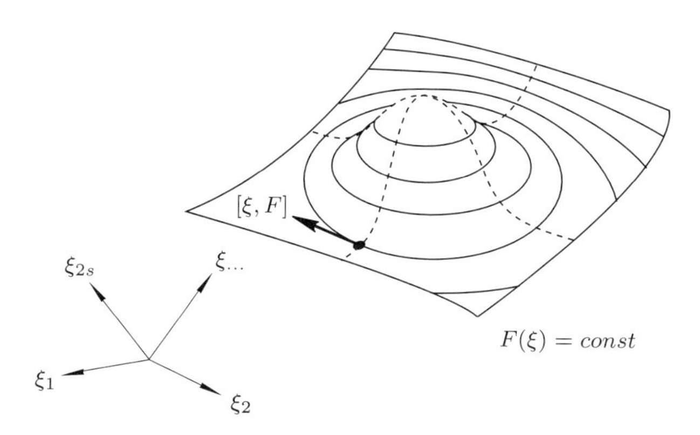

Figure 3-7. Flow Streamlines on the Level Surface of a Mechanical Quantity. The Poisson bracket $[\xi, F]$ generates the streamlines lying upon the level surfaces of the mechanical quantity $F(\xi)$.

a tangent equal to $[\xi, F]$. Such a curve is the solution to the differential equations (3.23) for a space curve lying in phase space. These curves are the streamlines of the flow as exhibited in Fig. 3-7. Since the "velocity" vectors $[\xi, F]$ are always tangent to the surface, the streamlines never escape the surface but lie wholly within it. The Poisson bracket thus shows that mechanical quantities $F(\xi)$ exist in phase space as interleaved sheets, each sheet covered with a streaming flow of points known as the *flow* of that mechanical quantity.

There is one mechanical quantity for which the streamline parameter τ

is indeed the time t and the Poisson bracket is indeed the velocity field $\dot{\xi}$ of the motion. That function is the Hamiltonian; and the Hamiltonian flow is the preeminent flow in phase space. Hamilton's equations induce a flow with a velocity equal to the Hamiltonian flow velocity (3.22) at each point, $\dot{\xi} = [\xi, H]$, and this flow is that of the trajectories of the particles in phase space.

Let us now find the significance of the Poisson bracket of any two quantities $F(\xi)$ and $G(\xi)$. Their Poisson bracket may be expressed as

$$[F, G] = \frac{\partial F}{\partial \xi} J \frac{\partial G}{\partial \xi} = \frac{\partial F}{\partial \xi} \cdot [\xi, G] = -\frac{\partial G}{\partial \xi} \cdot [\xi, F].$$

One now finds a more general interpretation of the Poisson bracket: the Poisson bracket of any two functions is the projection of the normals of the level surfaces of the one upon the tangents of the level surfaces of the other. Since the projections of the normals upon the tangents of the same surface vanish, one obtains a natural interpretation of the vanishing Poisson bracket of the same function.

The Poisson bracket shows how mechanical quantities intersect in phase space. If $[F, G] \neq 0$, then the flow of F does not stay on level surfaces of G but cuts across them. Thus, G is not constant on the flow of F. The same is true of the flow of G with respect to the level surfaces of F. But if $[F, G] = 0$, then the flow of F stays not only on its own level surface but also on the level surface of G. The level surfaces of these two quantities have become one.

A simple illustration shows schematically how this can occur. Suppose the quantities F and G each have one-dimensional level surfaces which are concentric circles. Each of the flows is a closed circular streamline on its respective circle. In general, these families of circles will intersect as shown in Fig. 3-8 *(a)* and the functions F and G will not be constant on each other's flows. But if $[F, G] = 0$, then these two sets of level surfaces are merged into the torus as shown in Fig. 3-8 *(b)* and both flows lie upon the merged surface of F and G.

A vanishing Poisson bracket signals a significant simplification in the flow. Although mechanical quantities generally fill phase space with level surfaces which entangle and intersect in complex ways, when the Poisson bracket of two quantities vanishes, the level surfaces of these two quantities become one common surface to which both flows are confined. A partial disentangling of the flow has occurred. These two mechanical quantities now possess a simple relationship. Their flows are bound to the same sheet, however convoluted the lay of that sheet in phase space and however complicated its relationship to the flows lying on the sheets of other mechanical quantities.

In summary, every mechanical quantity $F(\xi)$ has an image in phase space as sheets of level surfaces filled with streamlines generated by $[\xi, F]$ called its flow. The flow of a mechanical quantity necessarily lies upon its own

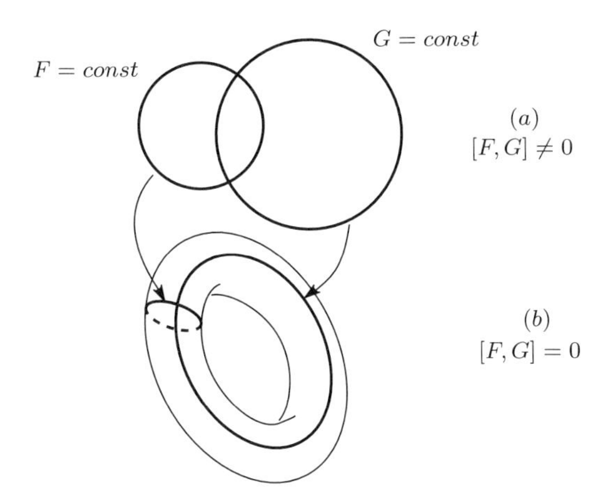

Figure 3-8. Two Flows with Vanishing Poisson Brackets Share a Common Level Surface. Each flow of the mechanical quantities F and G lies upon a level surface which is a circle. In general, $[F, G] \neq 0$ and the flows cut across one another in phase space *(a)*; but if $[F, G] = 0$, the flows no longer cross one another but arrange themselves upon a common level surface which is a torus *(b)*.

level surfaces; but it need not lie upon the level surfaces of another quantity. The Poisson bracket of any pair of mechanical quantities $[F, G]$ describes the intersection of their flows. If the Poisson bracket of any two quantities vanishes, their level surfaces are common and the mechanical quantities they represent are constant along one another's flows. Phase space has a correspondingly simple organization with respect to these quantities. A common surface contains both flows.

Poincaré Recurrence

Little is known about the general motion of many bodies in phase space; but one powerful theorem does exist and it has important implications: the phase flow is incompressible. An element of phase space containing a fixed number of points may be endlessly stretched, filamented, and dispersed by the flow. But it cannot be compressed or expanded. The volume of the same cluster of points, however distorted and smeared over phase space by the flow, is always the same as shown in Fig. 3-9.

The deformation of a fluid element is fully described by the deformation of the flow field in which it resides. An element of fluid must grow in volume if it flows through a region in which the streamlines are diverging as shown in Fig. 3-10. The divergence of the streamlines is described by gradients of the flow in the direction of the flow. For example, for the Hamiltonian flow in rectangular coordinates one has $\mathrm{div}(\dot{\xi}) = \partial\dot{\xi}_i/\partial\xi_i$, and it is this quantity

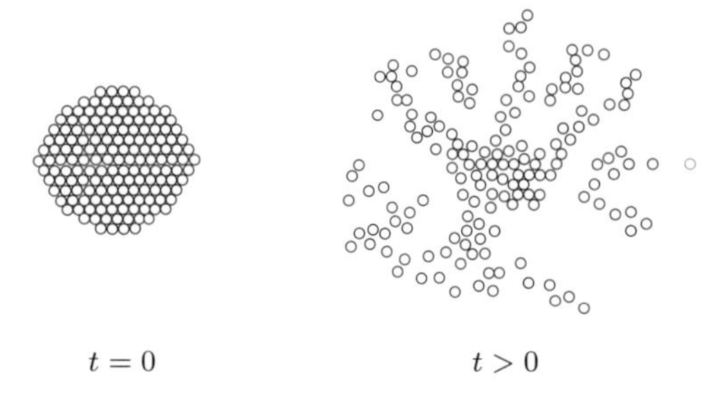

Figure 3-9. Volume Preservation in Phase Space. An element of phase space containing a fixed number of points at time $t = 0$ may be smeared throughout phase space by the flow for $t > 0$; but it cannot be compressed or expanded. The volume of the element, however dispersed, is always the same.

alone that controls the rate of volume change of a fluid element. If V is the volume of a fluid element, its rate of change is

$$\frac{1}{V}\frac{dV}{dt} = \operatorname{div}(\dot{\xi}) = \frac{\partial \dot{\xi}_i}{\partial \xi_i}.$$

Phase flows are incompressible ($dV/dt \equiv 0$) because their velocity fields are inherently divergence free. Again using the Hamiltonian flow as the illustration, one finds from Eq. (3.22)

$$\operatorname{div}(\dot{\xi}) = \frac{\partial}{\partial \xi} J \frac{\partial}{\partial \xi} H = 0.$$

Because the symplectic scalar product of any vector (in this case $\partial/\partial\xi$) with itself vanishes, the phase velocity field is a divergence-free field and the flow is incompressible. The Hamiltonian flow illustrates the point; but all flows $[\xi, F]$ are incompressible and preserve volume in phase space.

The incompressibility of the phase flow is the basis of the only generic theorems known for the general motion of many bodies. The most important of these concerns the recurrence of states of motions that are bound in a finite region of phase space and is due to Poincaré. A cluster of points singled out at any instant of time subsequently courses through phase space as tubes of flow sweeping out volume. Since the volume of the space is finite and incompressible but the time for sweeping is unbounded, the flow must eventually sweep out a volume it has already once swept through. The tubes of flow therefore intersect themselves and the flow returns to the vicinity of phase points through which it has already passed. States of a Hamiltonian flow in a compact region therefore recur over the course of the motion.

The Poincaré recurrence theorem says nothing about the time required for a particular state to recur; it may be absurdly long compared to other important time scales in the motion. States cannot always be expected to recur in meaningful finite times.

Invariants and the Elementary Flow

The motion of the world is imaged in phase space as a flow; but this flow only becomes known to us when the integrals of the Hamilton equations become known. Remarkably, the Hamilton equations, even those of the most complex motions, admit elementary integrals; and we are now going to find them. These integrals generate the elementary flow from which the flows of all possible motions may be built.

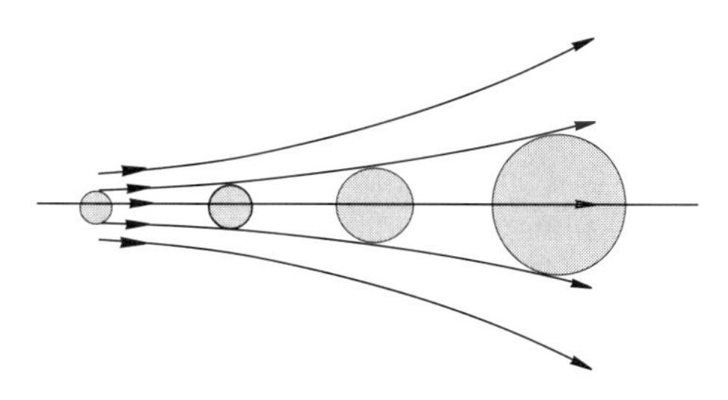

Figure 3-10. Flow Divergence and Volume Change. The divergence or convergence of streamlines corresponds to the expansion or compression of the volume of fluid elements.

There is an analogy here with complex matter and elementary particles. As all matter is built up from elementary particles, all motion may be built up from elementary flows. As elementary particles possess invariant properties such as charge and mass, elementary flows possesses the invariant properties of a motion. As the elementary particle invariants are carried over to the complex matter which they build up, so too are the elementary flow invariants carried over to the complex flows which they build. And just as matter may be a bewilderingly complex assemblage of elementary particles, so too may complex motions be an equally bewildering composition of elementary flows.

To discover the elementary flow and the invariants of motion with which it is endowed, one must first appreciate the power in the generality of the canonical equations. Because the motion of the world stands outside the particular quantities we use to apprehend it, there is a corresponding freedom in the choice of mechanical quantities in which we may cast our description. The law of motion as it is expressed in the canonical equations reflects this freedom. The mathematical forms of the action principle, the Hamilton equations, and the Poisson bracket are independent of the particular coordinates in which they are expressed.

Beginning with the most primitive physical quantities—position and velocity—we have worked our way up to the Hamiltonian summit to discover that the primitive coordinates that brought us here are not unique.

The law of motion allows other coordinate choices. Quantities built from the positions and velocity such as the momentum, angular momentum, energy, and other mechanical quantities we shall yet encounter (and, one might add, from which position and velocity may be built) serve with equal validity as canonical coordinates. What is required is that for motion with s degrees of freedom one has $2s$ canonical coordinates which form s conjugate pairs. It is not the coordinate strands themselves but their conjugate pattern which is universally woven into the fabric of motion by the action principle.

If the law of motion is to have the same form in all canonical coordinates, these coordinates must transform into one another in a particular way. Any set of canonical coordinates is related to another set by transformations that preserve the form of the action principle, the Hamilton equations, and the Poisson bracket. These transformations are naturally enough called canonical transformations and were worked out by the versatile nineteenth century mathematician C. G. J. Jacobi.

Consider another possible set of canonical coordinates (Q, P) in which one chooses to cast the same motion. These coordinates must be $2s$ in number divided between a "position-like" set, $Q = (Q_1, Q_2, \ldots, Q_s)$, and a conjugate "momentum-like" set, $P = (P_1, P_2, \ldots, P_s)$. The notation (Q, P) is misleading to the extent that it suggests that Q and P are necessarily an actual position and momentum. The coordinates Q and P need not at all be position or momentum. They may represent any mechanical quantity. Moreover, there is no significance to "position-like" and "momentum-like." It is the conjugal relationship between the coordinate pair rather than the individual member that is significant to the law of motion. This is easily seen by interchanging the position-like and momentum-like coordinates:

$$Q_i \to P_i, \qquad P_i \to Q_i.$$

Hamilton's equations

$$\dot{Q}_i = \partial H/\partial P_i, \qquad \dot{P}_i = -\partial H/\partial Q_i, \tag{3.24}$$

maintain the same form under such a change.

The connection between a set of canonical coordinates (q, p) with Hamiltonian $H = H(q, p)$ and another set of canonical coordinates (Q, P) with Hamiltonian $H' = H'(Q, P)$ is determined by requiring that they satisfy the same law of motion. This means that the action variations of both sets must be identical:

$$\delta S = \delta \int_{t_1}^{t_2} (p_i\, dq_i - H\, dt) = \delta \int_{t_1}^{t_2} (P_i\, dQ_i - H'\, dt)\,. \tag{3.25}$$

It is obvious that the variational problems appearing on each side of the right equality of this equation will lead to identical Hamiltonian and Poisson bracket structures, each expressed in its own coordinate language, that of Eqs. (3.15) and (3.17) for the (q, p) language, and Eqs. (3.24) for the (Q, P) language with Poisson bracket,

$$[F, G] = \left(\frac{\partial F}{\partial Q_i} \frac{\partial G}{\partial P_i} - \frac{\partial G}{\partial Q_i} \frac{\partial F}{\partial P_i} \right),$$

where $F(Q, P)$ and $G(Q, P)$ are any functions of the new coordinates. In particular, all canonical coordinates satisfy

$$[Q_i, Q_j] = 0, \qquad [Q_i, P_j] = \delta_{ij}, \qquad [P_i, P_j] = 0. \tag{3.26}$$

These observations in themselves, however, do not tell how the coordinates transform into one another. That connection is provided by the insight that the two integrals that appear under the variation in Eq. (3.25) need not be identical. It is their variations which must be identical. The integrals may differ by any function F that has a vanishing variation and still lead to the same variational problem for (q, p) and (Q, P). The difference between the integrands in Eq. (3.25) must therefore be the total differential of a function F,

$$dF = p_i \, dq_i - P_i \, dQ_i - (H - H') \, dt. \tag{3.27}$$

The requirement that the two variational problems be the same introduces a new function into the picture. This function, known as the generating function of the transformation, is an arbitrary function of one of the new and one of the old canonical coordinates and time. In the form (3.27) it is the function $F = F(q, Q, t)$ and the relationship of the coordinates to F may be read off Eqs. (3.27) as

$$p_i = \partial F / \partial q_i, \qquad P_i = -\partial F / \partial Q_i, \qquad H - H' = -\partial F / \partial t. \tag{3.28}$$

This generating function depends upon the old and new position-like coordinates. But it is only required that the generating function be a total differential of pairs of the two sets of coordinates. Another possible generating function can be formed from (3.27) as

$$dG \equiv d\Big(F + P_i Q_i\Big) = p_i \, dq_i + Q_i \, dP_i - (H - H') \, dt. \tag{3.29}$$

In this case the generating function is $G = G(q, P, t)$ and the coordinates are

$$p_i = \partial G / \partial q_i, \qquad Q_i = \partial G / \partial P_i, \qquad H - H' = -\partial G / \partial t. \tag{3.30}$$

The generating function $G(q, P, t)$ depends upon the old position-like coordinates and the new momentum-like coordinates. It is obvious that two other generating functions can be formed for the other pairs of old and new positions and momenta.

The generating function is the key actor in canonical transformations, transforming the state (q, p) into the state (Q, P). Two things are noteworthy about canonical transformations. First, the generating function incorporates one coordinate from the old pair and one from the new pair. It ties the two states together with one coordinate from each. Second, the canonical transformation presents one of the new coordinates explicitly and one of them implicitly. For example, in the transformation of the state (q, p) to the state (Q, P) by the generating function $G(q, P, t)$, the implicit–explicit presentation may be made graphic by writing the first two of Eqs. (3.30) as

$$\frac{\partial}{\partial q_i} G(q, P, t) = p_i, \qquad Q_i = \frac{\partial}{\partial P_i} G(q, P, t). \tag{3.31}$$

The new coordinates $Q(q, p)$ are given explicitly by the second of Eqs. (3.31) while their canonical mates $P(q, p)$ are given implicitly by the first.

As elementary illustrations, the generating function

$$G(q, P, t) = q_i P_i$$

is the identity transformation

$$Q_i = q_i, \qquad P_i = p_i, \qquad H' = H,$$

while the generating function

$$F(q, Q, t) = q_i Q_i$$

yields the inversion

$$Q_i = p_i, \qquad P_i = -q_i, \qquad H' = H.$$

The canonical structure of motion tells us that there are coordinates other than the primitive coordinates in which we may cast a motion which satisfy the same law of motion: they are governed by the Hamilton equations and possess the same invariants. The nineteenth century astronomer C. E. Delaunay saw the way to use this connection to good advantage. Since the only requirement of coordinates is that they be canonical, why not choose coordinates in which the integrals of the Hamilton equations $Q_i(t)$, $P_i(t)$ yield a flow of maximum simplicity, a simplicity not usually apparent in primitive coordinates? This flow is the *elementary flow* and occurs when one-half the coordinates, say the P_i, are selected as constants $P_i \equiv I_i$, with $\dot{I}_i = 0$. The other half, $Q_i \equiv \alpha_i$, are the canonical mates of these constants.

The Hamilton equations (3.24) now simplify dramatically, taking the form

$$\dot{\alpha}_i = \partial H'/\partial I_i, \qquad \dot{I}_i = -\partial H'/\partial \alpha_i = 0.$$

While the second Hamilton equation has been reduced to its maximum simplicity, the first has also become simple. This is because a condition on one of the canonical coordinates impacts the other due to their symplectic union. In addition to showing that the coordinates I are constants, the second set of Hamilton equations shows that the Hamiltonian is independent of the α coordinates: $H' = H'(I)$ only. The Hamiltonian derivative $\partial H'/\partial I_i$ is likewise a function only of I which may be expressed as $\omega_i(I) \equiv \partial H'/\partial I_i$. The first set of Hamilton equations thus has the simple integral by quadrature:

$$\alpha_i = \int (\partial H'/\partial I_i)\, dt = \omega_i t + \beta_i,$$

where $\beta = (\beta_1, \beta_2, \ldots, \beta_s)$ are its integration constants.

The Hamilton equations possess the elementary flow integrals

$$\alpha_i = \omega_i t + \beta_i, \qquad I_i = const. \tag{3.32}$$

The elementary flow has the property that it "flows" only along the α coordinates. Its phase velocity has no components in the invariant coordinate I directions since $\dot{I} = 0$. The integrals of the elementary flow depend upon constants of integration that are the two sets of s quantities β and I.

The elementary integrals of the Hamilton equations (3.32) may be gathered together as $2s$-dimensional symplectic vectors that are the phase vector, its phase velocity, and the integration constants,

$$\Xi = (\alpha, I), \qquad \Omega = (\omega, 0), \qquad \mathcal{I} = (\beta, I).$$

The elementary flow (3.32) underlying all motions is

$$\Xi(t) = \Omega t + \mathcal{I}$$

and its phase velocity is $\dot{\Xi} = \Omega$. The elementary flow consists of s symplectic pairs $\Xi_i = (\alpha_i, I_i)$. These are the fundamental flows from which all motions may be constructed by canonical transformation from the elementary phase space $\Xi = (\alpha, I)$ to the phase space $\xi = (q, p)$.

The elementary flow rests upon the constants $\mathcal{I} = (\beta, I)$. These constants are not only invariant on the elementary flow; they are invariant on the flow in the primitive coordinate phase space (q, p) of the motion built from it. Quantities invariant on the flow in one set of coordinates are invariant on the image of this flow in all other canonical coordinates. The constants $\mathcal{I} = (\beta, I)$ are therefore singularly important quantities: they are the $2s$ invariants of the motion. Invariants are significant because the elementary phase space coordinates $\Xi = (\alpha, I)$ are not connected in any simple way with their image phase space coordinates $\xi = (q, p)$ (the canonical transformation that connects them is generally not simple). But the invariants $\mathcal{I} = (\beta, I)$ are the same in both phase spaces. They are the invariant link between the two spaces.

An invariant $\mathcal{I}$ satisfies a conservation law,

$$\Delta(\mathcal{I}) = 0,$$

like the Galilean invariants in Eqs. (3.11). An invariant also satisfies the condition that its rate of change along the flow, or its total time derivative given by Eq. (3.16), vanishes:

$$\frac{d\mathcal{I}}{dt} = \frac{\partial \mathcal{I}}{\partial t} + [\mathcal{I}, H] = 0. \tag{3.33}$$

This condition can be seen to be the same as the conservation law $\Delta(\mathcal{I}) = 0$ by considering two points on the flow separated by an infinitesimal time interval Δt. The difference in the quantity $\mathcal{I}$ between these two points is $\Delta(\mathcal{I}) = (d\mathcal{I}/dt)\Delta t$. Since $\Delta(\mathcal{I})$ must vanish for an invariant and the time interval Δt need not vanish, one is led to the condition (3.33).

It frequently occurs that an invariant is not an explicit function of time ($\partial \mathcal{I}/\partial t = 0$) but depends only upon the canonical coordinates $\mathcal{I} = \mathcal{I}(q, p)$. Equation (3.33) for time-independent invariants thus reduces to the condition

$$[\mathcal{I}, H] = 0$$

showing that time-independent invariants have level surfaces in phase space that are coincident with the level surfaces of the Hamiltonian.

In addition to obeying a conservation law, invariants also possess canonical structure. One readily shows from Eqs. (3.32) and (3.26) that the invariant set $\mathcal{I} = (\beta, I)$ is a canonical pair satisfying the Poisson brackets

$$[\beta_i, \beta_j] = 0, \qquad [\beta_i, I_j] = \delta_{ij}, \qquad [I_i, I_j] = 0. \tag{3.34}$$

Invariants of the elementary flow, like the invariants of elementary particles, persist into the complex flows built from them. Ten of these $2s$ invariants are the Galilean invariants created by the fundamental symmetries of mechanics and are known in all motions; but most of the invariants of general motion are unknown to us as functions of the primitive coordinates $\mathcal{I}(q, p)$.

Explicit knowledge of a single invariant $I_i(q, p) = const$ certainly allows one to algebraically remove one of the $2s$ canonical coordinates leaving $2s-1$ integrals of the Hamilton equations to be found. Such a reduction applies to any system of differential equations; but the fact that the equations of mechanics are Hamiltonian gives mechanical invariants a greater power of reduction. Knowledge of one invariant produces *two* integrals of Hamilton's equations. This is because one invariant not only provides the integral of one Hamilton equation algebraically; it reduces the Hamilton equation of its mate to quadrature providing its integral as well as illustrated by the integrals (3.32). The power of one invariant to produce two Hamiltonian integrals is a unique and remarkable feature of the equations of motion. It sharply distinguishes Hamiltonian mechanics from other systems of differential equations.

Every invariant produces two Hamiltonian integrals. It is therefore sufficient to possess s invariants to find the complete $2s$ integrals of a Hamiltonian system. But there is a catch. The invariants I are canonical coordinates; they have vanishing Poisson brackets (3.34) and share a common manifold. It is this simplicity in the flow which reduces its Hamilton equations to quadratures. The collection of s invariants must be of one canonical half (say the Is); one cannot mix the Is and βs together. A number of integrals of a Hamiltonian system equal to $2m$ may be found if $m \leqslant s$ invariants with collectively vanishing Poisson brackets,

$$(I_1, I_2, \ldots, I_m), \qquad [I_i, I_j] = 0,$$

are known.

Action Redux

Given the elementary flow of the Hamilton equations, one is naturally interested in canonically transforming it into the flow in the primitive coordinates in which one observes the motion. What is the generating function that accomplishes this task? It is the action itself; and the existence of invariants brings us full circle to the action. We began our quest for the trajectories by minimizing the action. This procedure led to equations of motion and canonical coordinates while the action receded into the background. The invariants and the elementary flow now reunite the action S with the canonical coordinates (q, p) in the following way.

The momenta and energy are directly obtainable from the action. They are its derivatives with respect to the configuration coordinates and time. This follows from the fact that the total time derivative of the action is the Lagrangian, $dS/dt = L$. Since the energy H and the Lagrangian L are related as $L = -H + p_i\dot{q}_i$, the action time derivative is $dS/dt = -H + p_i\dot{q}_i$ and the differential of the action is

$$dS = -H\,dt + p_i\,dq_i. \tag{3.35}$$

Equation (3.35) shows that the action is a function of the configuration coordinates and time $S = S(q, t)$:

$$dS = \frac{\partial S}{\partial t}\,dt + \frac{\partial S}{\partial q_i}\,dq_i. \tag{3.36}$$

Comparison of Eq. (3.35) with Eq. (3.36) reveals that the energy and momenta are

$$H = -\partial S/\partial t, \qquad p_i = \partial S/\partial q_i. \tag{3.37}$$

In this form the law of motion may be regarded as a first order partial differential equation for the action S itself. For since $H = H(q,p)$ and the momenta are given by $p_i = \partial S/\partial q_i$ by the second of Eqs. (3.37), the first becomes

$$\partial S/\partial t + H(q, \partial S/\partial q) = 0. \tag{3.38}$$

Equation (3.38) is the Hamilton–Jacobi equation governing the action S. Its explicit form for a system of particles in rectangular coordinates $q = (\mathbf{x}_\alpha)$, $p = (\mathbf{p}_\alpha)$ with $\mathbf{p}_\alpha = \boldsymbol{\nabla}_\alpha S$ is

$$\partial S/\partial t + \sum_\alpha (1/2m_\alpha) \boldsymbol{\nabla}_\alpha S \cdot \boldsymbol{\nabla}_\alpha S + V(\mathbf{x}_\alpha) = 0.$$

The action $S(q,t)$ is the solution to the Hamilton–Jacobi equation (3.38), a first order partial differential equation in the s position coordinates q and the time t. The solution of such a partial differential depends upon $s+1$ constants of integration. One of these constants is purely additive; that is, if $S(q,t)$ is a solution of the Hamilton–Jacobi equation, so too is $S(q,t) + A$ where A is the additive constant.

The s remaining constants clearly have to be invariants. They may be selected as one-half of the invariant set, say the invariants I. Moreover, since the total energy is an invariant, say $H = \mathcal{E} = -\partial S/\partial t$, the time dependence of the action is simple; it is $-\mathcal{E}t$. The action must therefore be of the form

$$S(q, I, t) = -\mathcal{E}t + S_0(q, I), \tag{3.39}$$

where $S_0(q, I)$ is the time-independent part of the action. It is important to note that the energy $\mathcal{E}$ which appears in the first term on the right of Eq. (3.39) is one of the invariants $I = (I_1, I_2, \ldots, I_s)$.

The generating function of the elementary flow turns out to be the time-independent action $S_0(q, I)$. If one bases this generating function upon the old positions q and the new momentum-like invariants I according to the template $G(q, P, t)$ of Eqs. (3.29) and (3.30), then the invariants I in $S_0(q, I)$ must be identified with the new momentum-like coordinates $P \equiv I$ in $G(q, P, t) \equiv S_0(q, I)$. The remaining canonical coordinates are determined from the first two of Eqs. (3.30) in the form

$$p_i = \frac{\partial}{\partial q_i} S_0(q, I), \qquad \alpha_i = \frac{\partial}{\partial I_i} S_0(q, I).$$

Since $\alpha_i = \omega_i t + \beta_i$ and $\omega_i = \partial H'(I)/\partial I_i$, this result may also be expressed in terms of the full action $S(q, I, t)$ as

$$p_i = \frac{\partial}{\partial q_i} S(q, I, t), \qquad \frac{\partial}{\partial I_i} S(q, I, t) = \beta_i. \tag{3.40}$$

Equations (3.40) show how, given the invariants (β, I) and the action $S(q, I, t)$, the primitive coordinates (q, p) are obtained. The position coordinates q are contained implicitly in the second of Eqs. (3.40); the momenta p are contained explicitly in the first. Further, since $\partial G/\partial t \equiv \partial S_0/\partial t = 0$, the third of Eqs. (3.30) shows that the Hamiltonians in the two sets of coordinates are the same; they are just expressed in different coordinate language:

$$H'(I) = H(q, p).$$

The Hamilton equations and the Hamilton–Jacobi equation are different portraits of the same motion. In the Hamiltonian description motion is cast in canonical coordinates (q, p) and a system of $2s$ ordinary differential equations—the Hamilton equations—may be taken as the master equations of motion. One need not deal directly with the action itself [one can always find it by integrating Eq. (3.35) along the trajectories once the coordinates $q(t), p(t)$ are known as integrals of the Hamilton equations]. In the Hamilton–Jacobi description motion is cast in the action $S(q, I, t)$ and the Hamilton–Jacobi equation, a single partial-differential equation in s degrees of freedom plus the time and which depends upon s constants of integration (plus one additive constant) is taken as the master equation of motion. From the solution of the Hamilton–Jacobi equation $S(q, I, t)$, one can find the canonical coordinates (q, p) by differentiation as in Eqs. (3.40).

Poincaré Invariant

The action is the central mechanical quantity in the orchestration of motion. Invariants are also basic quantities upon which motion rests. Is there a direct connection between action and invariants? There is: one of the invariants is always directly related to the action. It is the action on a closed contour in phase space and is known as the integral invariant of Poincaré.

From Eq. (3.35), the action can always be expressed as an integral over the canonical coordinates:

$$S = \int (p_i \, dq_i - H \, dt). \tag{3.41}$$

The action in the indefinite integral representation (3.41) is clearly not an invariant; but if instead of the indefinite integral (3.41), one considers this integral over any closed contour γ lying in phase space,

$$\mathcal{S} = \oint_\gamma (p_i \, dq_i - H \, dt), \tag{3.42}$$

one finds it is invariant. The contour γ itself is not invariant; it is deformed as it is swept through phase space by the flow; but the integral over this moving contour is always the same. This closed integral is the Poincaré invariant.

One should be aware that although the Poincaré invariant exists for all motion, it is an integral of the canonical coordinates. It is inaccessible unless the canonical coordinates are known functions and the integrals can be performed. A known solution of the Hamilton or Hamilton–Jacobi equations is a prerequisite to an explicit representation of the Poincaré invariant.

Illustration: Hooke Motion

Let us now illustrate invariants, elementary flows, and the canonical transformation linking a particular flow to its elementary flow. We do so for the motion of a particle moving in one dimension and bound by a law of force proposed by Robert Hooke, Isaac Newton's most prominent contemporary. The Hooke law of force corresponds to the potential

$$V(q) = \kappa q^2/2, \tag{3.43}$$

where κ is a constant. It gives rise to an attractive force which is linear in the displacement, $f = -\partial V/\partial q = -\kappa q$.

The Hooke force does not correspond to any fundamental force in nature; but it does approximate the force between two bodies bound together by an elastic material such as a spring. (The actual forces in the spring material are electrical inverse-square forces between the nuclei and electrons of the atoms of which it is composed. The forces are inverse-square between any *two* charged nuclei or electrons; but the average force over myriads of charged nuclei and electrons for small displacements from their equilibrium positions turns out to be a linear force. This is because of the enormous cancellation of forces in a large ensemble of particles.)

Hooke's law of force leads to the Hamiltonian

$$H(q,p) = p^2/2m + \kappa q^2/2. \tag{3.44}$$

This is the Hamiltonian of a harmonic oscillator whose natural frequency is $\omega_0 = \sqrt{\kappa/m}$. The canonical invariant I of this motion turns out to be

$$I = H/\omega_0 \tag{3.45}$$

and its elementary flow, like all elementary flows, is given by Eqs. (3.32). The Hamiltonian in the elementary flow coordinates (α, I) is $H'(I) = \omega_0 I$, and the phase velocity of the elementary flow is

$$\omega = \partial H'/\partial I = \omega_0.$$

To construct the flow in primitive coordinates one must have the generating function of a canonical transformation from the elementary flow to the primitive coordinate flow. This generating function is the action $S(q, I, t) = -\mathcal{E}t + S_0(q, I)$; and it is obtained by solving the Hamilton–Jacobi equation of the motion. This can readily be done; for the Hamilton–Jacobi equation (3.38) for one degree of freedom reduces to the ordinary differential equation $dS_0/dq = p(q)$. The Hamilton–Jacobi equation of all one degree of freedom motions is integrable as the quadrature $S_0 = \int p\, dq$. The momentum $p(q)$ may be taken from Eq. (3.44) with H eliminated in favor of I according to Eq. (3.45) thereby giving the action

$$S_0(q, I) = \int \sqrt{2m\omega_0(I - m\omega_0 q^2/2)}\, dq.$$

The Poincaré invariant is just this integral over a closed contour in the (q, p) plane: $\mathcal{S} = \oint_\gamma p\, dq$. This contour integral can be shown to be

$$\mathcal{S} = 2\pi I.$$

The action $S_0(q, I)$ generates a canonical transformation in the $G(q, P, t)$ format of Eqs. (3.29) and (3.30) with $G(q, P, t) \equiv S(q, I, t) = -\mathcal{E}t + S_0(q, I)$ and with coordinates $P(q, p) \equiv I$ and $Q(q, p) \equiv \alpha$. From Eq. (3.30) one finds

$$p = \frac{\partial S}{\partial q} = \sqrt{2m\omega_0(I - m\omega_0 q^2/2)}, \qquad \alpha = \frac{\partial S}{\partial I} = \cos^{-1}\left(\sqrt{m\omega_0/2I}q\right).$$

These relationships show that the primitive coordinates (q, p) are obtained from the elementary flow (α, I) as

$$q(\alpha, I) = \sqrt{2I/m\omega_0}\, \cos\alpha, \qquad p(\alpha, I) = \sqrt{2Im\omega_0}\, \sin\alpha,$$

with the inverse relationships

$$\alpha(q, p) = \tan^{-1}(p/m\omega_0 q), \qquad I(q, p) = (p^2/2m + \kappa q^2/2)/\omega_0.$$

The invariant $I = H/\omega_0$ is a single-valued function of (q, p). The other half of the invariant set is $\beta = \alpha - \omega t$ or

$$\beta(q, p) = \tan^{-1}(p/m\omega_0 q) - \omega_0 t, \tag{3.46}$$

and is seen to be a many-valued but periodic function of (q, p) with period 2π.

The harmonic oscillator of Hooke motion has a unique phase space—it is Euclidean as shown by the metric condition embodied in its Hamiltonian (3.44) and Fig. 3-11. The canonical transformation of the elementary

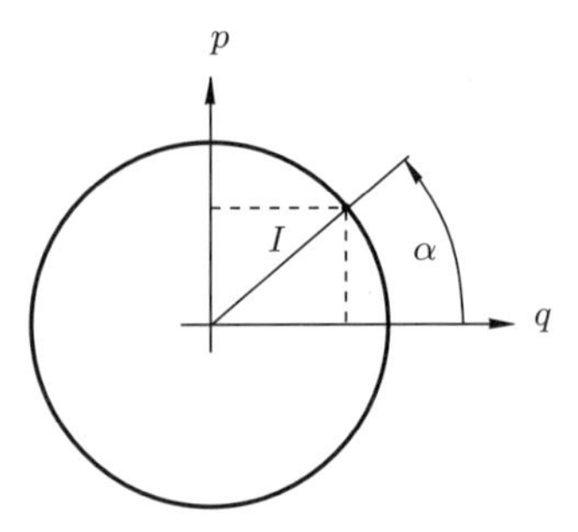

Figure 3-11. Phase Space of the Hooke Oscillator. The elementary flow consists of concentric circles whose radii are the invariants I and whose angular configurations are fixed by the coordinate α. The canonical transformation of the elementary flow to primitive coordinates consists of projections of the invariant I upon the primitive coordinate axes q, p—a cylindrical to rectangular coordinate transformation in this Euclidean space.

flow can be seen to be a transformation from rectangular to cylindrical coordinates in this Euclidean space.

The flow of Hooke motion in phase space consists of concentric circles in the (q, p) plane whose radii are the invariants I. The projections of the invariant I on the q and p axes give the flow in primitive coordinates. These projections are specified by the angle coordinate α whose phase is fixed by the invariant β.

The Complexity of Motion

The elementary flow is simple. It removes the complexity of motion from the Hamilton equations and relocates it in the action of the canonical transformation which builds the motion in primitive coordinates. But whence comes the action $S(q, I, t)$? The action is the solution of the Hamilton–Jacobi equation—an equation which contains all the complexity of the Hamilton equations and whose solution is obtainable with a degree of difficulty equal to that of solving the Hamilton equations in primitive coordinates.

Although nature expresses the trajectories of motion from their underlying Hamilton or Hamilton–Jacobi equations with ease, we who seek to form an image of motion have a more difficult task expressing this image. Finding the integrals of the Hamilton equations in primitive coordinates or the solution of the Hamilton–Jacobi equation for the generic motion of many bodies turns out to be a forbidding task.

The motion of many bodies is complex. This means it is impossible to exhibit the solutions of their Hamilton or Hamilton–Jacobi equations even as transcendental functions of the primitive coordinates—functions of great mathematical generality consisting of infinite series of algebraic polynomi-

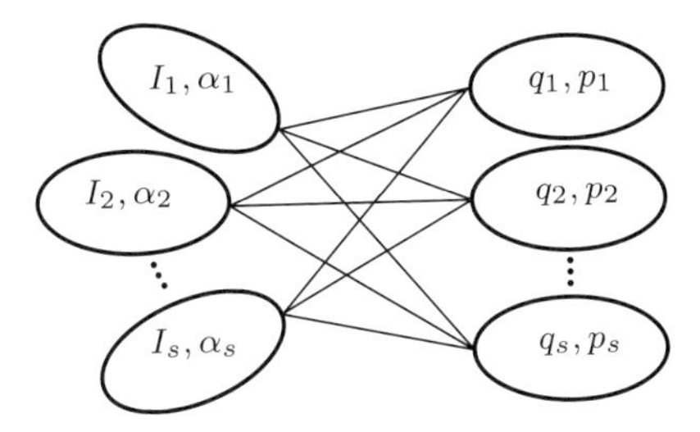

Figure 3-12. Complex Motion. Many primitive coordinate pairs (q, p) are entangled within each invariant manifold $I_i(q, p)$ of a complex motion.

als. (We shall see why this is so in Chap. 7.) Numerical integration is the only generally available method for finding these integrals.

Since each of the s elementary flow pairs (α_i, I_i) is simple, the complexity of motion arises from the action $S(q, I, t)$ which links them to the primitive coordinates (q, p). A typical invariant I_i of a complex motion is linked to not one but all s pairs of the primitive flow coordinates (q, p) in its canonical transformation as indicated in Fig. 3-12. The fact that many primitive coordinate pairs (q, p) are entangled within *each* invariant manifold $I_i(q, p)$ makes the generic motion of many bodies complex.

Motions so structured by their symmetries that each invariant I_i can be built from only *one* corresponding primitive coordinate pair (q_i, p_i) are freed of their complexity. The primitive coordinates do not entangle within the invariants and are said to be *separable* as in Fig. 3-13. When this happens, the Hamilton equations in primitive coordinates will be found to be integrable by quadrature, as was the case for the Hooke oscillator (we shall shortly describe how this happens more generally). Separable motions are also known as integrable motions and the two notions are interchangeable. Integrable motions lack complexity because the simplicity of the elementary flow is preserved in the canonical transformation to primitive coordinates.

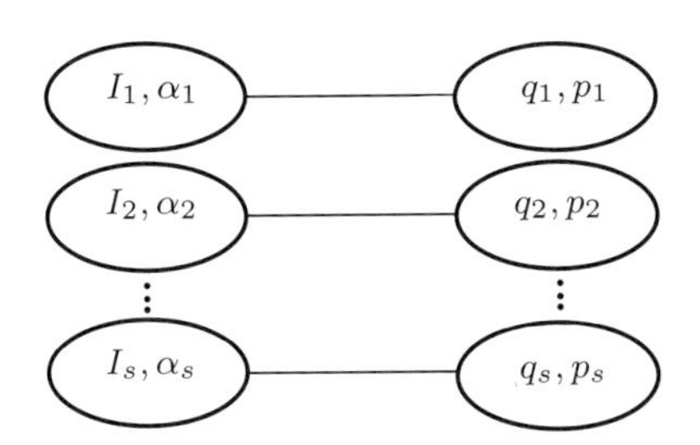

Figure 3-13. Separable Motion. In contrast to complex motion, only one primitive coordinate pair (q_i, p_i) is linked to an invariant manifold $I_i(q_i, p_i)$ of a separable motion.

Symmetry competes with the degrees of freedom of a motion to determine its integrability. Symmetry imposes order which separates the coordinates among the invariants; increasing degrees of freedom dilute that order allowing some of the coordinates to entangle. Two-body motion is so strongly structured by its symmetries that it is an integrable motion whereas the motion of many bodies is not. Though subject to Galilean symmetries, the motion of more than two bodies has sufficient degrees of freedom to evade the order imposed by the Galilean symmetries.

The separation of the coordinates by symmetry has a corresponding impact on the way in which the flow fills the phase space. Symmetric flows possess *sameness*; they are made of repetitive patterns. Less information is therefore required for the description of a symmetric flow than for a nonsymmetric flow. One can think of each point of phase space as a repository of information about the flow. (The position coordinates tell which points presently experience a flow and the momentum coordinates point out the adjoining points which experience the same flow.) The maximum information that can be packed into phase space occurs when every point is used by the flow. A symmetric flow which contains repetitive patterns does not require the full information storage capability of phase space. Some points are surplus and need never be visited by the flow. A highly symmetric flow need arrange itself over very few points of phase space. A flow lacking symmetry is so laden with information that it consumes the entire phase space.

One is accustomed to the Galilean invariants which are single-valued functions of the primitive coordinates; but generic motions possess a full range of invariants that include but go beyond the Galilean invariants. The additional invariants beyond the Galilean invariants for complex motions are many-valued functions of the primitive coordinates. Complex flows have many-valued invariants because they wander over large numbers of points of phase space. Since each flow is identified by an invariant, the invariant $I = const_1$ of a complex flow is visited over large volumes of points (q, p) as the flow wanders about. Additional flows defined by $I = const_2, const_3, \ldots$ also distribute their invariant values over large volumes of (q, p).

In a finite phase space with incompressible volume it is inevitable that flows defined by different invariant values $I = const_1, const_2, \ldots$ which each occupy large volumes of points in a finite region of phase space will intersect. (This is another implication of the incompressibility of the phase flow.) Points (q, p) within the intersections correspond to many values of I. Flows which consume large regions of phase space intersecting one another form complex tangled webs.

If a motion is sufficiently lacking in symmetry, a given point (q, p) may be swept by an infinite number of flow surfaces defined by the invariants $I = const_1, const_2, \ldots$. The invariants of motion whose flow consumes every point of a region of phase space (it is said to densely fill the phase space) are infinitely many-valued functions of (q, p). In contrast to complex flows whose invariants are many-valued, an integrable motion has invariants

$I(q,p)$ which are single-valued functions of the primitive coordinates.

The Simplicity of Motion

The counterpoint to complex motion is integrable motion. Integrable motion possesses symmetries which separate the coordinates: each invariant I_i is related to only one primitive coordinate pair (q_i, p_i) as in Fig. 3-13. Symmetry renders motions separable and separability in turn leads directly to integrability.

A separable symmetry splits the Hamilton–Jacobi equation into parts, each of which is an invariant depending upon a single primitive coordinate pair. This occurs when a symmetric motion is cast in a set of coordinates $q_i = q_i(\mathbf{x}_\alpha)$ which possess that very symmetry. Any coordinate system formed by independent lengths or angles may be utilized to specify a point in space. Each coordinate system also embodies particular symmetries. As the most elementary example, the rectangular coordinate system has planar symmetries formed by mutually orthogonal sets of planes $x_1 = const$, $x_2 = const$, $x_3 = const$. Other coordinate systems also exist which possess particular symmetries such as spherical, cylindrical, conical, and others.

In a coordinate system whose symmetries match the symmetries of the motion, the Hamilton–Jacobi equation—a partial differential equation in $s = 3n$ dimensions—splits into s ordinary differential equations whose integrals are quadratures. The whole action of the motion $S(q,t)$ separates into a sum of "partial actions" $S_i(q_i)$, each of which depends upon only one position coordinate,

$$S_0(q) = \sum_i S_i(q_i).$$

The momenta are given by $p_i = \partial S/\partial q_i = \partial S_0/\partial q_i$ or

$$p_i = \partial/\partial q_i \sum_j S_j(q_j) = dS_i(q_i)/dq_i$$

and each momentum coordinate is expressible as a function of only its canonical mate $p_i(q_i)$. The partial actions are therefore integrable by quadrature,

$$S_i(q_i) = \int p_i(q_i)\, dq_i \tag{3.47}$$

(no implied sums are intended in the above expressions).

For bound motion a particular coordinate pair (q_i, p_i) can only take values in a finite range; but the time for the motion is not bounded. The coordinate pair must, over the run of time, eventually retrace the same points again and again. Each action integral $S_i(q_i)$ therefore consists of repetitive increments, increasing by the same amount ΔS_i every time a complete cycle is completed by the coordinate pair (q_i, p_i). The action integrals (3.47) for bound integrable motion are therefore infinitely many-valued functions of the primitive coordinates: any $S_i(q) + n\Delta S_i$ for $n = 0, \pm 1, \pm 2, \ldots$ corresponds to the same value of q.

The $S_i(q_i)$ are almost invariants, but not quite; they do not satisfy the conservation law (3.33) because of their continual increase with time. Moreover, they are many-valued (actually, infintely many-valued) functions of the primitive coordinates. Nonetheless, in a single stroke one both turns the S_i into invariants and makes them single-valued. One does so by defining a single-valued quantity that contains all the information contributed by that coordinate pair as

$$J_i = (2\pi)^{-1}\Delta S_i = (2\pi)^{-1} \oint p_i dq_i, \tag{3.48}$$

where $\oint$ indicates integration over one cycle of the motion and the factor 2π is inserted for later convenience.

The quantities J_i are not only invariant; they are single-valued invariants known as action invariants. The invariants $I = (J_1, J_2, \ldots, J_s)$ are the invariant elementary flow coordinates made single-valued functions of the primitive coordinates.* The α of Eqs. (3.32) are their canonical mates and the trajectories (3.32) for integrable motion are

$$\alpha_i = \omega_i t + \beta_i, \qquad J_i = const,$$

with $\omega_i = \partial H/\partial J_i$.

A close connection exists between the Poincaré invariant $\mathcal{S} = \oint \sum_i p_i\, dq_i$ and the action invariants $J_i = (2\pi)^{-1} \oint p_i\, dq_i$ (no sum); and it is important to recognize their similarities and differences. The Poincaré invariant is a sum over all the coordinates; each action invariant involves only one coordinate pair. The Poincaré invariant exists for all motion, integrable or not; the action invariants exist only for integrable motions. Both invariants are integrals over closed contours; but the contour for the Poincaré invariant can be *any* closed contour in phase space whereas the contour for the action invariants is not arbitrary. It lies on a phase trajectory over a complete cycle of periodic motion. The contour of the Poincaré invariant need not be oriented in any special way just as long as it is closed. It therefore will generally be swept through phase space by the flow. But the contour for an action invariant lies *along* the flow. It remains stationary in phase space. If the motion is integrable and one chooses identical contours for the Poincaré

---

* To see that the J_i are invariant, recognize that the contour of integration lies along the flow and therefore remains stationary. One can then calculate $\dot{J}_i = \oint(\dot{p}_i dq_i + p_i d\dot{q}_i)$ (no sum), integrate the term $\oint p_i d\dot{q}_i$ by parts with the result $\dot{J}_i = \oint(\dot{p}_i dq_i - \dot{q}_i dp_i)$. This result may be arranged to $\Delta J_i = \dot{J}_i dt = \oint(dp_i dq_i - dq_i dp_i) \equiv 0$.

invariant and the action invariants, then the Poincaré invariant is seen to be a sum of the action invariants:

$$\mathcal{S} = 2\pi(J_1 + J_2 + \cdots + J_s).$$

The reduction of the many-valued invariants generic to nonintegrable motion to single-valued invariants is a *sine qua non* of integrable motion. The invariants of a complex motion are many-valued functions of the primitive coordinates. A many-valued function of a complex motion illustrated in Fig. 3-14 *(a)* is steadfastly many-valued. Its many-values occur without order and there is no way one can construct a non-trivial single-valued function from it. The lack of order in the many-values makes such a function complex.

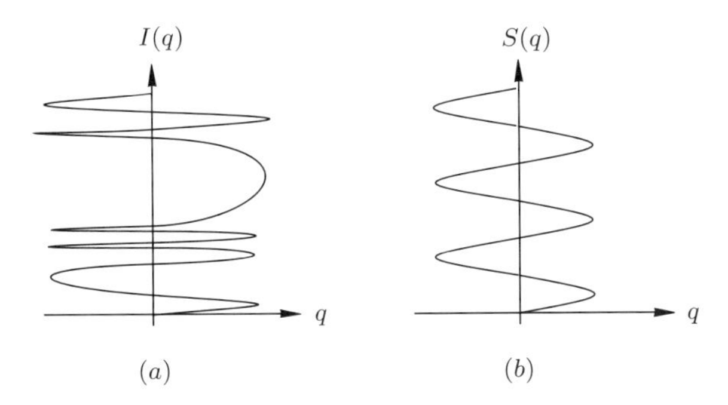

Figure 3-14. Complex and Simple Many-Valued Functions. A complex many-valued function $I(q)$ lacks order in the occurrence of its many-values *(a)*. A periodic many-valued function $S(q)$ is not complex because its many-values occur in a simple order *(b)*.

But if a many-valued function like the partial action $S_i(q)$ happens to be periodic, its many-values will occur in an ordered fashion as shown in Fig. 3-14 *(b)*. A periodic many-valued function is not complex. It is periodicity that distinguishes the many-valuedness of integrable motion from the many-valuedness of complex motion. Each many-value is simply an integer multiple of a single-valued function. It may therefore be made single-valued, as in the method of Eq. (3.48). Many-valuedness that is periodic takes motion from the realm of complexity and makes it simple.

Since the form of the action is known in Eqs. (3.47) and (3.48) for integrable motion, the topology of the elementary flow of bound integrable motion is known. Each action J_i is a "radius" of its circle (one-torus) and the canonical mate α is an angular coordinate on this circle. The torus of the complete motion is a Cartesian product of all these circles. Moreover, since the coordinates are separable, each pair of elementary flow coordinates $\Xi_i = (\alpha_i, J_i)$ and its torus are linked to only *one* primitive coordinate pair $\xi_i = (q_i, p_i)$ and its two-dimensional manifold (rather than to all of them as in the nonintegrable case). The submanifolds are therefore not glued up into complex topologies in building up the complete phase flow but are rather "tori all the way up."

The invariant manifold for one integrable mode is a circle for which the amplitude is the invariant radius J_i and the angle spins at an invariant frequency ω_i as shown in Fig. 3-15 *(a)*. A second integrable mode conflated with the first spins the first circle around a second center with the radius of the second action invariant and with an angle that spins at the second frequency creating the 2-torus as shown in Fig. 3-15 *(b)*. Conflations with the third and higher modes create higher-dimensional tori which cannot be visualized in three-dimensional space [imagine the 2-torus being spun round like the circle to create a 3-torus as indicated in Fig. 3-15 *(c)*; the 3-torus spun again to create a 4-torus, and so on]. The trajectory flow swirls upon the surfaces of these tori.

The flow lines of the composite flow may or may not close upon themselves. If they close, then there are points on the surface never visited by the flow. If the streamlines are open, the flow covers every point on the torus. (Although the flow densely covers the torus, it does not densely cover the phase space because, being the flow of an integrable motion, it is confined to the surface of the torus.)

Integrable motions are exceptional. A phase space of invariant tori, the signature of integrable motion, is likewise exceptional. Two-body motion is one of these exceptional motions. All two-body motion (no matter the force law) is integrable. Aside from the motion of a free particle, whose separability is trivial, two-body motion is the *only* many-body motion that is rigorously integrable.

There exist idealized many-body situations in which integrable motions can also be found. The most important of these is the rigid body—an ensemble of arbitrary numbers of particles rigidly linked to one another. Rigidity is the key to separability; removing all the degrees of freedom of the motion except six which separate into six of the Galilean invariants.

Integrable motion is highly specialized. One is therefore delightfully surprised to find that integrable motion describes much of the essential features of the world including the solar system and the atom. Moreover, integrable motion offers more insight into the generic motion of many bodies than one might expect from its specialized nature. This is because most many-body motion may be represented as a perturbation of many isolated two-body motions. In this sense, all many-body motion lies close to a collection of integrable two-body motions whose invariant tori have been distorted or broken by one another's presence. Integrable motions and their invariant tori subjected to small perturbations are therefore windows into many-body motions which reveal properties which lie well beyond those of two-bodies. We shall see how this occurs in Chap. 7.

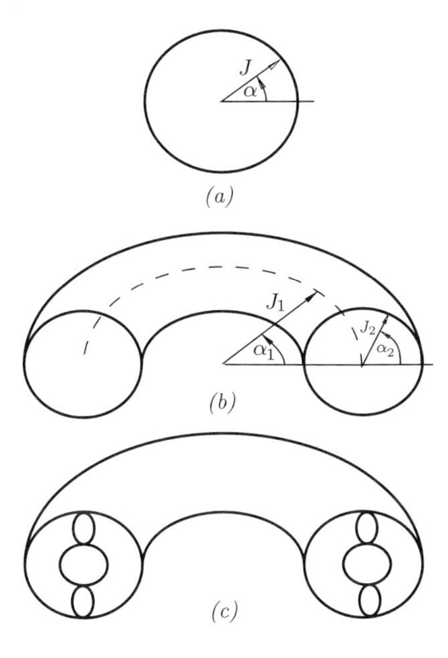

Figure 3-15. Tori as the Invariant Manifolds of Integrable Motion. A single integrable mode generates a circle whose radius is the action invariant J and whose angle α spins at the mode frequency *(a)*. A second mode conflated with the first rotates the first circle on a radius equal to a second action invariant thereby creating the torus *(b)*. Conflation with the third mode creates a torus with a torus [it cannot be visualized in three-dimensional space; Fig. 3-15 *(c)* is merely suggestive]. The complete manifold is an $s = 3n$-dimensional torus.

Ultimate Simplicity

Integrable motions consist of an ensemble of independent oscillating modes, each mode defined by a pair of canonical invariants, $\mathcal{I}_i = (\beta_i, J_i)$ which fix its amplitude and phase. Each mode has a phase velocity ("frequency") $\omega_i = \partial H / \partial J_i$ and its motion is an identity transformation of the elementary flow $\alpha_i = \omega_i(J_i)t + \beta_i$, $J_i = const.$

Although integrable motions lack complexity, they nonetheless can exhibit structure rich enough to be superficially mistaken for complex motions. This is because although each oscillator of action S_i supports a periodic motion with frequency ω_i, the ensemble action $S_0 = \sum_i S_i$ built from them is in general aperiodic: the separable modes of the motion are completely unsynchronized and the motion as a whole never repeats itself. The fact that an ensemble of periodic motions is not generally itself periodic is an important generic property of integrable motion; and it is this aperiodicity that gives such motions an illusory semblance of complexity.

This property is illustrated in Fig. 3-16 where one regards each separable mode as the motion on a clock face, each clock hand rotating at the mode frequency. If all the motions are initiated with the hands pointing at high noon, the ensemble of hands will generally never simultaneously return to this position (or any other) for arbitrarily given frequencies ω_i (although each individual clock will return to its initial position).

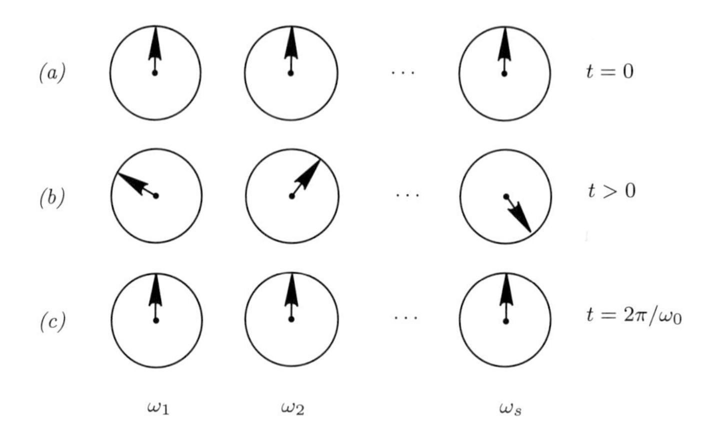

Figure 3-16. Aperiodic Motion from Periodic Motion. Each mode of an integrable motion may be represented as a clock face. Motion initiated with all modes at high noon *(a)* will generally never return to this state *(b)*. When the frequencies are rationally commensurate $\omega_j = l_j\omega_0$ with l_j an integer and ω_0 the lowest frequency, the motion becomes periodic. The initial state is repeated after intervals of time $t = 2\pi/\omega_0$ *(c)*.

A significant change in the character of integrable motion occurs when two or more of its mode frequencies ω_i are integer multiples of some fundamental frequency ω_0. (The frequencies are then said to be rationally commensurate.) When this happens, the motion is said to possess a degeneracy. If all frequencies are integer multiples of some fundamental frequency, the motion is fully degenerate and becomes periodic. (For s degrees of freedom, an ensemble may possess up to $s - 1$ degeneracies, the limit at which it is fully degenerate.) There exist times at which the hands of rationally commensurate clocks all return simultaneously to the same position as illustrated in Fig. 3-16 *(c)*. The difference between two such successive times is the period of the ensemble.

An integrable motion is generally aperiodic with structure rich enough to support flows which densely cover its invariant tori. But if an integrable motion possesses a degeneracy, its structure is further simplified and the flow no longer densely covers all its tori. Portions of the motion now become repetitive. If the motion is fully degenerate, the flow acquires its maximum simplicity and consists of streamlines which close upon themselves.

Degeneracy is a critical distinction in integrable motion. If no degeneracy exists, the motion is aperiodic; but when degeneracy is present, corresponding parts of the motion become periodic. At full degeneracy, the motion becomes fully periodic. Generic integrable motion is therefore said to be *quasiperiodic*: it can be either aperiodic or periodic depending upon its degeneracy.

The role of degeneracy in integrable motion may be further illuminated by representing any mechanical quantity $F(q,p)$ as a Fourier series in the angle variables $\alpha_i = \omega_i t + \beta_i$:

$$F(q,p) = \sum_{m_1,m_2,\ldots,m_s=-\infty}^{\infty} F(J)_{m_1,m_2,\ldots,m_s} e^{i(m_1\alpha_1 + m_2\alpha_2 + \cdots + m_s\alpha_s)}. \quad (3.49)$$

The Fourier coefficients $F(J)_{m_1,m_2,\ldots,m_s}$ describe the way in which the amplitudes of each integrable mode are combined together so as to represent any particular motion. The dependence of F on (q,p) is given through the canonical transformations (3.40) of $\beta = \beta(q,p)$, $J = J(q,p)$.

Let us use the shorthand notation $m \equiv (m_1, m_2, \ldots, m_s)$ for the s sets of integers in the Fourier expansion as well as $F_m(J) \equiv F(J)_{m_1,m_2,\ldots,m_s}$ for the Fourier amplitudes. The ensemble frequency corresponding to any set of integers m is

$$\Omega_m(J) \equiv m_1\omega_1(J_1) + m_2\omega_2(J_2) + \cdots + m_s\omega_s(J_s).$$

The corresponding phase of the ensemble is

$$\phi_m(\beta) \equiv m_1\beta_1 + m_2\beta_2 + \cdots + m_s\beta_s.$$

One may therefore express the angle variables in terms of their frequencies and phases in Eqs. (3.49) with the result

$$F(q,p) = \sum_{m=-\infty}^{\infty} F_m(J) e^{i[\Omega_m(J)t + \phi_m(\beta)]}.$$

The mechanical quantities of an integrable motion are a sum of oscillations of frequencies $\Omega_m(J)$ and phases $\phi_m(\beta)$. The canonical pair of invariants thus enter in distinctly different ways into the motion. The action invariants J fix the frequencies $\Omega_m(J)$ while the invariants β fix its phases $\phi_m(\beta)$ (hence the name "phase invariant").

One can now observe that if ratios of the frequencies ω_i are not rational, the ensemble frequency Ω_m will not be a rational multiple of some fundamental frequency ω_0. The quantity $F(q,p)$ will then be aperiodic. This is the generic case of integrable motion.

Suppose now two frequencies are rationally commensurate in the form

$$\omega_1 = l_1\omega_0, \qquad \omega_2 = l_2\omega_0, \quad (3.50)$$

where l_1 and l_2 are integers. These two frequencies now appear in the ensemble frequency Ω_m in the form

$$m_1\omega_1 + m_2\omega_2 = (m_1 l_1 + m_2 l_2)\omega_0.$$

But now a remarkable thing has happened. The coefficient $(m_1 l_1 + m_2 l_2)$ is itself an integer. Since the sums in Eq. (3.49) run over all the integers, one can just as well write

$$m' \equiv m_1 l_1 + m_2 l_2$$

and sum over the single set of integers m' rather than the two sets m_1 and m_2. The existence of a degeneracy has reduced the number of Fourier sums and independent frequencies of the ensemble from s to $s-1$:

$$\Omega_m = m' \omega_0 + \Omega'_m,$$

where Ω'_m consists of the $s-2$ nondegenerate frequencies. The collapse of two infinite sums into one is the origin of the term "degeneracy."

The degeneracy of the frequencies also shows

$$\partial H / \partial (l_1 J_1) = \partial H / \partial (l_2 J_2) = \omega_0.$$

When a degeneracy exists between two frequencies such as ω_1 and ω_2, the actions J_1 and J_2 appear in the Hamiltonian $H(J)$ in the rationally commensurate form $(l_1 J_1 + l_2 J_2)$. The converse is also true. If a Hamiltonian depends upon the action invariants in the form $(l_1 J_1 + l_2 J_2)$, then the motion possesses a degeneracy and has one less independent frequency.

Each additional appearance of a frequency in the form $\omega_k = l_k \omega_0$ where l_k is an integer signals the appearance of an additional degeneracy. When all frequencies are rationally commensurate the motion is fully degenerate with only a single fundamental frequency ω_0 and a single set of integers, $m' \equiv l_1 m_1 + l_2 m_2 + \cdots + l_s m_s$. The Fourier representation degenerates to the single sum

$$F(q,p) = \sum_{m'=-\infty}^{\infty} F_{m'}(\beta, J) e^{i m' \omega_0 (J) t}.$$

The existence of degeneracy also has an important impact on the phase invariants. The phase invariants β are generally many-valued. Although many-valued, the phase invariants are periodic. Each β_j increases by 2π every time a coordinate pair (q_j, p_j) returns to the same state as in the example of the Hooke oscillator, Eq. (3.46). The many-valuedness is therefore not complex; one may construct a single-valued function by taking a trigonometric function of the phase. Thus, the many-valued periodic invariant β_j may be made into a single-valued invariant:

$$B_j = e^{i\beta_j}.$$

Although each B_j, considered in isolation, is a single-valued function of its phase β_j, the various β_j are functions of (q,p). Hence, the B_j considered as functions of (q, p) cannot in general be single-valued because cycles of (q, p) induce cycles in β which are not in synchrony. If the motion possesses no degeneracies, one may construct only one additional single-valued invariant, say $B_1 = e^{i\beta_1}$. One single-valued phase invariant always exists.

On the other hand, if a degeneracy (3.50) exists between ω_1 and ω_2, then the cycles of β_1 and β_2 come into synchrony. The invariant B_2 is then also single-valued if B_1 is single-valued. For each degeneracy given by $\omega_k = l_k\omega_0$, an additional single-valued invariant B_k comes into being.

To sum up, the action invariants J_i of an integrable motion are always single-valued functions of the coordinates; the phase invariants $B_j = e^{i\beta_j}$ need not be. But if the motion is degenerate, the phase invariants also become single-valued.

Isolating Invariants

The flow of a generic, nonintegrable motion fills the entire manifold of phase space. When motion possesses a symmetry that allows it to be integrable, a single-valued action invariant J_i exists for the integrable mode corresponding to that symmetry (its corresponding phase invariant B_i generally remains many-valued). The flow no longer fills the full manifold of phase space of dimension $2s = 6n$. It now lies upon a submanifold—the union of a one-dimensional torus upon which the invariant is constant and the remaining $2(s-1)$-dimensional phase space of the nonintegrable coordinates.

Single-valued invariants are called *isolating invariants.* They confine the flow to smaller submanifolds thereby isolating it from the full manifold of phase space. If the motion is fully integrable, then s single-valued invariants exist and the flow lies upon an s-dimensional torus. The flow of a fully integrable motion lies upon a manifold half the size of the manifold of its phase space.

Can the flow be isolated to a yet smaller manifold? It can. When a fully integrable motion is also degenerate, the manifold upon which the flow lies will be further reduced. This is because the phase invariants also become single-valued when degeneracy exists; and it is the single-valuedness of invariants that leads to isolation. In addition to the s single-valued invariants J a degenerate motion now includes the single-valued invariants B_i, one for each degeneracy. The flow correspondingly occupies a manifold of dimension less than s, a decrease of one dimension for each degeneracy. If the motion is fully degenerate (a maximum of $s - 1$ degeneracies), all invariants are single-valued and the flow closes upon itself on a one-dimensional torus—the knotted circle (it is knotted because the streamline may wind more than once around the torus before closing).

Flows may occupy a manifold of dimension as great as $2s$ or as small as one dimension depending upon their symmetries and the corresponding single-valued invariants they possess. It is this vast range in the dimension of the flow manifold that spans the most simple to the most complex motions.

Only s single-valued invariants are required to integrate the Hamilton equations (3.15). Integrable motions have at least this number of single-valued invariants constructed in Eq. (3.48). But when a motion is degenerate, more single-valued invariants exist and the set exceeds s. One may therefore assemble more than one unique set of s invariants (the actions J) in which the motion is integrable. This means there exists more than one set of coordinates in which the Hamilton–Jacobi equation is integrable. This is the case for two-body motion with inverse-square forces.

In summary, integrable motion lacks complexity; nonetheless it exhibits a richness of structure that ranges from simple one-loop closed orbits to open orbits so dense and seemingly irregular they can be superficially mistaken for those of nonintegrable motion. Isolating invariants and the degeneracy they induce distinguish the range of simplicity in integrable motion. A fully degenerate motion is periodic and its orbits are closed curves. A nondegenerate motion has the maximum richness of structure possible in integrable motion. It is aperiodic and its orbits wind densely over its torus without closing.

The Algebra of Symmetry

The Galilean symmetries of mechanics are transformations of space and time to which mechanical quantities which are functions of space and time such as the energy H, momentum $\boldsymbol{P}$, mass-center $\boldsymbol{N}$, and angular momentum $\boldsymbol{J}$ are invariant. The configuration space in which the particles reside, however, is only half of the wider world of phase space in which motion fully exists. The full symmetries of motion are the symmetries of phase space.

Phase space naturally possesses Galilean symmetries because it contains configuration space. But phase space also possesses another fundamental symmetry flowing from the action principle itself: symplectic symmetry. The fundamental symmetries of phase space are the Galilean and symplectic symmetries. Additional symmetries come into being with specific motions. They are embodied in the particular Hamiltonians of those motions. They represent additional transformations of phase space into itself to which the mechanical quantities defined on it are invariant.

How are mechanical quantities imprinted by a particular symmetry? Remarkably, symmetries reveal themselves in the algebraic behavior of these quantities. This mechanical algebra, inspired by the Norwegian mathematician Sophus Lie at the end of the nineteenth century, differs from ordinary algebra in the definition of the product operation. It was Lie's ingenious insight that an algebra with powerful and revealing properties would result if the product of any two quantities is taken as the Poisson bracket of these quantities. It is appropriate that the behavior of mechanical quantities under the Poisson bracket forms a portrait of their underlying symmetry in phase space. The Poisson bracket is the symplectic template—an antisymmetric product of derivatives with respect to the phase space coordinates themselves.

Products of mechanical quantities of interest may be arrayed in a multiplication table. This multiplication table is a portrait of the corresponding symmetry. By examining its patterns, one discerns the structure of the symmetry. For the elementary Galilean symmetries of mechanics, the mechanical quantities of interest are H, $\boldsymbol{P}$, $\boldsymbol{N}$, and $\boldsymbol{J}$ and their multiplication table is

$$\begin{array}{lcccc} & H] & P_j] & N_j] & J_j] \\ \\ [H, & 0 & 0 & P_i & 0 \\ [P_i, & 0 & 0 & \delta_{ij}m & \epsilon_{ijk}P_k \\ [N_i, & -P_i & -\delta_{ij}m & 0 & \epsilon_{ijk}N_k \\ [J_i, & 0 & -\epsilon_{ijk}P_k & -\epsilon_{ijk}N_k & \epsilon_{ijk}J_k \end{array} . \tag{3.51}$$

Four symmetries are contained in this table, each symmetry carried by a particular invariant. They are translations in time (H), translations in configuration space ($\boldsymbol{P}$), translations in velocity space ($\boldsymbol{N}$), and rotations ($\boldsymbol{J}$). Products of the components of an individual invariant vector, such as $[J_i, J_j]$, describe the symmetry carried by that invariant and lie on the diagonal of the table. Products of the components of invariant vectors carrying different symmetries, such as $[P_i, J_j]$, describe the joint action of the two symmetries when they are combined and form the off-diagonal products.

Time translational symmetries described by H are simple. The products of all quantities with H are zero except for $\boldsymbol{N}$ for which it is $-\boldsymbol{P}$. This is because $\boldsymbol{N} = \boldsymbol{P}t - m\boldsymbol{X}$ is an explicit function of time. The Poisson bracket $[\boldsymbol{N}, H] = -\boldsymbol{P}$ must cancel the time derivative $\partial\boldsymbol{N}/\partial t = \boldsymbol{P}$ so as to yield $d\boldsymbol{N}/dt = 0$ according to Eq. (3.33).

Translational symmetries in configuration space and in velocity space described by $\boldsymbol{P}$ and $\boldsymbol{N}$ are also simple. Their self-Poisson brackets on the diagonal vanish while their off-diagonal product vanishes unless both are components in the same direction in which case their product yields the total mass of the ensemble. Translations in configuration space couple with translations in velocity space like canonical invariants. The invariants $\boldsymbol{P}$ and $-\boldsymbol{N}/m$ have a Poisson bracket behavior that is identical to that of the invariants I and β.

Rotational symmetry and its algebra are described by that portion of the multiplication table containing the elements J_i and are richer in structure than translational symmetry. Quantities possessing rotational symmetry, such as the angular momentum, have the cyclic Poisson bracket which, in contrast to the brackets for translations, does not vanish. A product of any two of the three elements J_1, J_2, J_3 is the third. Moreover, the magnitude of the elements of rotational symmetry has a vanishing Poisson bracket with all elements, $[J^2, J_i] = 0$.

Translations and rotations are distinct symmetries. Nonetheless, they do couple in a general frame of reference as indicated by the nonvanishing off-diagonal products $[P_i, J_j]$ and $[N_i, J_j]$. This coupling is superficial. It is eliminated in a frame of reference translating with the center of mass and known as the center of mass frame of motion.

All motion may be decomposed into a center of mass motion described by $\boldsymbol{P}$ and $\boldsymbol{N}$ whose angular momentum is

$$\boldsymbol{X} \times \boldsymbol{P} = \boldsymbol{P} \times \mathbf{N}/m$$

and a relative motion observed in a frame of reference moving with the center of mass. The center of mass motion is free of forces and its total energy is the kinetic energy $P^2/2m$. The energy $\mathcal{H}$ and the angular momentum $\boldsymbol{M}$ of the relative motion are therefore

$$\mathcal{H} \equiv H - P^2/2m, \qquad \boldsymbol{M} \equiv \boldsymbol{J} - \boldsymbol{P} \times \mathbf{N}/m.$$

The translations and rotations coupled in the Poisson bracket algebra of H, $\boldsymbol{P}$, $\boldsymbol{N}$, and $\boldsymbol{J}$ in a general frame of reference uncouple into two pieces: a center of mass component described by $\boldsymbol{P}$ and $\boldsymbol{N}$ and a relative component described by $\mathcal{H}$ and $\boldsymbol{M}$. The center of mass components have the Poisson bracket algebra

$$[N_i, N_j] = 0, \qquad [N_i, P_j] = -\delta_{ij} m, \qquad [P_i, P_j] = 0.$$

The center of mass motion is integrable and constitutes a six-dimensional phase space with elementary flow identical to the flow in primitive coordinates,

$$X_i = (P_i/m)t - N_i/m, \qquad P_i = const.$$

Its invariants are $\mathcal{I} = (\beta, I) = (-\mathbf{N}/m, \boldsymbol{P})$. The action of the center of mass motion is

$$S = -\mathcal{E}t + P_i X_i$$

and is an identity canonical transformation of the elementary flow.

The center of mass motion of all motions is always known. All the complexity of general motion rests in the remaining $6(n-1)$-dimensional relative motion whose Galilean invariants are $\mathcal{H}$ and $\boldsymbol{M}$ with the Poisson brackets

$$[\mathcal{H}, \mathcal{H}] = 0, \qquad [\mathcal{H}, M_i] = 0, \qquad [M_i, M_j] = \epsilon_{ijk} M_k.$$

Symmetries create invariants; and the invariant of symplectic symmetry is the zero element of the Poisson bracket algebra. All mechanical quantities manifest the symplectic symmetry of phase space in the zero element: the product of a mechanical quantity with itself (i.e., Poisson bracket of a quantity with itself) vanishes. Although the components of quantities like $\boldsymbol{M}$ do not have vanishing Poisson brackets among themselves, quantities may be constructed from them which have a vanishing Poisson bracket with every element of the algebra. If one restricts attention to the angular momentum describing three-dimensional rotational symmetry, the magnitude M^2 or M is such a quantity with vanishing Poisson brackets with all components as indicated by Eq. (3.19). Such a quantity is the key to the organization of the phase flow by symmetry and is known as a Casimir invariant after the twentieth century physicist H. B. G. Casimir who first distinguished them in his classic work on the mechanics of rigid body rotations.

A Casimir invariant is a quantity that has a vanishing Poisson bracket with every element of the symmetry. For translations in configuration space, all the elements (P_1, P_2, P_3) are Casimir invariants as are (N_1, N_2, N_3) for translations in velocity space.

In the case of rotations none of the elements of the symmetry algebra is a Casimir invariant. The Casimir invariant of rotational symmetry is the relative angular momentum magnitude M. In contrast to J, it is also a Casimir invariant of translations since it has vanishing Poisson brackets with $\boldsymbol{P}$ and $\boldsymbol{N}$. The relative energy $\mathcal{H}$ is easily seen to be a Casimir invariant not only of time translations but of all the Galilean symmetries.

When all the Galilean symmetries of mechanics are combined, only two invariants survive with vanishing Poisson brackets with all the elements H, $\boldsymbol{P}$, $\boldsymbol{N}$, and $\boldsymbol{J}$. These Casimir invariants of the complete Galilean symmetries are the relative energy $\mathcal{H}$ and the magnitude of the relative angular momentum M; and they are possessed by all motions. Particular motions described by particular Hamiltonians possess additional symmetries beyond the Galilean symmetries. The Casimir invariants expand accordingly.

In summary, symmetries create an algebra through the Poisson bracket as illustrated by the fundamental translational and rotational symmetries of mechanics. The algebra of rotations may be uncoupled from that of translations by viewing the motion in the center of mass frame. The center of mass motion is integrable; it is an identity transformation of the elementary flow. The remainder of the motion which contains all its possible complexity is contained in the $(s-3)$-dimensional relative motion. For a given symmetry, special quantities exist that have vanishing Poisson brackets with every element of the algebra. These are the Casimir invariants of the symmetry. The complete Galilean symmetries of mechanics have two Casimir invariants, the relative energy $\mathcal{H}$ and angular momentum magnitude M (the energy and the angular momentum magnitude in the frame of motion moving with the center of mass). All motion possesses these Casimir invariants.

A Summary of Symmetry in Invariant Sets

Symmetry is captured in the Poisson bracket behavior of its invariants. Sets of quantities whose Poisson brackets collectively vanish summarize the symmetry. This is because they form one-half of a symplectic pair of coordinates as indicated by Eqs. (3.34) which allow the Hamilton equations to be integrable. We shall call sets that contain all the elements of a symmetry with collectively vanishing Poisson brackets *invariant* or *Casimir sets*. Casimir invariants are always members of this distinguished set.

A Casimir set need not only consist of Casimir invariants. In the case of rotational symmetry, one may always select one of the three quantities M_1, M_2, M_3 to place with the Casimir invariants M and $\mathcal{H}$ to form a set whose Poisson brackets collectively vanish such as $(\mathcal{H}, M, M_3)$. The elements $\mathcal{H}$ and M of this Casimir set are Casimir invariants; but the element M_3 is not.

A given symmetry may have more than one independent Casimir set. The rotational symmetry of the relative motion turns out to have two. The independent invariants of the relative motion are $(\mathcal{H}, M_1, M_2, M_3)$. They may also be expressed in terms of the angular momentum magnitude M and any two components as $(\mathcal{H}, M, M_3, M_1)$. It can be easily seen that one may combine these invariants together in the two independent Casimir sets

$$(\mathcal{H}, M, M_3), \qquad (\mathcal{H}, M, M_1).$$

If one restricts attention to translations, Casimir sets consist of the two complete sets

$$(P_1, P_2, P_3), \qquad (N_1, N_2, N_3).$$

They form the pair of three-dimensional invariants of the ensemble translational motion $\mathcal{I} = (-\mathbf{N}/m, \boldsymbol{P})$.

Casimir sets contain all the contributions of a given symmetry to the invariant set I. The sets (P_1, P_2, P_3) and (N_1, N_2, N_3) are each Casimir sets of the translational symmetry possessed by the center of mass motion. The sets $(\mathcal{H}, M, M_3)$ and $(\mathcal{H}, M, M_1)$ are the Casimir sets of the three-dimensional rotational and time-translational symmetry of the relative motion. These Casimir sets for translations and rotations may be combined into Casimir sets for the complete Galilean symmetries, such as

$$[(P_1, P_2, P_3), (\mathcal{H}, M, M_3)], \qquad [(P_1, P_2, P_3), (\mathcal{H}, M, M_1)]. \tag{3.52}$$

The full Galilean group allows the Casimir sets (3.52). All the Casimir sets one may form for the complete Galilean symmetries contain a maximum of six invariants, a number less than the full number of Galilean invariants which is ten.

The Hamilton equations have the unique property that they may be reduced to quadratures in *pairs* for *each* single-valued invariant which is known thereby allowing a reduction of two in the dimension of the phase space for each single-valued invariant. One might be led to conclude that since there are ten Galilean invariants, the dimension of the phase space of all motions may be reduced by twenty; but this is not so. The invariants that effect the reduction must form a set with collectively vanishing Poisson brackets. These are the Casimir sets; and they contain only six rather than the full ten of the Galilean invariants. The reduction in the dimension of the phase space of all motions is twelve rather than twenty.

The fact that there exists more than one Casimir set for a symmetry means that one has a choice as an observer of viewing the same motion through different sets of mechanical quantities. It also means that the same symmetry is natural to more than one geometry. Rotational symmetry has two independent Casimir sets. It is natural to both the geometry of the sphere and the geometry of the paraboloid as we shall find.

Casimir sets show how symmetries organize the canonical structure of motion. Symmetries create invariants; and the invariants of mechanics $\mathcal{I} = (\beta, I)$ are a symplectic pair, each half of which is a Casimir set. The flows of a Casimir set all lie on a common manifold of phase space. Casimir sets are the building blocks for constructing all integrable motions. If they had no other significance than this, Casimir sets would be amply distinguished quantities in mechanics. But in the extension of motion into the quantum realm, Casimir sets take on deeper significance. Casimir sets are those quantities that can be simultaneously observed without quantum uncertainty. It can well be argued that the most important quantities in mechanics are the Casimir sets of the symmetries of nature.

Reprise

Our journey with the law of motion began with space and time and culminated in the principle of least action. The principle of least action blossoms with implications: conservation laws, invariants, the Poisson bracket and its Hamiltonian flow, and a revealing algebra of symmetry and invariants. We shall follow these implications in a detailed exposition of the two-body motions of heavenly bodies and the hydrogen atom. These are nature's integrable motions whose behavior is completely open to us. And we shall find that the law of motion reflects the beauty of the physical world in the mathematical beauty of its own unfolding form.

Chapter 4

Classical Mechanics: The Heavens

The universe is in motion. Least action is the principle underlying that motion. Let us look upon the revelations of this principle for the most elementary celestial motion, that of two heavenly bodies such as a planet and the sun. The trajectories of the bodies are the final flowering of the mathematical description. The action principle has provided their equations of motion; however the trajectories do not become apparent until the integrals of the equations of motion are found.

But the action principle provides more than equations of motion. It also offers conservation laws created by the symmetries of nature. The conservation laws present us with the invariants of mechanics; and these invariants provide direct access to the trajectories of two bodies. The motion of two bodies may be decomposed into a center of mass component and a component relative to the center of mass. The invariants $\boldsymbol{N}$ and $\boldsymbol{P}$ tell us that no matter what complex motion each body executes, the center of mass of the two bodies translates at a constant velocity through the celestial space. The invariants $\boldsymbol{N}$ and $\boldsymbol{P}$ each constitute an appropriate Casimir set for the center of mass motion, each contributing three invariants. The trajectory flow of the center of mass consists of straight lines given by

$$\boldsymbol{X} = (\boldsymbol{P}/m)t - \boldsymbol{N}/m, \qquad \boldsymbol{P} = const.$$

Now turn to the relative motion. The energy of the relative motion $\mathcal{H}$ is an invariant. We shall find that for energies $\mathcal{H} < 0$, the two bodies are bound together—they must always remain within a finite distance of one another. For energies $\mathcal{H} > 0$, the motion is unbound—the bodies may move infinitely far from one another.

The angular momentum $\boldsymbol{M}$ of two bodies is also an invariant vector forever pointing in a fixed direction as the two companions stroll through the heavens. By definition, the position and the momentum vectors of the relative motion are both perpendicular to the angular momentum; they define a plane. The result is exceptional: the entire relative motion is confined to a plane perpendicular to the angular momentum vector.

Bodies in arbitrary motion may be expected to explore all values of position and momentum in phase space. This is not so in two-body motion. The angular momentum is an isolating invariant. It defines a plane permanently fixed in configuration space to which the relative motion is restricted. No matter how the initial conditions of the bodies may be varied for a given angular momentum, the trajectory flow always lies in the same plane. The isolating power of the angular momentum is reflected in the fact that it contributes not just one but two Casimir sets to the relative motion:

$$(\mathcal{H}, M, M_3), \qquad (\mathcal{H}, M, M_1). \tag{4.1}$$

The phase space of the relative motion is a six-dimensional manifold. The existence of one Casimir set with three single-valued invariants immediately shows the motion is integrable and its phase flow confined to a submanifold of three rather than six dimensions. But the fact that the angular momentum contributes two Casimir sets rather than one means the motion is degenerate and restricted to an even smaller submanifold of two dimensions. When the motion is bound, this submanifold is a two-torus. Angular momentum in concert with the energy isolates the trajectory flow within a manifold of six dimensions to one of two dimensions—a powerful reduction.

The trajectories lie on a plane in configuration space and upon a two-torus in phase space; but they are still quite open. The trajectories form generally wandering orbits, never closing upon themselves as shown in Fig. 4-1 *(a)*. A closed orbit for which the bodies retrace the same streamline as shown in Fig. 4-1 *(b)* represents another degeneracy of the motion. Such a trajectory is bound to an even more severely restricted region of space. Not only is the motion confined to the plane; the trajectory is closed restricting the motion from the full space of the plane.

If the trajectories of two-body motion are to be closed, there must be an additional single-valued invariant beyond $\mathcal{H}$ and M underlying this degeneracy that fully isolates the trajectory. Is there such an isolating invariant? In general there is not. Additional isolating invariants that close the orbit do not exist for arbitrary force laws and the orbits are generally open. There is one, and only one, inverse-power force law that endows two-body motion with another isolating invariant that closes the orbit (**Note 2**). That force is the *inverse-square* force.

This coincidence is one of the compelling correspondences we shall find in our journey with the two-body problem. Inverse-square forces are special forces indeed. They are precisely the forces of gravitation and electricity for which bodies move slowly compared to the speed of light. Inverse-square force orbits, the orbits of two particles bound by electrical and gravitational forces, are closed; the motion is completely degenerate.

The trajectories of two bodies bound by the inverse-square force are closed orbits lying in a plane. What of their shape? One is drawn by a

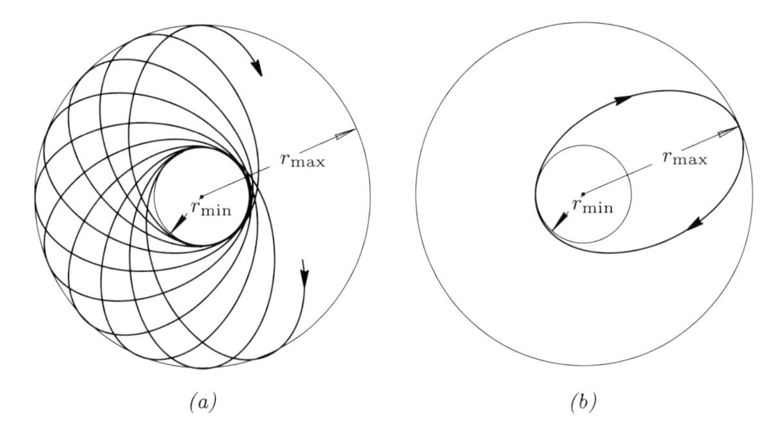

Figure 4-1. Closed and Open Trajectories of Two-Body Motion. The trajectory is confined to the torus in phase space and the annulus $r_{\min} \leqslant r \leqslant r_{\max}$ in configuration space. In *(a)*, the orbit never closes upon itself. Over the course of time the trajectory densely covers the annulus. In *(b)* the trajectory is closed. Most of the surface of the annulus is untouched by the trajectory.

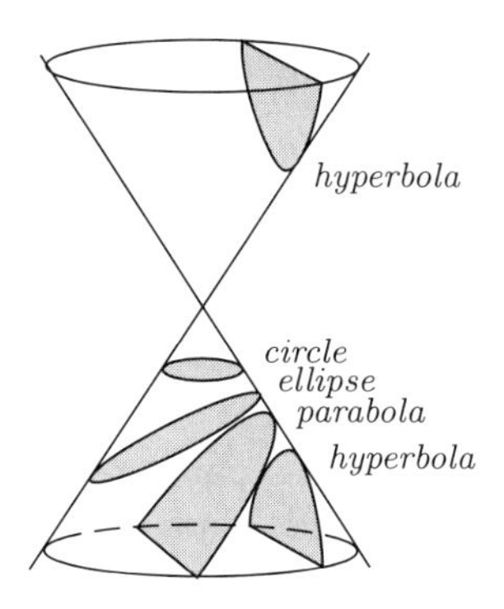

Figure 4-2. The Family of Conic Sections. A plane cuts a cone in the shapes of circle, ellipse, parabola, and hyperbola.

classical sense of perfection to the circle. But the general trajectories of the relative motion in the plane cannot be circles; for once one has fixed the orientation of the plane of motion by the angular momentum vector, two Galilean invariants, the magnitude M and the energy $\mathcal{H}$, remain. A circle possesses only one invariant, its radius.

Enter the ellipse. It is a closed, smooth curve one step beyond the circle. It has two invariants that match the physical invariants $\mathcal{H}$ and M, a major axis and a minor axis. But let us not rush into ellipses just yet. There are other shapes as well—the full set of *conic sections*—that satisfy the laws of two-body, inverse-square force motion. Conic sections are the curves that form the intersections of a cutting plane with a cone as shown in Fig. 4-2.

Two-Body Coordinates

Before proceeding further, it is useful to introduce a natural set of coordinates for the relative motion of two bodies. The total mass of the bodies m and their *reduced mass* μ are

$$m = m_1 + m_2, \qquad \mu^{-1} = m_1^{-1} + m_2^{-1},$$

where m_1 and m_2 are the masses of the two bodies. The relative position vector $\boldsymbol{r}$ of body 2 with respect to body 1 and its corresponding relative momentum are

$$\boldsymbol{r} = \mathbf{x}_2 - \mathbf{x}_1, \qquad \boldsymbol{p} = \mu\dot{\boldsymbol{r}}.$$

The relative angular momentum expressed in two-body coordinates is readily shown to be

$$\boldsymbol{M} = \boldsymbol{r} \times \boldsymbol{p}. \tag{4.2}$$

The energy in two-body coordinates is

$$\mathcal{H} = p^2/2\mu - k/r. \tag{4.3}$$

An unusual connection exists between the energy and the angular momentum in two-body motion, a direct result of the existence of the angular momentum as an isolating invariant. The momentum may be represented as a sum of the component in the radial direction of the position vector $\boldsymbol{p}_r = \mu\dot{r}(\boldsymbol{r}/r)$ and that in the direction perpendicular to it $\boldsymbol{p}_\perp$:

$$\boldsymbol{p} = \boldsymbol{p}_r + \boldsymbol{p}_\perp.$$

The radial direction momentum does not contribute to the angular momentum so the angular momentum is expressible as $\boldsymbol{M} = \boldsymbol{r} \times \boldsymbol{p}_\perp$. Since $\boldsymbol{r}$ and $\boldsymbol{p}_\perp$ are orthogonal, a simple identity shows that the perpendicular momentum is contained within the angular momentum:

$$\boldsymbol{p}_\perp = -\boldsymbol{r} \times \boldsymbol{M}/r^2. \tag{4.4}$$

Since $\boldsymbol{r}$ and $\boldsymbol{M}$ are perpendicular, the magnitude of the perpendicular momentum is given by

$$p_\perp = M/r.$$

This result provides two-body motion a sweeping simplification; for the perpendicular momentum need not explicitly appear in any mechanical expression. It is contained within the invariant $\boldsymbol{M}$. The perpendicular momentum can be represented as a pure function of $\boldsymbol{r}$ through Eq. (4.4). Thus, the total kinetic energy of the motion can be expressed in terms of the total angular momentum as

$$T = p_r^2/2\mu + M^2/2\mu r^2,$$

where $p_r = \mu\dot{r}$ is the radial momentum. This representation of T in terms of M^2 is a result of the isolating power of the angular momentum invariant. The total energy may be expressed in the compact form

$$\mathcal{H} = p_r^2/2\mu + W(r), \tag{4.5}$$

where $W(r)$ is the *effective potential* defined as

$$W(r) = M^2/2\mu r^2 + V(r) = M^2/2\mu r^2 - k/r. \tag{4.6}$$

The relative motion in the center of mass reference frame may be regarded as an equivalent *one-body* problem of mass μ, position coordinate $\mathbf{r}$, linear momentum $\boldsymbol{p} = \mu\dot{\mathbf{r}}$, and angular momentum $\boldsymbol{M} = \boldsymbol{r} \times \boldsymbol{p}$. In this equivalent one-body problem in the frame of relative motion translating with the center of mass, $\boldsymbol{M}$ and $\mathcal{H}$ are invariants; but the relative linear momentum $\boldsymbol{p} = \mu\dot{\mathbf{r}}$ *is not.*

The Hidden Invariant of Kepler Motion

The isolating invariant that closes the orbit of two-body motion with an inverse-square force, $\boldsymbol{f} = -\partial V/\partial \mathbf{r} = -k\mathbf{r}/r^3$, is the *eccentricity* $\mathbf{e}$:

$$\mathbf{e} = (\boldsymbol{p} \times \boldsymbol{M})/k\mu - \mathbf{r}/r. \tag{4.7}$$

One may verify that the eccentricity obeys the conservation law $d\mathbf{e}/dt = 0$. This is done in **Note 3**. There it is shown that the Poisson bracket $[\mathbf{e}, \mathcal{H}]$ vanishes. Since $\mathbf{e}$ does not depend explicitly upon time, it is an invariant by Eq. (3.33).

The eccentricity invariant was first recognized in the early eighteenth century by Jacob Hermann, a student and protege of Jacob Bernoulli and a favorite colleague of Leibniz. It was elaborated by Simon Pierre Marquis de Laplace in the *Traité de méchanique céleste* of 1799 and discovered, apparently independently, by William Rowan Hamilton again in 1845.*
The eccentricity invariant was introduced into this century by Carl Runge in his *Vektoranalysis* of 1919. It has also been referred to as the *Laplace vector* and the *Runge–Lenz vector.*

How is the eccentricity vector oriented with respect to the other fundamental vectors of the motion: $\boldsymbol{r}$, $\boldsymbol{p}$, and $\boldsymbol{M}$? Since $\boldsymbol{r}$ and $\boldsymbol{p} \times \boldsymbol{M}$ are

---

* Hermann's invariant may be found in "Extrait d'une lettre de M. Herman à M. Bernoulli, datée de Padoue le 12. Juillet 1710" published in *Histoires de L'Academie Royale des Sciences avec les Mémoires de Mathematique et Physique*, 519–523, Paris (1712). [For a brief survey of Hermann's life and works, see F. Nagel, *Historia Mathematica*, **18**, 36–54 (1991).] For an historical account of the eccentricity invariant see H. Goldstein, *Am. J. Phys.* **43**, 8 (1975) and **44**, 11 (1976) which includes P. S. Laplace, *Traité de méchanique céleste*, Tome I, Premiere Partie, Livre II, 165ff, Paris (1798–1799); W. R. Hamilton, *Proc. R. Irish Acad.* **3**, Appendix No. III, xxxviff (1847).

each perpendicular to $\boldsymbol{M}$, the eccentricity vector is also perpendicular to $\boldsymbol{M}$ and shares the plane of $\boldsymbol{r}$ and $\boldsymbol{p}$ as shown in Fig. 4-11. As an invariant, $\boldsymbol{e}$ remains permanently perpendicular to the invariant $\boldsymbol{M}$; but the angles $\boldsymbol{r}$ and $\boldsymbol{p}$ make with $\boldsymbol{e}$ in the plane that they share will vary because these quantities are not invariant.

It may seem that the three-vector $\boldsymbol{e}$ brings three additional invariants to the motion; but this is not so. Since $\boldsymbol{e}$ is expressed in terms of $\boldsymbol{M}$, these two vectors are not independent. The magnitude of $\boldsymbol{e}$ is completely expressible in terms of the other invariants $\mathcal{H}$ and M. In addition, the vectors $\boldsymbol{M}$ and $\boldsymbol{e}$ are orthogonal. They by nature satisfy the subsidiary conditions

$$e^2 = 1 + \frac{2\mathcal{H}}{k^2\mu} M^2, \qquad \boldsymbol{M} \cdot \boldsymbol{e} = 0. \tag{4.8}$$

Because of the two conditions (4.8), the eccentricity contributes only one additional independent invariant to the motion.

The Symmetries of Kepler Motion

The only symmetries present in all classical motions are the elementary symmetries of translations and rotations of space and time. The only invariants present in all classical motion are the Galilean invariants that spring from these elementary symmetries. However, each configuration of particles and force law introduces additional symmetries and accompanying invariants. The inverse-square nature of gravitational and electrical forces for two bodies expands the symmetries of the Galilean invariants in a remarkable manner: the three-dimensional rotational symmetry generated by central forces (and its corresponding invariant, the angular momentum) is expanded to a four-dimensional rotational symmetry by the additional action of the inverse-square nature of the force (and its corresponding invariant, the eccentricity). It need not be so and, indeed, would not be so if the force were not inverse-square.

Put differently, the symmetries of two-body motion that exist independently of the law of force are three-dimensional rotations and their invariant is the angular momentum, the three-vector $\boldsymbol{M}$. When one further specifies the force is inverse-square, a second three-vector, the eccentricity $\boldsymbol{e}$, comes into being also bearing rotational symmetry; but the rotations represented by these two vectors in concert is in four rather than three dimensions. (This does not mean the three-dimensional configuration space of the motion has become four-dimensional. The space in question is the phase space of the relative motion.)

To see how four-dimensional rotations are the symmetries of two-body motion with inverse-square forces one must first grasp the nature of rotations in spaces of various dimension. The rotation of a vector $\xi =$

$(\xi_1, \xi_2, \ldots, \xi_n)$ in n-dimensional space consists of transformations

$$\xi' = R\xi$$

that leave the magnitude of the vector unchanged:

$$\xi'^2 = \xi^2.$$

The matrix R that effects the rotations is an $n \times n$ matrix. The unique properties that make it a rotation matrix may be established by forming the magnitude

$$\xi'^2 = R\xi R\xi.$$

The product $R\xi$ may be rearranged as $R\xi = \xi R^\dagger$ where $R^\dagger$ is the transpose of R obtained by interchanging its rows and columns. The invariance of the magnitudes then becomes

$$\xi'^2 = \xi R^\dagger R\xi = \xi^2$$

showing that the product of the rotation matrix and its transpose must be the identity*

$$R^\dagger R = I. \tag{4.9}$$

The condition (4.9) is the fundamental condition that all rotation matrices must satisfy. In particular, one finds that the identity $R = I$ satisfies Eq. (4.9) and is itself a rotation matrix.

One may further draw out the structure of a rotation by considering rotations which depart from the identity by an infinitesimal amount:

$$R = I + \delta w,$$

where $|\delta w| \ll 1$. Substitution of the infinitesimal rotation into Eq. (4.9) shows that to first order, $\delta w + \delta w^\dagger = 0$, or δw is an antisymmetric matrix ("skew-symmetric" if one prefers a more refined term):

$$\delta w_{ij} + \delta w_{ji} = 0.$$

Antisymmetry makes less than half the n^2 components of R independent. The diagonal consists of unit elements and the off-diagonal components are reflections across the diagonal. The number of independent components is therefore equal to the number of components lying either above or below

---

* The argument in subscript notation: $\xi'_i = R_{ij}\xi_j$ gives $\xi'^2_i = R_{ij}\xi_j R_{ik}\xi_k$; the transpose $R_{ij} = R^\dagger_{ji}$ allows the product of the matrix and the vector to be rearranged to $R_{ij}\xi_j = \xi_j R^\dagger_{ji}$ from which follow $\xi'^2_i = \xi_j R^\dagger_{ji} R_{ik}\xi_k = \xi^2_i$ and $R^\dagger_{ji} R_{ik} = \delta_{jk}$.

the diagonal which is $n(n-1)/2$. One finds an n-dimensional rotation is specified by $n(n-1)/2$ parameters.

Rotations do not exist in spaces of one dimension. In two-dimensional space one parameter specifies a rotation. In three dimensions there are three parameters. In four dimensions six parameters are required to specify a rotation. The correspondence between the parameters which describe a rotation and the dimension of the space is neither simple nor intuitive. Only in three dimensions is the number of parameters equal to the dimension of the space while in four dimensions the number of parameters is double the number of those in three dimensions. The fact that the number of parameters in four dimensions is twice those in three dimensions is a manifestation of a remarkable correspondence between three- and four-dimensional space we will soon encounter.

The invariant of a rotation in n-dimensional space,

$$\xi_1{}^2 + \xi_2{}^2 + \cdots + \xi_n{}^2 = \xi^2,$$

is also the equation of an $(n-1)$-sphere embedded in n-dimensional space. Spheres and rotations are therefore intimately related. The symmetry of the sphere is a static, geometric thing: it is the locus of all points equidistant (in the Euclidean sense) from an origin. The symmetry of rotations is a dynamic that lies upon a sphere. The rotation $\xi' = R\xi$ sends the point ξ' over the surface of a sphere of radius $|\xi|$ as the parameters within R are varied. A ray connecting the origin at the center of the sphere to the moving point ξ' performs rotations, its tip constrained to the surface of this sphere.

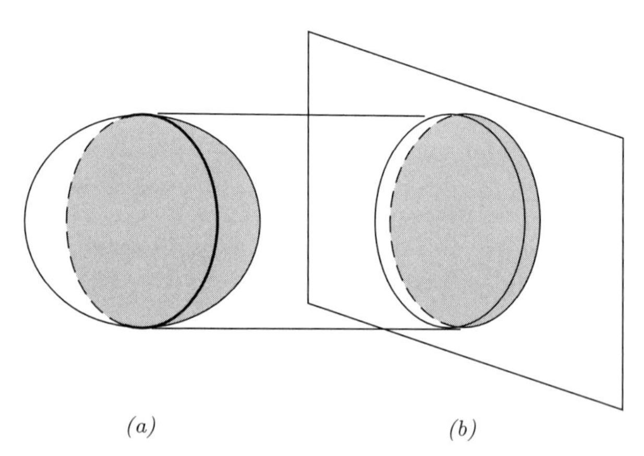

Figure 4-3. Projection of a Two-Sphere onto the Two-Dimensional Plane. A two-sphere *(a)* projected onto the two-dimensional plane reveals a pair of superimposed disks *(b)*, each disk filled or *foliated* with a family of circles.

The most elementary rotation is that in two dimensions. The geometric object which is the basis of rotational symmetry in the plane is a one-sphere:

the circle. Consistent with the existence of one rotation parameter in two dimensions, points lying on a circle in *two*-dimensional space are described by *one* coordinate (such as the distance along its circumference or an angle of rotation); hence the name *one*-sphere.

The basis of rotational symmetry in three dimensions is the familiar two-sphere (basketball, balloon, map-globe). It has two degrees of freedom to locate a point upon it. In four dimensions the three-sphere is the basis of rotational symmetry. All spheres exhibit rotational symmetry in that when points on the surface are subjected to a rotation, the sphere does not change form. The angles specifying the location of a point on the surface of the sphere change; but the radius—the invariant of rotational symmetry—does not.

Human beings can visualize one-spheres (circles) and two-spheres (balloons) embedded in three-dimensional Euclidean space; but we have difficulty visualizing higher-dimensional spheres. As an aid to visualization and insight into the underlying connections between rotations in spaces of various dimension, one can examine the way in which higher-dimensional spheres project into more familiar spaces like the two-dimensional plane and three-dimensional Euclidean space.

The two-sphere projects onto the two-dimensional plane as a pair of superimposed disks as shown in Fig. 4-3. Imagine a semi-transparent map-globe whose image falls on the plane. Each of the hemispheres of the globes projects as a disk. The superimposed disks may be thought of as the two sides of an infinitely thin coin, each side a projection of one of the hemispheres. The two-sphere may be defined by

$$\xi_1{}^2 + \xi_2{}^2 + \xi_3{}^2 = const$$

and the projection achieved by setting the coordinate perpendicular to the plane, say ξ_3, constant. This leads to the family of circles $\xi_1{}^2 + \xi_2{}^2 = const$, each circle a projection from the two-sphere for each value of ξ_3. Each disk is built up with circles (it is said to be *foliated* by circles). There are two families of circles because it takes two disks to completely cover both hemispheres of the two-sphere. One disk covers the region $\xi_3 < 0$, the other covers $\xi_3 > 0$ and the circular boundary (great circle) joining both covers $\xi_3 = 0$. Although the projective points are infinitely dense within one another, points on different disks are disconnected by direct paths. They can only be connected by a path which extends to and crosses over the circular boundary joining them at the edges of the disks.

The three-sphere is the basis of rotational symmetry in four dimensions. It projects into three-dimensional Euclidean space as two three-dimensional spherical bodies superimposed *within* one another as shown in Fig. 4-4. Each of these spherical bodies is foliated with a family of nested two-spheres within it. If the three-sphere is defined by

$$\xi_1{}^2 + \xi_2{}^2 + \xi_3{}^2 + \xi_4{}^2 = const,$$

and the projection is along the ξ_4 axis, one of the bodies covers the hemisphere $\xi_4 < 0$, the other covers $\xi_4 > 0$, and the bounding two-sphere ("great two-sphere") which connects both covers $\xi_4 = 0$. There are two families because it takes two spherical bodies filled with two-spheres to completely cover the three-sphere much like it takes two disks filled with circles to completely cover the two-sphere. The points lying within the projection are again "doubly infinite" because they are contributed by two distinct regions of the three-sphere just as points on the disks were contributed by two different hemispheres of the two-sphere. Points within different spherical bodies, like those of the disks, are disconnected by direct paths. Points within one spherical body are linked to points within the other by paths which cross the great two-sphere joining both.

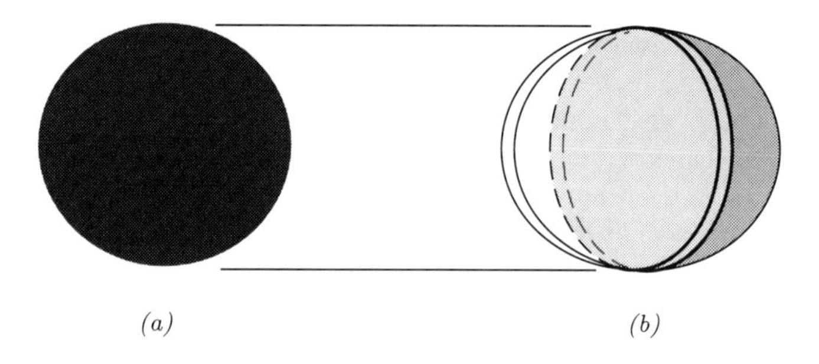

Figure 4-4. Projection of a Three-Sphere into Three-Dimensional Euclidean Space. The projection of a three-sphere *(a)* into three-dimensional Euclidean space reveals a pair of interpenetrating spherical bodies *(b)*, each body foliated by a family of two-spheres.

This property of four-dimensional space in which the three-sphere underlying its rotational symmetry may be covered by two families of two-spheres is more than an aid to visualization. It is exceptional. One might suppose that all higher-dimensional spheres could be projected into foliations of lower-dimensional spheres; but this turns out to be impossible except for the case of four dimensions. Four-dimensional space is unique; and its topological and geometric properties are quite surprising.* The remarkable family of relationships between one-, two-, and three-dimensional spheres was discovered by the topologist Heinz Hopf in the first half of this century. These properties, which Hopf established as pure mathematical entities, turn out to be the fundamental topological properties of Kepler motion. The beautiful mapping of the three-sphere into a pair of families of two-spheres may be found in **Note 6**.

For our purposes it is sufficient to recognize that the unique properties of four-space allow a three-sphere embedded in it to be covered by a pair of families of nested two-spheres. Each family of two-spheres has three-

---

* See Ian Stewart, *The Problems of Mathematics*, Oxford University Press, Chaps. 9 and 17 (1987).

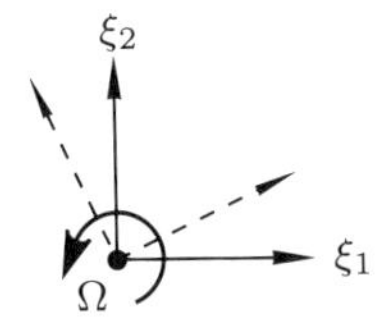

Figure 4-5. Rotation in Two-Space. A single piece of information such as the angle Ω through which the two coordinate axes are turned specifies a rotation in the two-dimensional plane.

dimensional rotations as a symmetry. We therefore arrive at a defining moment for the Kepler symmetries: a four-dimensional rotation may be represented as a pair of three-dimensional rotations. A pair of three-dimensional rotational symmetry algebras like that of the angular momentum (3.18) therefore describes the symmetries of two-body motion with inverse-square forces. Let us see how this happens.

In two dimensions with coordinates ξ_1, ξ_2 the rotation lies in the only two-dimensional subspace possible, the plane (ξ_1, ξ_2). A rotation in two dimensions possesses one free parameter, such as the angle Ω through which the coordinates are rotated in Fig. 4-5.

In three dimensions with coordinates ξ_1, ξ_2, ξ_3, three independent planes are defined by the pairs of components (ξ_2, ξ_3), (ξ_3, ξ_1), (ξ_1, ξ_2). These planes correspond to the three independent parameters required to describe a three-dimensional rotation. It turns out that any rotation in three-space may be realized as a sequence of three two-dimensional rotations, one in each of these planes. In each of these elementary rotations, one of the coordinates is held fixed while the other two are rotated. Each of these two-dimensional rotations is specified by a rotation angle which fixes the rotation in each plane as shown in Fig. 4-6.

Three pieces of information are naturally carried by a three-vector. A rotation in three dimensions may therefore be described by a three-vector such as $\boldsymbol{\Omega} = (\Omega_1, \Omega_2, \Omega_3)$ as was done in Chap. 3. Each symmetry parameter has a corresponding invariant; and the invariant which corresponds to $\boldsymbol{\Omega}$ is the angular momentum $\boldsymbol{M}$ satisfying the cyclic Poisson bracket (3.18) of rotational symmetry.*

Now leap to four dimensions. Six parameters are required to specify a rotation in four-space. They match the six independent planes that exist in four-space: the three planes of three-space plus three new planes (ξ_1, ξ_4), (ξ_2, ξ_4), (ξ_3, ξ_4), formed by the fourth dimension paired one at a time with the three other dimensions as illustrated in Fig. 4-7. The six planes are the property of the pair of families of two-spheres with which the three-sphere

---

* The rotation matrix in three dimensions may be explicitly exhibited through the alternating tensor ϵ_{ijk} as $R_{ik} = e^{\epsilon_{ijk}\Omega_j}$; but its form in four and higher dimensions is more complicated.

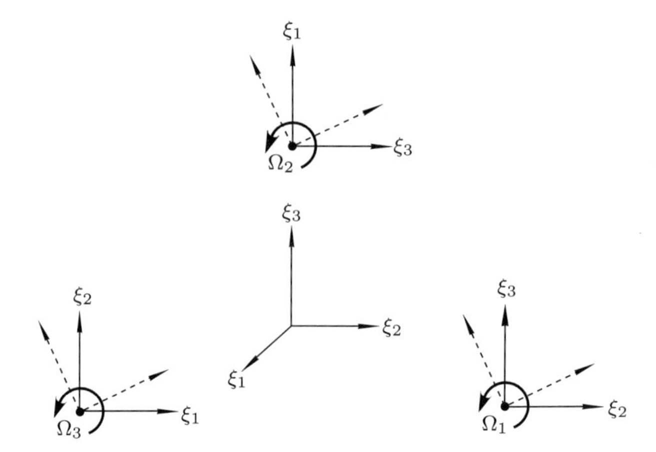

Figure 4-6. Rotation in Three-Space. A rotation in three dimensions may be realized as three successive two-dimensional rotations, one in each of the three independent planes that exist in three-space. Each of these two-dimensional rotations is specified by a rotation angle. The three angles naturally form a three-vector, $\boldsymbol{\Omega} = (\Omega_1, \Omega_2, \Omega_3)$.

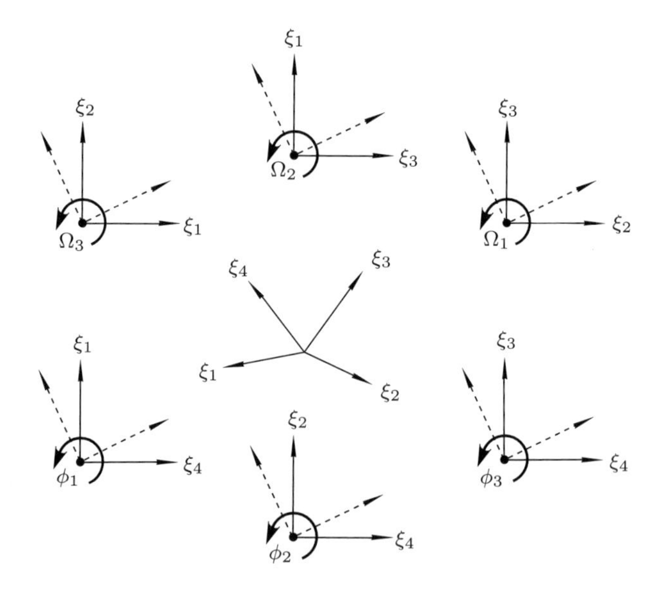

Figure 4-7. Rotation in Four-Space. Six independent planes exist in four-space. A rotation in four-space may be realized as six successive two-dimensional rotations, one in each of these planes. The six angles specifying the rotations form two three-vectors, $\boldsymbol{\Omega} = (\Omega_1, \Omega_2, \Omega_3)$ and $\boldsymbol{\phi} = (\phi_1, \phi_2, \phi_3)$.

of rotational symmetry in four-space is foliated. A rotation in four-space may therefore be realized as six successive two-dimensional rotations, one in each of the six planes as shown in Fig. 4-7.

Six pieces of information are required to describe a rotation in four-space. Two three-vectors carry six pieces of information and are the natural vehicles to describe a four-dimensional rotation. One of these is again the vector $\boldsymbol{\Omega}$. It describes the rotations in the planes (ξ_2, ξ_3), (ξ_3, ξ_1), (ξ_1, ξ_2) and its invariant is the angular momentum $\boldsymbol{M}$. The other which we symbolize as $\boldsymbol{\phi} = (\phi_1, \phi_2, \phi_3)$ describes the rotations in the planes (ξ_1, ξ_4), (ξ_2, ξ_4), (ξ_3, ξ_4) as indicated in Fig. 4-7. Its invariant turns out to be the eccentricity $\boldsymbol{e}$.

In summary, a single set of three-dimensional rotations possesses a single invariant three-vector. Two sets of three-dimensional rotations possess two three-vector invariants; and such a pair represents* four-dimensional rotations. The symmetry parameter for one set of three-dimensional rotations is $\boldsymbol{\Omega}$ and has associated with it the invariant $\boldsymbol{M}$. The additional set of rotations described by the symmetry parameter $\boldsymbol{\phi}$ possesses the additional invariant $\boldsymbol{e}$. These vectors cannot be arbitrary three-vectors. They must possess rotational character. One must show they satisfy a rotational Poisson bracket algebra.

Begin by observing that the quantity

$$h \equiv \sqrt{k^2\mu/(-2\mathcal{H})}$$

which appears as the coefficient of M^2 in the first of Eqs. (4.8) has the dimensions of action or angular momentum. This quantity is a key actor in the classical motion of two bodies; among its many roles, we shall see it as the Poincaré invariant. Because of its recurring importance, we shall call h the *Kepler constant.* It is intimately connected to an equally important quantity in the quantum motion. The Kepler constant h is the classical counterpart of the quantum Planck constant $\hbar$.

The eccentricity and the angular momentum invariants may be put on the same footing by using a slightly different dimensional form of the eccentricity created through the Kepler constant:

$$\boldsymbol{E} \equiv h\boldsymbol{e}.$$

Since $\boldsymbol{E}$ is represented as a function of the invariants $\boldsymbol{e}$ and $\mathcal{H}$ (through h), it is similarly invariant and serves equally well in favor of $\boldsymbol{e}$ in the complete set of invariants for two-body, inverse-square force motion.

Let us dare to call $\boldsymbol{E}$ the *eccentrum.* The eccentrum has the same dimensions as the angular momentum; indeed, it is the intimate connection between $\boldsymbol{E}$ and $\boldsymbol{M}$ that underlies the symmetry. For bound motions ($\mathcal{H} < 0$) the components of the eccentrum are real quantities (the components of the eccentricity $\boldsymbol{e}$ are always real); but for free motions ($\mathcal{H} > 0$), the eccentrum components—and the Kepler constant h—are imaginary (**Note 3**).

---

* A representation of a symmetry is a set of vectors that satisfy its symmetry algebra.

The Poisson bracket of the angular momentum $\boldsymbol{M}$ cyclically reproduces $\boldsymbol{M}$ itself. A straightforward calculation that is carried out in **Note 3** shows that the Poisson bracket algebra for the pair of vectors $\boldsymbol{M}$ and $\boldsymbol{E}$ is

$$[M_i, M_j] = \epsilon_{ijk} M_k, \qquad [E_i, M_j] = \epsilon_{ijk} E_k, \qquad [E_i, E_j] = \epsilon_{ijk} M_k. \quad (4.10)$$

From the algebra of the individual components one readily finds the following algebra of the magnitudes:

$$[M^2, E_i] = -[E^2, E_i] = 2\epsilon_{ijk} E_j M_k$$

and all scalars have vanishing Poisson brackets with the angular momentum,

$$[M^2, M_i] = 0, \qquad [E^2, M_i] = 0, \qquad [M^2, E^2] = 0.$$

The algebra of the eccentrum in the third of Eqs. (4.10) is not the same as the paradigmatic algebra of three-dimensional rotations typified by the angular momentum in the first of Eqs. (4.10). The algebra of $\boldsymbol{E}$, rather than cyclically reproducing $\boldsymbol{E}$ itself in the manner of angular momentum, reproduces $\boldsymbol{M}$! The reason for this twist is that the invariants $\boldsymbol{M}$ and $\boldsymbol{E}$ are not themselves the two invariants belonging to the pair of spheres into which four-dimensional rotations project, each of which is a rotationally symmetric vector with algebra like that of the angular momentum. Rather, they are sums and differences of these invariants as we shall soon see. The Poisson brackets of $\boldsymbol{E}$ are not confined to $\boldsymbol{E}$ but instead incorporate $\boldsymbol{M}$. The elements of the symmetry algebra are therefore six in number: M_1, M_2, M_3, E_1, E_2, E_3 and correspond to the six components of the two three-vectors required to represent a four-dimensional rotation.

The three-dimensional rotational symmetry underlying the invariant $\boldsymbol{M}$ is a fundamental symmetry of mechanics, no matter the force law; and the symmetry algebra reflects this. It can be seen that the algebra for $\boldsymbol{M}$ alone involves only its three components as indicated by the first of Eqs. (4.10). The algebra of $\boldsymbol{E}$ mixes in $\boldsymbol{M}$ but that of $\boldsymbol{M}$ does not mix in $\boldsymbol{E}$. This means that three-dimensional rotations (represented by $\boldsymbol{M}$ alone) form a subgroup of the full group of four-dimensional rotations (represented by both $\boldsymbol{M}$ and $\boldsymbol{E}$). Three-dimensional rotational symmetry persists in the presence of the wider symmetry in four dimensions induced by the inverse-square nature of the force. This symmetry is described by the angular momentum $\boldsymbol{M}$ and possesses a single Casimir invariant, the magnitude M. (Recall that a Casimir invariant is a quantity that has a vanishing Poisson bracket with all elements of the symmetry algebra).

Since four-dimensional rotations are described by two three-vectors, they possess two Casimir invariants. It is not difficult to show from the first of the equations following Eqs. (4.10) that the Casimir invariants of four-dimensional rotations described by $\boldsymbol{M}$ and $\boldsymbol{E}$ are $M^2 + E^2$ and $\boldsymbol{M} \cdot \boldsymbol{E}$. In this light the subsidiary conditions (4.8) natural to the eccentricity invariant come into their own. When $\boldsymbol{e}$ is expressed as $\boldsymbol{e} = \boldsymbol{E}/h$, they take the simple form

$$M^2 + E^2 = h^2, \qquad \boldsymbol{M} \cdot \boldsymbol{E} = 0. \tag{4.11}$$

The subsidiary conditions of the two-body problem for gravity and electricity turn out to be conditions on the Casimir invariants of four-dimensional rotations!

The four-dimensional rotational symmetry may be made fully transparent as two sets of three-dimensional rotations appropriate to a pair of two-spheres by using the invariant vectors of these spheres out of which $\boldsymbol{M}$ and $\boldsymbol{E}$ are constructed. Let $\boldsymbol{S}$ and $\boldsymbol{D}$ be the invariants of these two sets of three-dimensional rotations. The vectors $\boldsymbol{M}$ and $\boldsymbol{E}$ turn out to be sums and differences of $\boldsymbol{S}$ and $\boldsymbol{D}$:

$$\boldsymbol{M} = \boldsymbol{S} + \boldsymbol{D}, \qquad \boldsymbol{E} = \boldsymbol{S} - \boldsymbol{D},$$

with the inverse relationships

$$\boldsymbol{S} = (\boldsymbol{M} + \boldsymbol{E})/2, \qquad \boldsymbol{D} = (\boldsymbol{M} - \boldsymbol{E})/2.$$

It is an easy matter to show from Eqs. (4.10) that $\boldsymbol{S}$ and $\boldsymbol{D}$ are each representations of purely three-dimensional rotations just like the angular momentum, each reproducing only itself according to the cyclic paradigm of rotational symmetry:

$$[S_i, S_j] = \epsilon_{ijk} S_k, \qquad [S_i, D_j] = 0, \qquad [D_i, D_j] = \epsilon_{ijk} D_k. \tag{4.12}$$

The vectors $\boldsymbol{S}$ and $\boldsymbol{D}$ each represent rotations in three dimensions. Each corresponds to one of the families of two-spheres into which the three-sphere of four-dimensional rotational symmetry projects. Each vector accordingly has the identical algebra of rotations in three dimensions and each vector's algebra is independent of the other's.

One could therefore have begun with two copies of three-dimensional rotational symmetry corresponding to a pair of two-spheres, the vectors $\boldsymbol{S}$ and $\boldsymbol{D}$. Putting this pair of three-dimensional representations together as the sums and differences $\boldsymbol{M} = \boldsymbol{S} + \boldsymbol{D}$ and $\boldsymbol{E} = \boldsymbol{S} - \boldsymbol{D}$, one would then have found that $\boldsymbol{M}$ and $\boldsymbol{E}$ obey the Poisson bracket algebra (4.10) and turn out to be the angular momentum and eccentrum of Kepler motion.

Although $\boldsymbol{M}$ and $\boldsymbol{E}$ are more familiar, the vectors $\boldsymbol{S}$ and $\boldsymbol{D}$ are an equally valid set of invariants of the symmetry equivalent to $\boldsymbol{M}$ and $\boldsymbol{E}$ in the sense that if one specifies the vectors $\boldsymbol{M}$ and $\boldsymbol{E}$, the vectors $\boldsymbol{S}$ and $\boldsymbol{D}$ are uniquely specified. The Casimir invariants $E^2 + M^2$ and $\boldsymbol{M} \cdot \boldsymbol{E}$ in the two subsidiary conditions (4.11) become in terms of $\boldsymbol{S}$ and $\boldsymbol{D}$:

$$S^2 + D^2 = h^2/2, \qquad S^2 - D^2 = 0, \tag{4.13}$$

from which it follows that S and D are the same Casimir invariant and this Casimir invariant, aside from a factor of one-half, is none other than the Kepler constant:

$$S = D = h/2.$$

The radii of the spheres of four-dimensional rotational symmetry are Kepler constants and depend only upon the energy. The spheres have real radii for bound motions ($\mathcal{H} < 0$) and imaginary radii for free motions ($\mathcal{H} > 0$). The phase space of the invariants for free motions is therefore non-Euclidean (**Note 3**).

The Two Faces of Rotational Symmetry

Four-dimensional rotations are represented by a pair of three-vectors; but there exist two alternative representations of these three-vectors, one in $\boldsymbol{M}$ and $\boldsymbol{E}$, the other in $\boldsymbol{S}$ and $\boldsymbol{D}$. The representations are distinct because the Poisson bracket algebras (4.10) and (4.12) for each representation are different; yet the states of motion corresponding to these different representations are the same because given $\boldsymbol{M}$ and $\boldsymbol{E}$, the vectors $\boldsymbol{S}$ and $\boldsymbol{D}$ are uniquely fixed.

Why are there two different representations of the motion of two bodies with inverse-square forces? They are the manifestation of a fundamental feature of three-dimensional rotations: three-dimensional rotations come in two mathematical forms, one based on three-vectors with real components, the other based on two-vectors with complex components known as spinors. The four-dimensional rotational symmetry of the motion is equivalent to a pair of three-dimensional rotations; and these three-dimensional rotations have both a vector and a spinor face.

The significance of this duality for the two-body problem is that both the classical motion of planetary bodies and the quantum motion of the hydrogen atom have two distinct faces: one vector, the other spinor. Each of these faces has a corresponding geometric form. The vector face is exhibited in spheres and cones, the spinor face in paraboloids. The representations in $\boldsymbol{S}$ and $\boldsymbol{D}$ on the one hand and $\boldsymbol{M}$ and $\boldsymbol{E}$ on the other correspond to these two faces. We shall explore these two faces of rotational symmetry algebraically. Then we shall show their different geometrical forms.

The familiar face of rotational symmetry is the vector face as exhibited by a three-vector such as the angular momentum,

$$\boldsymbol{M} = \begin{pmatrix} M_1 \\ M_2 \\ M_3 \end{pmatrix},$$

which satisfies the Euclidean metric

$${M_1}^2 + {M_2}^2 + {M_3}^2 = M^2.$$

Whereas a vector consists of real elements, a spinor consists of complex elements. A spinor is built from complex two-vectors, such as

$$\boldsymbol{\eta} = \begin{pmatrix} \eta_1 \\ \eta_2 \end{pmatrix}. \tag{4.14}$$

Since the two-vector is complex, four unique quantities may be formed from products of its two components η_1, η_2 and their conjugates η_1^*, η_2^*. Let us call these four quantities M^+, M^-, M_+, M_-. These four quantities create a two-matrix; and this two-matrix turns out to be the spinor (complex two-matrix) equivalent of a three-vector:

$$\begin{pmatrix} M^+ & M_+ \\ M_- & M^- \end{pmatrix} = \begin{pmatrix} \eta_1\eta_1^* & \eta_1\eta_2^* \\ \eta_2\eta_1^* & \eta_2\eta_2^* \end{pmatrix}. \tag{4.15}$$

One can easily see that this spinor possesses a vanishing determinant, a fundamental property of all two-matrices built from a single two-vector:

$$M^+M^- - M_+M_- = 0. \tag{4.16}$$

The spinor equivalent of a three-vector has a quadratic form which is an invariant vanishing identically. Equation (4.16) is the "spinor metric" condition which creates an invariant spinor magnitude (it is always zero) just as the Euclidean metric creates an invariant vector magnitude.

The coincidence between the vector and the spinor is therefore this: a collection of four real things obeying the Euclidean metric is equivalent to a collection of two complex things obeying the spinor metric (the four spinor components M^+, M^-, M_+, M_- are built from a single two-vector with two complex components):

$$\boldsymbol{M} = \begin{pmatrix} M_1 \\ M_2 \\ M_3 \end{pmatrix} \Longleftrightarrow \begin{pmatrix} M^+ & M_+ \\ M_- & M^- \end{pmatrix}.$$

It may seem incongruous that there are four spinor components to represent a three-vector; but the magnitude of the vector is encoded in the spinor along with its three components. The four real things in the case of the vector are its three components M_1, M_2, M_3, and its magnitude M.

One therefore has two sets of four quantities (M, M_1, M_2, M_3) for the vector and (M^+, M^-, M_+, M_-) for the spinor, each of which is a representation of the same physical quantity. The relationships between the two forms were first worked out by Wolfgang Pauli in the nineteen twenties. The spinor components are given in terms of the vector components by

$$\begin{aligned} M^+ &= M + M_3, \qquad & M_+ &= M_1 + iM_2, \\ M^- &= M - M_3, \qquad & M_- &= M_1 - iM_2 \end{aligned} \tag{4.17}$$

(note that M_+ and M_- are complex conjugates). The inverse relationships of the vector components to the spinor components are

$$\begin{aligned} M &= (M^+ + M^-)/2, \qquad & M_1 &= (M_+ + M_-)/2, \\ M_3 &= (M^+ - M^-)/2, \qquad & M_2 &= (M_+ - M_-)/2i. \end{aligned} \tag{4.18}$$

Equations (4.17) and (4.18) reveal that the spinor neatly divides into two parts, the part contained on the left-hand sides of Eqs. (4.17) and (4.18) pairs (M, M_3) with (M^+, M^-) while that on the right-hand sides pairs (M_1, M_2) with (M_+, M_-). These two parts are "hinged" by the common product $M^+M^- = M_+M_-$ which also shows how the spinor metric and Euclidean metric imply one another. The first product is $M^+M^- = M^2 - {M_3}^2$ and the second is $M_+M_- = {M_1}^2 + {M_2}^2$. These two products, according to the spinor metric (4.16), are the same. They yield the vector metric condition

$$M^+M^- = M_+M_- = M^2 - {M_3}^2 = {M_1}^2 + {M_2}^2. \tag{4.19}$$

Each of the spinor parts has significance in its own right. Readers have perhaps already taken note of the fact that (M, M_3) which appears in one of the pairings constitutes the Casimir set of the vector. This means one part of the spinor is naturally the Casimir set. The spinor representation of the Casimir set of rotational symmetry is (M^+, M^-).

Consider now the part involving (M_1, M_2) and (M_+, M_-). Unlike M^+ and M^- which have a vanishing Poisson bracket, the quantities M_+ and M_- have the same non-vanishing Poisson bracket as M_1 and M_2 save for a factor:

$$[M_+, M_-] = -2i[M_1, M_2] = -2iM_3.$$

Although M_+ and M_- do not have a vanishing Poisson bracket, their bracket bears a decisive relationship to the component M_3. The Poisson bracket algebra of M_+ and M_- produces M_3 while their brackets with M_3 result in *self-* rather than cyclic-reproduction:

$$[M_\pm, M_3] = \pm iM_\pm.$$

While the elements of a rotationally symmetric vector like $\boldsymbol{M}$ cyclically reproduce one another, the spinors (M_+, M_-) constructed from them reproduce only themselves. Self-reproduction gives the spinor pair (M_+, M_-) the power to generate the values of a given component of the angular momentum such as M_3. This is because when one forms Poisson bracket products of $M_\pm$ and M_3, the algebra remains confined to $M_\pm$. Increasing (or decreasing) quantities of M_3 are the outcomes. For this reason M_+ and M_- are called creation and annihilation operators: they create and annihilate values of M_3. This property is exploited in quantum motion where M_+

and M_- take the central role in the quantum description of rotational symmetry. They generate all the values of the angular momentum which are possible in quantum mechanics. We shall see the creation and annihilation operators M_+ and M_- in action in Chap. 5.

Each pair of spinor components of a rotationally symmetric vector carries an additional significance beyond that of simply keeping book on the components and magnitude of the vector. The spinor pair (M^+, M^-) is the Casimir set of this vector while the pair (M_+, M_-) contains its creation and annihilation operators.

The two representations of rotational symmetry have two different but closely related geometrical forms. To see this, turn to the fundamental condition that all vectors in Euclidean space obey: ${M_1}^2 + {M_2}^2 + {M_3}^2 = M^2$. This is a second order algebraic expression, the general family of which describes spheres, cones, paraboloids, and hyperboloids. The vector specification of the Casimir set (M, M_3) describes spheres $M = const$ and cones $M_3/M = const$. The quantities M and M_3 are therefore said to be spherical-polar. The angular momentum vector must lie on the cone of half-angle $\tan^{-1}(M_3/M)$ centered on the M_3 axis. The spherical-polar surfaces are shown in Fig. 4-8 *(a)*.

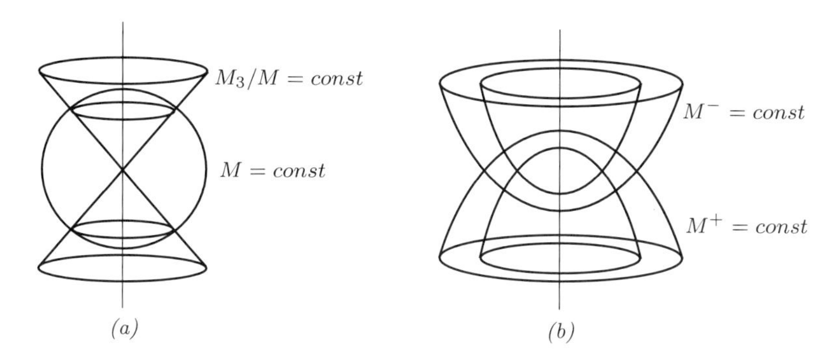

Figure 4-8. The Two Faces of Rotational Symmetry. In *(a)* the Casimir set has the vector specification (M, M_3) and the surfaces which correspond to constant values of M are spheres of radius M while those which correspond to constant values of M_3/M are cones. In *(b)* the Casimir set has the spinor representation (M^+, M^-). The surfaces which correspond to constant values of M^+ and M^- are downward and upward opening paraboloids of revolution with foci at the origin.

To see the spinor face of rotational symmetry, use $M = M^+ - M_3$ to eliminate M in favor of M^+ in the metric condition ${M_1}^2+{M_2}^2+{M_3}^2 = M^2$ with the result

$${M_1}^2 + {M_2}^2 + 2M^+M_3 = (M^+)^2.$$

This is the equation of a paraboloid. The surface generated by $M^+ = const$ is a paraboloid of revolution with focus at the origin and which opens downward in the negative M_3 direction. In like fashion the metric condition may be re-written in terms of M^- rather than M as

$$M_1{}^2 + M_2{}^2 - 2M^-M_3 = (M^-)^2.$$

The surface generated by $M^- = const$ is a paraboloid of revolution with focus at the origin and which opens upward in the positive M_3 direction. These surfaces are shown in Fig. 4-8 *(b)*. The spinor representation M^+ and M^- is therefore said to be parabolic.

Let us summarize the duality of rotational symmetry. Two representations of a rotationally symmetric quantity like the angular momentum are possible: one vector, the other spinor. Corresponding to each representation is a family of surfaces appropriate to the representation. The vector representation manifests itself in spherical-polar geometry while the spinor representation is parabolic. In the spherical-polar expression the Casimir set (M, M_3) is specified and the components are given in terms of it and the free angle $\tan\phi = M_2/M_1$ as

$$\begin{aligned} M_1 &= \sqrt{M^2 - M_3{}^2}\cos\phi, \\ M_2 &= \sqrt{M^2 - M_3{}^2}\sin\phi, \\ M_3 &= M_3. \end{aligned} \tag{4.20}$$

This system describes spheres ($M = const$) and cones ($M_3/M = const$). The total magnitude of the components M_1, M_2 is specified by the Casimir set (M, M_3); but their angular position given by $\tan\phi = M_2/M_1$ is not. It is the free member of the three independent quantities (M_1, M_2, M_3) or (M, M_3, ϕ) or (M^+, M^-, ϕ) required to specify a vector in three-dimensional space.

In the parabolic expression the Casimir set has the spinor representation (M^+, M^-) and the vector components are given in terms of it by

$$\begin{aligned} M_1 &= \sqrt{M^+M^-}\cos\phi, \\ M_2 &= \sqrt{M^+M^-}\sin\phi, \\ M_3 &= (M^+ - M^-)/2. \end{aligned} \tag{4.21}$$

This system describes downward ($M^+ = const$) and upward ($M^- = const$) opening paraboloids of revolution.

Invariant Sets of Kepler Motion

Three-dimensional rotations are described by the pair of Casimir sets $(\mathcal{H}, M, M_3)$, $(\mathcal{H}, M, M_1)$. What are the Casimir sets for four-dimensional rotations? Since three-dimensional rotations are a subgroup of the full symmetry, four-dimensional rotations also possess the sets $(\mathcal{H}, M, M_3)$, $(\mathcal{H}, M, M_1)$. By studying all the combinations of Casimir invariants and components in the light of Eqs. (4.12), one finds the four quantities (S, S_3, D, D_3) are the only others with collectively vanishing Poisson brackets. But since the subsidiary conditions (4.13) require $S = D = h/2$, only three are independent. The inverse-square force symmetry therefore contributes one additional Casimir set which, since h, S, and D are all fixed by the energy $\mathcal{H}$, may be expressed as $(\mathcal{H}, S_3, D_3)$.

Because there are two representations of four-dimensional rotations, one in $\boldsymbol{M}$ and $\boldsymbol{E}$, the other in $\boldsymbol{S}$ and $\boldsymbol{D}$, the additional Casimir set of four-dimensional rotations may also be expressed in terms of the angular momentum and eccentrum components M_3 and E_3 rather than S_3 and D_3 as $(\mathcal{H}, E_3, M_3)$. There are therefore two representations of the third Casimir set of four-dimensional rotations: $(\mathcal{H}, S_3, D_3)$ and $(\mathcal{H}, E_3, M_3)$. These two alternatives are the sign of the underlying dual representation of rotations in either vector or spinor form because four-dimensional rotations consist of a pair of three-dimensional rotations.

Both vector and spinor representations therefore occur in four dimensions as well as three. The Casimir sets $(\mathcal{H}, M, M_3)$ and $(\mathcal{H}, M, M_1)$ are clearly vector representations; but let us look more carefully at the set $(\mathcal{H}, S_3, D_3)$ and its alternative $(\mathcal{H}, E_3, M_3)$. The set $(\mathcal{H}, S_3, D_3)$ is a vector representation combining the two rotationally symmetric vector sets (S, S_3) and (D, D_3) (recall that specifying $\mathcal{H}$ is equivalent to specifying S and D). On the other hand, the set $(\mathcal{H}, E_3, M_3)$ turns out to be the spinor representation corresponding to $(\mathcal{H}, S_3, D_3)$.

The spinor nature of $(\mathcal{H}, E_3, M_3)$ can be seen by observing that the spinor representations of $\boldsymbol{S}$ and $\boldsymbol{D}$ are

$$S^{\pm} = S \pm S_3, \qquad D^{\pm} = D \pm D_3.$$

Since $S = D = h/2$, $S_3 = (M_3 + E_3)/2$, and $D_3 = (M_3 - E_3)/2$, the spinors $S^{\pm}$ and $D^{\pm}$ can be written

$$S^{\pm} = h/2 \pm (M_3 + E_3)/2, \qquad D^{\pm} = h/2 \pm (M_3 - E_3)/2. \tag{4.22}$$

Eliminating h, one finds M_3 and E_3 are completely expressible in the spinor components $S^{\pm}$ and $D^{\pm}$:

$$M_3 = \pm(S^{\pm} - D^{\mp}), \qquad E_3 = \pm(S^{\pm} - D^{\pm}).$$

A representation of the motion in the invariants $(\mathcal{H}, E_3, M_3)$ is thus a spinor representation incorporating both the spinors $S^{\pm}$ and $D^{\pm}$. Each of these spinors again corresponds to one of the two-spheres into which the three-sphere of four-dimensional rotational symmetry projects. The invariants M_3 and E_3 are a composite of these two spinors, one from each two-sphere.

To sum up, the third Casimir set of Kepler motion when expressed in terms of the rotational invariants of the individual two-spheres, $(\mathcal{H}, S_3, D_3)$, is a vector representation. This Casimir set has an alternative representation in terms of the components of the eccentrum and angular momentum, $(\mathcal{H}, E_3, M_3)$ which are sums and differences of the individual two-sphere invariants. In this form it is a spinor representation. When expressed in terms of these spinor invariants, the motion will be viewed in parabolic coordinates.

Three independent Casimir sets of all the symmetries of the relative motion such as

$$(\mathcal{H}, M, M_3), \qquad (\mathcal{H}, M, M_1), \qquad (\mathcal{H}, E_3, M_3)$$

are capital quantities. The number of single-valued invariants in each Casimir set is three. This means the three-dimensional relative motion is completely integrable. There are three Casimir sets. This means the motion is fully degenerate. The bound flow lies upon a one-dimensional torus and its trajectories are closed. These Casimir sets embody all the symmetries of the relative motion of two bodies with gravitational or electrical forces. They play the organizing role in both the classical and quantum motion. They are the invariants with which symmetry turns complexity into simplicity.

The Vector View of Celestial Mechanics

The Casimir sets $(\mathcal{H}, M, M_3)$ and $(\mathcal{H}, M, M_1)$ are both vector expressions of rotational symmetry. They give rise to the same motions, but with different labels since all one need do is interchange M_1 and M_3 to obtain one motion from the other. It is sufficient therefore to consider only one of them, say $(\mathcal{H}, M, M_3)$.

The symmetric coordinates of the vector specification $(\mathcal{H}, M, M_3)$ are spherical-polar. The position vector in the center of mass frame of reference is $\mathbf{r} = (x_1, x_2, x_3)$. (The symbols x_1, x_2, x_3 are now to be understood as the coordinates in the center of mass frame of reference.) In spherical-polar coordinates ($q_1 = r$, $q_2 = \phi$, $q_3 = \theta$) shown in Fig. 4-9, a point in space is specified by the length r of the position vector and two angular coordinates, a polar angle ϕ which fixes the rotation of the position vector about the x_3 or polar axis and a latitudinal angle θ which fixes a cone (centered about the x_3 axis) upon which the position vector must lie. (One requires a single axis for designation as the "polar" axis. The x_3 direction is arbitrarily selected.)

The coordinate grid of the system is formed by the intersections of spheres, planes, and cones. The sphere $r = const$ corresponds to the purely radial dependence of the total energy $\mathcal{H}$. The cone $\theta = const$ corresponds to the central force symmetry. It is the surface upon which the angular momentum vector of magnitude M must lie. The plane $\phi = const$ also corresponds to the central force symmetry and is the surface upon which both the M_3 component and the total angular momentum vector must lie. The symmetric surfaces of spherical-polar coordinates therefore correspond to constant values of the three invariants of the Casimir set $(\mathcal{H}, M, M_3)$. The motion of two-bodies bound by central forces is naturally matched to the symmetries of these surfaces; and the Hamilton–Jacobi equation will separate in their coordinates.

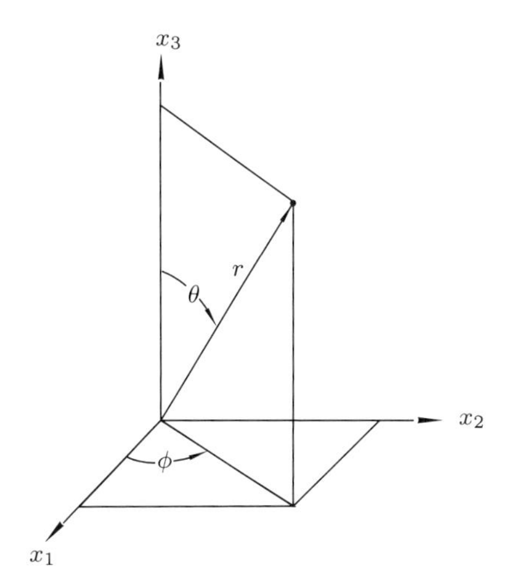

Figure 4-9. Spherical-Polar Coordinates. A point in space is specified by the length of the position vector r and two angular coordinates: a polar angle ϕ which fixes the rotation of the position vector about the x_3 axis and a latitudinal angle θ which fixes a cone (centered about the x_3 axis) upon which the position vector must lie. The coordinate grid of this system is formed by the intersections of spheres (r = *const*), planes (ϕ = *const*), and cones (θ = *const*).

The integrals for the central force symmetry may be determined by separating the Hamilton–Jacobi equation in spherical-polar coordinates. The spherical-polar coordinates r, ϕ, θ are related to the rectangular coordinates x_1, x_2, x_3 by

$$x_1 = r \sin\theta \cos\phi, \qquad x_2 = r \sin\theta \sin\phi, \qquad x_3 = r \cos\theta. \tag{4.23}$$

The corresponding velocities are

$$\begin{aligned}
\dot{x}_1 &= \sin\theta \cos\phi \dot{r} - \sin\theta \sin\phi r\dot{\phi} + \cos\theta \cos\phi r\dot{\theta}, \\
\dot{x}_2 &= \sin\theta \sin\phi \dot{r} + \sin\theta \cos\phi r\dot{\phi} + \cos\theta \sin\phi r\dot{\theta}, \\
\dot{x}_3 &= \cos\theta \dot{r} - \sin\theta r\dot{\theta}.
\end{aligned} \tag{4.24}$$

The Lagrangian is $L = T - V = \mu \dot{x}_i^2/2 - V$, or using Eq. (4.24),

$$L = \mu(\dot{r}^2 + r^2 \sin^2\theta \dot{\phi}^2 + r^2\dot{\theta}^2)/2 - V. \tag{4.25}$$

The three momentum components are $p_r = \partial L/\partial \dot{r}$, $p_\phi = \partial L/\partial \dot{\phi}$, $p_\theta = \partial L/\partial \dot{\theta}$, or

$$p_r = \mu \dot{r}, \qquad p_\phi = \mu r^2 \sin^2\theta \dot{\phi}, \qquad p_\theta = \mu r^2 \dot{\theta}. \tag{4.26}$$

It should be noted that the momenta in these coordinates have different dimensions. The momenta p_ϕ and p_θ have the same dimensions as the angular momentum.

The angular momentum takes a natural form in spherical-polar coordinates. The polar angle coordinate ϕ describes rotations about the x_3 or polar axis. Since the angular momentum describes the rotational aspects of motion, the polar angular momentum component M_3 must bear a direct relationship to the angular coordinate ϕ. A calculation using the relationships (4.23) and (4.26) shows that the polar angular momentum component M_3 is identical to the linear momentum p_ϕ conjugate to ϕ:

$$M_3 = \mu(x_1\dot{x}_2 - x_2\dot{x}_1) = \mu r^2 \sin^2\theta \dot{\phi} = p_\phi. \tag{4.27}$$

By similar calculation, the total angular momentum in spherical-polar coordinates has the form

$$M^2 = \mu^2 r^4 \left(\sin^2\theta \dot{\phi}^2 + \dot{\theta}^2\right).$$

Using the second and third of Eqs. (4.26), the total angular momentum can also be expressed as

$$M^2 = {M_3}^2/\sin^2\theta + p_\theta^2. \tag{4.28}$$

The Hamiltonian $\mathcal{H} = p_i\dot{q}_i - L$ follows from the Lagrangian (4.25). The Hamiltonian and Hamilton–Jacobi equation for $\mathcal{H} = \mathcal{E}$ are then

$$\mathcal{H} = p_r^2/2\mu + p_\phi^2/2\mu r^2 \sin^2\theta + p_\theta^2/2\mu r^2 + V(r) = \mathcal{E}. \tag{4.29}$$

Since each momentum component is a function of only its conjugate position coordinate, the potential and kinetic energy terms corresponding to each coordinate may be grouped together and the Hamilton–Jacobi equation separated. The potential energy must be of such a form that this is possible. For a central force potential $V = V(r)$ separation is indeed possible and the Hamilton–Jacobi equation (4.29) may be partitioned as

$$r^2 \sin^2\theta \left[p_r^2/2\mu + p_\theta^2/2\mu r^2 + V(r) - \mathcal{E}\right] + p_\phi^2/2\mu = 0. \tag{4.30}$$

The Hamilton–Jacobi equation now splits into three equations. The first term on the left of Eq. (4.30) is a pure function of r and θ while the second is a pure function of ϕ. Since they must sum to zero for arbitrary r, θ, and ϕ, this can only occur if both terms are the same constant with opposite sign. The Hamilton–Jacobi equation for a central force thus reaffirms that p_ϕ is the invariant angular momentum component $p_\phi = M_3$.

The invariance of the angular momentum leads directly to the second of Kepler's laws: *a planetary body sweeps out equal areas of its orbit in equal times.* The differential of a wedge-shaped area of the plane formed by the angle $d\phi$ is

$$dA = r^2 d\phi/2 = r^2 \dot{\phi}\, dt/2.$$

The polar momentum component is given by $p_\phi = M_3 = \mu r^2 \dot{\phi}$. Eliminate $r^2\dot{\phi}$ in the area differential in favor of the angular momentum and obtain the desired proportionality:

$$dA = (M_3/2\mu)\, dt.$$

One learns more than that the area swept out is proportional to time. One also learns that the factor of proportionality is the ratio of the angular momentum to the reduced mass. Kepler's second law rests strictly on the fact that the force is central. It is not restricted to inverse-square forces. It holds for arbitrary central force fields.

Equation (4.30) may be further separated by re-expressing it in terms of $p_\phi = M_3$ and rearranging as

$$r^2 \left[p_r^2/2\mu + V(r) - \mathcal{E}\right] + \left[{M_3}^2/2\mu \sin^2\theta + p_\theta^2/2\mu\right] = 0. \tag{4.31}$$

The two terms on the left of Eq. (4.31) are pure functions of r and θ, respectively; hence each of them must be a constant which, from Eq. (4.28), may be recognized as the total angular momentum M^2. Two final separated equations thus split from the original Hamilton–Jacobi equation:

$$p_\theta^2 + {M_3}^2/\sin^2\theta = M^2 \tag{4.32}$$

and

$$p_r^2 + 2\mu\left[V(r) - \mathcal{E}\right] = -M^2/r^2. \tag{4.33}$$

The Hamilton–Jacobi equation splits into three equations and the three elements of the Casimir set turn out to be the separation constants of these three equations.

The action in spherical-polar coordinates may be expressed as

$$S(r, \theta, \phi, t) = -\mathcal{E}t + S_r(r) + S_\phi(\phi) + S_\theta(\theta).$$

The complete solution for the action is now reduced to quadratures of the separated Hamilton–Jacobi equations. For the partial action $dS_\phi/d\phi = p_\phi = M_3$, the integral is

$$S_\phi(\phi) = \int p_\phi\, d\phi = M_3\phi. \tag{4.34}$$

The partial action $S_\theta(\theta) = \int p_\theta \, d\theta$ is formed with p_θ taken from Eq. (4.32):

$$S_\theta(\theta) = \int \sqrt{M^2 - {M_3}^2/\sin^2\theta} \, d\theta. \tag{4.35}$$

The radial momentum p_r may be isolated from Eq. (4.33) for the radial action $S_r(r) = \int p_r \, dr$:

$$S_r(r) = \int \sqrt{2\mu \left[\mathcal{E} - V(r) - M^2/2\mu r^2\right]} \, dr. \tag{4.36}$$

The actions $S_r(r)$, $S_\phi(\phi)$, $S_\theta(\theta)$ constitute the solutions of the Hamilton–Jacobi equation. These solutions are expressed in terms of constants of integration appropriate to the vector representation of rotational symmetry, $I = (\mathcal{H}, M, M_3)$ where $\mathcal{H} = \mathcal{E}$.

The actions $S_\phi(\phi)$ and $S_\theta(\theta)$ are fully determined; but the action $S_r(r)$ is not determined until the potential $V(r)$ is specified. This is because the separation is a result of the rotational symmetry of the central force. An infinite family of generally nonisolating actions and orbits satisfy this symmetry. They do not yield a closed trajectory. The motion is separable; but the integrals are not fully isolating.

For the *particular* central force potential $V = -k/r$ of gravity and electricity, the action $S_r(r)$ becomes fully determined as the quadrature

$$S_r(r) = \int \sqrt{2\mu \left[\mathcal{E} + k/r - M^2/2\mu r^2\right]} \, dr \tag{4.37}$$

and this particular action integral *is isolating* (see **Note 2**).

Symmetry permits separation of the Hamilton–Jacobi equation into pure functions of the separate coordinates which may be integrated by quadrature. The separation constants which appear in the integrals turn out to be the Casimir set of the symmetry. Since spherical-polar coordinates match the rotational symmetry of central forces whose corresponding invariant is the angular momentum, the separation constants are appropriately $(\mathcal{H}, M, M_3)$. The integrals for $S_r(r)$, $S_\theta(\theta)$, and $S_\phi(\phi)$ are achieved for any potential of the form $V = V(r)$.

The Spinor View of Celestial Mechanics

The Casimir set $(\mathcal{H}, E_3, M_3)$ is the spinor expression of rotational symmetry. This symmetry exists because of the inverse-square nature of the force and manifests itself in parabolic geometry. The parabolic coordinate system is shown in Fig. 4-10. The parabolic coordinates $q_1 = \xi$, $q_2 = \eta$, $q_3 = \phi$ are related to the rectangular coordinates x_1, x_2 as

$$x_1 = \rho \cos\phi, \qquad x_2 = \rho \sin\phi,$$

where ρ is the projection of the position vector onto the plane $x_3 = 0$ given by

$$\rho^2 = \xi\eta = {x_1}^2 + {x_2}^2.$$

The radial coordinate r and polar-axis coordinate x_3 are given by

$$r = (\xi + \eta)/2, \qquad x_3 = (\xi - \eta)/2,$$

with the corresponding relationships

$$\xi = r + x_3, \qquad \eta = r - x_3.$$

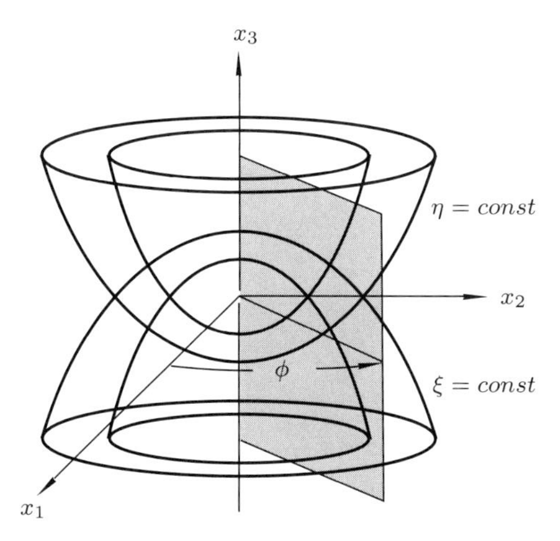

Figure 4-10. Parabolic Coordinates. The polar coordinate ϕ represents rotations about the x_3 axis. Surfaces formed by $\phi = const$ are planes. Surfaces formed by $\xi = const$ are paraboloids of revolution about the x_3 axis whose focus is at the origin and which open downward in the negative x_3 direction. Surfaces formed by $\eta = const$ are paraboloids of revolution which open upward in the positive x_3 direction.

The polar coordinate ϕ plays the same role in parabolic coordinates that it plays in spherical-polar coordinates representing rotations about the x_3 or polar axis. Surfaces formed by $\xi = const$ are paraboloids of revolution about the x_3 axis whose focus is at the origin and which open downward in the negative x_3 direction. Surfaces formed by $\eta = const$ are similar paraboloids of revolution about the x_3 axis with focus at the origin and which open upward in the positive x_3 direction. The coordinate grid consists of the intersections of downward and upward opening paraboloids of revolution $\xi = const$, $\eta = const$, and planes $\phi = const$.

Since the magnitude of the position vector in parabolic coordinates is $r = (\xi + \eta)/2$, the intersections of the two paraboloids of revolution $\xi = const$, $\eta = const$ are circles upon which the magnitude r of the position vector is constant. The energy $\mathcal{H}$, which is purely a function of r, is therefore constant on these circles. We shall find that the eccentricity component e_3 is constant on the paraboloidal surfaces $\xi = const$ and $\eta = const$. The angular momentum component M_3 is constant on surfaces $\phi = const$ just as in spherical-polar coordinates. Hence, the surfaces $\xi = const$, $\eta = const$, $\phi = const$ correspond to constant values of the invariants of the Casimir set $(\mathcal{H}, e_3, M_3)$. (Since $e_3 = E_3/h$ and h depends only upon $\mathcal{H}$, the set may be expressed either in terms of the eccentricity or the eccentrum.)

The velocities in parabolic coordinates are

$$\dot{x}_1 = \cos\phi\dot{\rho} - \sin\phi\rho\dot{\phi}, \qquad \dot{x}_2 = \sin\phi\dot{\rho} + \cos\phi\rho\dot{\phi}, \qquad \dot{x}_3 = (\dot{\xi} - \dot{\eta})/2,$$

where

$$\dot{\rho} = \rho\left(\dot{\xi}/2\xi + \dot{\eta}/2\eta\right).$$

The Lagrangian $L = \mu\dot{x}_i^2/2 - V$ is then

$$L = \mu\,(\xi + \eta)\left(\dot{\xi}^2/4\xi + \dot{\eta}^2/4\eta\right) + \mu\xi\eta\dot{\phi}^2/2 - V,$$

and the momenta $p_\xi = \partial L/\partial\dot{\xi}$, $p_\eta = \partial L/\partial\dot{\eta}$, $p_\phi = \partial L/\partial\dot{\phi}$ are

$$p_\xi = \mu(\xi + \eta)\dot{\xi}/4\xi, \qquad p_\eta = \mu(\xi + \eta)\dot{\eta}/4\eta, \qquad p_\phi = \mu\xi\eta\dot{\phi}.$$

The Hamiltonian and Hamilton–Jacobi equation then take the form

$$\mathcal{H} = \frac{4}{\xi + \eta}\left(\xi p_\xi^2/2\mu + \eta p_\eta^2/2\mu\right) + \frac{1}{\xi\eta}p_\phi^2/2\mu + V = \mathcal{E}. \tag{4.38}$$

To separate the Hamilton–Jacobi equation in parabolic coordinates first observe that since $V = V(r)$ and $r = (\xi + \eta)/2$, the polar momentum p_ϕ contains the only ϕ dependence; hence it is a constant, the polar angular momentum M_3.

An inverse power potential (corresponding to the inverse-square force) is a function only of ξ and η in $V = -k/r = -2k/(\xi + \eta)$. This structure matches the structure of the kinetic energy in Eq. (4.38) and it is therefore possible to completely separate the Hamilton–Jacobi equation in parabolic coordinates. To do so, eliminate p_ϕ in favor of the invariant M_3, substitute $V = -2k/(\xi + \eta)$, and re-express Eq. (4.38) as

$$\begin{aligned} &\xi\left(\mathcal{E} - 2p_\xi^2/\mu - {M_3}^2/2\mu\xi^2 + k/\xi\right) \\ &+ \eta\left(\mathcal{E} - 2p_\eta^2/\mu - {M_3}^2/2\mu\eta^2 + k/\eta\right) = 0. \end{aligned} \tag{4.39}$$

Since the two lines of Eq. (4.39) are each pure functions of different coordinates which sum to zero, they must each be the same constant with opposite sign. Appropriately, the separation constant for this symmetry is e_3, the polar component of the eccentricity.

This connection may be shown by directly calculating e_3 in parabolic coordinates. The term $\boldsymbol{p} \times \boldsymbol{M}$ which appears in the eccentricity (4.7) may be expressed in terms of $\boldsymbol{r}$ and $\mathbf{p}$ through the identity $\boldsymbol{p}\times\boldsymbol{M} = \mathbf{r}p^2 - \mathbf{p}(\mathbf{r}\cdot\mathbf{p})$. The polar component of the eccentricity is then

$$e_3 = \left(x_3 p^2 - p_3\, \mathbf{r}\cdot\mathbf{p}\right)/k\mu - x_3/r. \tag{4.40}$$

Using the parabolic coordinate relationships, find

$$p_3 = \frac{2}{\xi+\eta}(\xi p_\xi - \eta p_\eta), \qquad \mathbf{r}\cdot\mathbf{p} = \xi p_\xi + \eta p_\eta. \tag{4.41}$$

The total squared momentum may be taken from the kinetic energy in the first term of Eq. (4.38) as

$$p^2 = \frac{4}{\xi+\eta}\left(\xi p_\xi^2 + \eta p_\eta^2\right) + \frac{1}{\xi\eta}p_\phi^2. \tag{4.42}$$

Put together the pieces from Eqs. (4.41) and (4.42) and the eccentricity (4.40) has the parabolic coordinate form

$$e_3 = \frac{2\xi\eta}{\xi+\eta}\left(p_\eta^2 - p_\xi^2\right)/k\mu + \frac{\xi-\eta}{\xi\eta}p_\phi^2/k\mu - \frac{\xi-\eta}{\xi+\eta}. \tag{4.43}$$

The momentum component p_ϕ is the invariant M_3. The eccentricity may be separated into pure functions of ξ and η by partitioning Eq. (4.43) in the manner

$$\begin{aligned} &[(e_3 - 1) + 2\xi p_\xi^2/k\mu + {M_3}^2/2k\mu\xi]/\xi \\ &+ [(e_3 + 1) - 2\eta p_\eta^2/k\mu - {M_3}^2/2k\mu\eta]/\eta = 0. \end{aligned} \tag{4.44}$$

The terms on the first line of Eq. (4.44) are a pure function of ξ while those on the second line are a pure function of η. Each line must therefore be the same constant with opposite sign. This separation constant may be seen to be the total energy $\mathcal{H} = \mathcal{E}$ by comparing Eq. (4.44) with Eq. (4.39). The total energy is the separation constant for the eccentricity while the eccentricity is the separation constant for the total energy in the Hamilton–Jacobi equation (4.39).

The Hamilton–Jacobi equation therefore splits into the additional pair

$$\begin{aligned} \xi\left(\mathcal{E} - 2p_\xi^2/\mu - {M_3}^2/2\mu\xi^2\right)/k + 1 &= e_3, \\ \eta\left(\mathcal{E} - 2p_\eta^2/\mu - {M_3}^2/2\mu\eta^2\right)/k + 1 &= -e_3. \end{aligned} \tag{4.45}$$

The polar component of the eccentricity e_3 is given by the same function of the parabolic coordinates ξ and η. The $\eta = const$ family is the reflection of the $\xi = const$ family.

This symmetry in which ξ and η interchange roles when e_3 changes sign is a reflection symmetry. The eccentricity vector, unlike the angular momentum, changes sign when the coordinates are reflected: $\boldsymbol{r} \to -\boldsymbol{r}$. The radial coordinate $r = \sqrt{x_i x_i}$ is unchanged under reflection; however, the polar-axis coordinate x_3 is reflected. Since the parabolic coordinates are given by $\xi = r + x_3$ and $\eta = r - x_3$, coordinate reflection interchanges ξ and η as it reflects e_3.

The full description of the motion is contained in the action

$$S(\phi, \xi, \eta, t) = -\mathcal{E}t + S_\phi(\phi) + S_\xi(\xi) + S_\eta(\eta).$$

The partial actions follow as quadratures $S_\phi(\phi) = \int p_\phi \, d\phi$, $S_\xi(\xi) = \int p_\xi \, d\xi$, and $S_\eta(\eta) = \int p_\eta \, d\eta$ of the separated Hamilton–Jacobi equations. The action $S_\phi(\phi) = M_3\phi$ is identical to that in spherical-polar coordinates. The momenta p_ξ and p_η may be isolated from the separated Hamilton–Jacobi equations, Eqs. (4.45), to yield the action integrals

$$\begin{aligned} S_\xi(\xi) &= \frac{1}{2}\int \sqrt{2\mu\left[\mathcal{E} + (1-e_3)k/\xi - {M_3}^2/2\mu\xi^2\right]}\, d\xi, \\ S_\eta(\eta) &= \frac{1}{2}\int \sqrt{2\mu\left[\mathcal{E} + (1+e_3)k/\eta - {M_3}^2/2\mu\eta^2\right]}\, d\eta. \end{aligned} \tag{4.46}$$

The two action integrals $S_\xi(\xi)$, $S_\eta(\eta)$ are identical functions save for the sign of e_3. With $S_\xi(\xi)$ and $S_\eta(\eta)$ represented by quadratures according to Eq. (4.46), $S_\phi(\phi) = M_3\phi$, and $\mathcal{H} = \mathcal{E}$, the action is completely determined in terms of the spinor Casimir set $I = (\mathcal{H}, e_3, M_3)$.

The action integrals (4.46) for $S_\xi(\xi)$ and $S_\eta(\eta)$ should be compared with that for $S_r(r)$ given by Eq. (4.36). Whereas the action $S_r(r)$ is not determined because the central force potential is the general potential $V(r)$, the actions $S_\xi(\xi)$ and $S_\eta(\eta)$ are fully determined because they exist only for the solitary central force potential $V = -k/r = -2/(\xi + \eta)$. Separation in parabolic coordinates is fully degenerate whereas separation in spherical-polar coordinates is only partly degenerate. When $S_\xi(\xi)$ and $S_\eta(\eta)$ given by Eq. (4.46) are compared with $S_r(r)$ given by Eq. (4.37) for the particular potential $V = -k/r$, all integrals are found to have an identical structure: the arguments of the square roots are quadratic functions of r, ξ, or η. The spinor actions $S_\xi(\xi)$ and $S_\eta(\eta)$ also differ from the vector action $S_r(r)$ by the appearance of the factor 1/2.

Separation of the Hamilton–Jacobi equation yields all the invariants of motion, particularly those invariants which lie beyond the Galilean invariants. The *forte* of the method of separation of the Hamilton–Jacobi equation is the direct link it forges between a mechanical symmetry and a corresponding coordinate geometry. For two-body motion subject to gravitational and electrical forces these invariants are the energy, the angular momentum, and the eccentricity. The corresponding geometrical figures of these symmetries are spheres, cones, planes, and paraboloids of revolution.

Planar Symmetries—A Digression

One can scarcely avoid wondering if the most basic coordinates of all—rectangular coordinates—possess any symmetries of motion for two bodies. They do; but not for inverse-square forces. Rectangular coordinates, formed by the intersections of mutually perpendicular planes $x_1 = const$, $x_2 = const$, $x_3 = const$, are the symmetry coordinates for the potential $V = \kappa r^2/2$ of Hooke motion. This potential corresponds to the attractive linear force $\boldsymbol{f} = -\kappa \boldsymbol{r}$ and the Hooke Hamiltonian is

$$\mathcal{H} = p^2/2\mu + \kappa r^2/2.$$

The Hooke Hamiltonian in the six-dimensional phase space $\xi = (\boldsymbol{r}, \boldsymbol{p})$ of the relative motion can be seen to be the equation of a five-sphere embedded in this six-dimensional space. It shows that the phase space of Hooke motion is Euclidean.

The linear central force $f = -\kappa r$ and the inverse-square central force $f = -k/r^2$ are unique in mechanics. Whereas the inverse-square force is the only inverse power-law force for which an additional isolating invariant exists that endows the motion with a closed orbit, the linear force is the only direct power-law force for which the motion likewise possesses an additional isolating invariant which closes the orbit (**Note 2**). For the full range of power-law central forces $f = -\gamma r^{\beta}$, where β is any positive or negative number, isolating invariants corresponding to closed orbits of the motion may be found for only the two values $\beta = 1$ and $\beta = -2$. The intimate connection between the bound motion of two-bodies subject to the inverse-square force (Kepler motion) and to the linear force (Hooke motion) is described in **Note 6**.

The additional isolating invariant for the linear force $\boldsymbol{f} = -\kappa \boldsymbol{r}$ corresponding to the potential $V = \kappa r^2/2$ turns out to be the matrix A*:

$$A_{ij} = p_i p_j/2\mu + \kappa x_i x_j/2.$$

It is a straightforward matter to show that the Poisson bracket $[A, \mathcal{H}]$ vanishes and A is an invariant. Since A is a symmetric three-by-three matrix, it has only six independent components. But A, like the eccentricity $\mathbf{e}$, contributes only one new scalar invariant to the motion because it is not independent of the other invariants $\mathcal{H}$ and $\boldsymbol{M}$. For example, the sum of the diagonal components of A is the total energy of the motion:

$$A_{ii} = p^2/2\mu + kr^2/2 = \mathcal{H}.$$

As a final tribute to the intimate connection between the inverse-square and linear central force laws, it turns out that bound Kepler and Hooke

---

* See Jean Sivardière, "Laplace vectors for the harmonic oscillator," *Am. J. Phys.* **57** (6), (1989).

motions may be transformed into one another. Hence, bound Kepler motion also has a phase space that is Euclidean. This is a remarkable correspondence that originates in the four-dimensional rotational symmetry of Kepler motion. Since it must satisfy the equation of the three-sphere and this equation is the condition of a Euclidean space, Kepler motion may be transformed into coordinates in which the phase space is Euclidean.

The orbits of bound motions for both the Hooke force and inverse-square force are ellipses; but the center of mass for the Hooke force lies at the center of the ellipse while for inverse-square forces the center of mass lies at one of the foci. Readers may consult **Note 6** to see how the three-sphere upon which Kepler motion exists in phase space projects onto a pair of two-spheres and in so doing find that Kepler motion is transformed into Hooke motion.

The Orbits

The orbits for inverse-square forces may now be drawn out; they directly follow from the action. Since both a vector (spherical-polar) and spinor (parabolic) representation of the action exist, either may be used to produce the orbits. Let us use the spherical-polar representation. [Interested readers may want to develop the orbits using Eqs. (4.46) for the parabolic actions.]

The directions natural to two-body motion are the three orthogonal directions formed by the vectors $\boldsymbol{M}$, $\mathbf{e}$, and $\boldsymbol{M} \times \mathbf{e}$ of Fig. 4-11. These three vectors form an orthogonal set of basis vectors for the motion. The vector Casimir set $(\mathcal{H}, M, M_3)$ singles out the x_3 direction as the direction of the component M_3. It is therefore appropriate to align the x_3 axis with $\boldsymbol{M}$ as shown in Fig. 4-11 for the vector case. As a consequence, $e_3 \equiv 0$. (For the spinor representation $(\mathcal{H}, e_3, M_3)$, it is appropriate to align the x_3 axis with $\mathbf{e}$ in which case $M_3 \equiv 0$.) The angle θ then takes the permanent value $\theta = \pi/2$, the momenta p_3 and p_θ vanish, and the angular momentum possesses only a polar component for which $M = M_3$. Let the x_1 axis be aligned with $\mathbf{e}$. The invariant $\mathbf{e}$ is a vector drawn from the origin of a conic section at the focus to its center. Since x_3 is aligned with $\boldsymbol{M}$, the motion lies completely in the plane $\theta = \pi/2$. The eccentricity vector $\mathbf{e}$ lies in this plane with $e_3 = 0$. Choose the origin for the angle ϕ by aligning the position vector $\boldsymbol{r}$ with the eccentricity vector $\mathbf{e}$ at $\phi = 0$.

In this potential the radial motion is an equivalent one-dimensional motion in the potential $W(r)$ as shown in Fig. 4-12. The potential $W(r)$ has a minimum energy of magnitude

$$\mathcal{E}_0 = -\mu k^2/2M^2. \tag{4.47}$$

This is the lowest value or *ground state* of the energy available to the relative motion. The energy is bounded in the range $\mathcal{E}_0 \leqslant \mathcal{E} < \infty$.

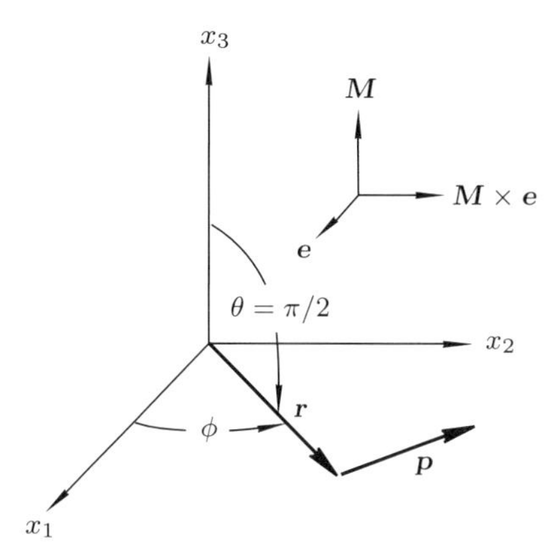

Figure 4-11. Coordinates for Two-Body Motion. The three natural orthogonal coordinate directions are $\boldsymbol{M}$, $\mathbf{e}$, and $\boldsymbol{M} \times \mathbf{e}$. For the vector representation in spherical-polar coordinates, the x_3 axis is aligned with the angular momentum vector; hence $p_3 = 0$ and $e_3 = 0$. (For the spinor representation in parabolic coordinates it is appropriate to align the x_3 axis with the eccentricity vector in which case $M_3 = 0$.) The motion lies completely in the plane $\theta = \pi/2$. The eccentricity vector lies in this plane and is aligned with the x_1 axis. The x_2 axis is aligned with $\boldsymbol{M} \times \mathbf{e}$.

A point at which the radial momentum p_r vanishes is a libration point (a point at which a coordinate reaches a maximum or minimum in the periodic motion of the trajectory). For a given energy $\mathcal{H} = \mathcal{E}$, this point is defined by the condition $p_r = 0$ in Eq. (4.5) which requires $\mathcal{E} = W(r)$ and hence, according to Eq. (4.6),

$$\mathcal{E} = W(r) = M^2/2\mu r^2 - k/r.$$

Since this condition is quadratic in r, there are two libration points corresponding to its two roots. The roots are neatly summarized by the quantities

$$r_0 \equiv M^2/k\mu, \qquad e^2 \equiv 1 - \mathcal{E}/\mathcal{E}_0. \tag{4.48}$$

The quantity r_0 is the value of r at the minimum energy point and e is the magnitude of the eccentricity vector $\mathbf{e}$ which may also be expressed in terms of the Kepler constant as

$$e^2 = 1 - M^2/h^2.$$

The two roots representing the libration points are then

$$r_{\min} = r_0/(1+e), \qquad r_{\max} = r_0/(1-e). \tag{4.49}$$

The libration points are the intersections of the lines $\mathcal{H} = \mathcal{E}$ with the potential curve $W(r)$ in Fig. 4-12. The distance of closest approach to the

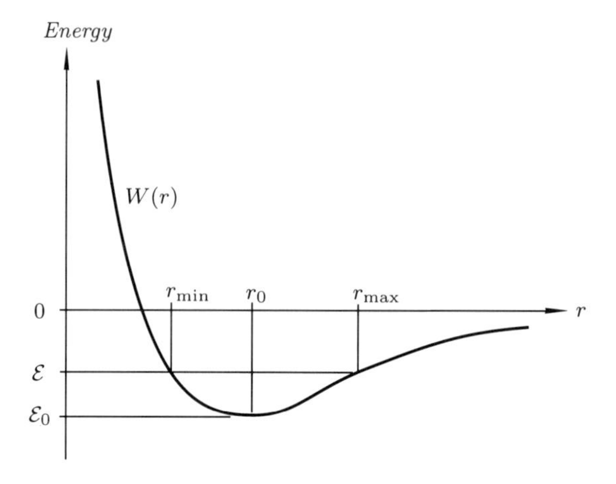

Figure 4-12. The Effective Potential Energy Function $W(r)$ of Two-Body, Inverse-Square Force Motion. Bound orbits exist for energy $\mathcal{E} < 0$. The ground state energy is $\mathcal{E}_0 = -\mu k^2/2M^2$.

origin, or *pericentron*, is r_{min} and this quantity exists for all values of the energy above the ground state energy $\mathcal{E}_0$. On the other hand, the distance of furthest excursion, the *apocentron*, only exists for energies $\mathcal{E} < 0$ which implies $0 < e < 1$. For $\mathcal{E} = \mathcal{E}_0$, there results $e = 0$, $r_{\text{min}} = r_{\text{max}} = r_0$, and the orbit is a circle of radius r_0. The motion is confined to an annulus of outer radius r_{max} and inner radius r_{min} as shown in Fig. 4-1.

How much of the total energy of two bodies is kinetic and how much is potential? From Eqs. (4.47) and (4.48) one sees that the ground state total energy is half the potential energy:

$$\mathcal{E}_0 = V/2 = -k/2r_0.$$

Since $\mathcal{E}_0 = T + V$, the kinetic energy in the ground state must be half the magnitude of the potential energy and of opposite sign. The ground state energies therefore bear the relationship $T = -\mathcal{E}_0$ and $V = 2\mathcal{E}_0$. For inverse-square forces, the potential energy has two-thirds of the total energy while the kinetic energy gets one-third. It is an interesting fact that these relationships are true for the average values of T and V over any orbit, not just the ground state orbit:

$$\mathcal{E} = \langle T \rangle + \langle V \rangle,$$
$$\langle T \rangle = -\mathcal{E}, \qquad \langle V \rangle = 2\mathcal{E}.$$

Only in the ground state are T and V as well as $\mathcal{E}$ constant over the orbit.

It can be seen that the energy determines the kind of trajectory permitted. Three classes exist depending upon whether $\mathcal{E}$ is negative, zero, or positive as shown in Fig. 4-13. The class corresponding to $\mathcal{E} < 0$ consists of bound orbits for which the bodies are forever within finite range of one another; they have insufficient kinetic energy to reach infinity. The class with $\mathcal{E} > 0$ consists of those trajectories that are free. In this class the bodies have an excess of kinetic energy at infinity. They approach from infinity, encounter one another, and then escape, never to encounter one another again. Bodies in motion at infinity are necessarily on free trajectories.

The trajectory $\mathcal{E} = 0$ separates the free trajectories from the bound orbits. Bound orbits with $\mathcal{E} \to 0$ are very large ellipses with apocentrons which approach infinity but nonetheless remain periodic. At the limit $\mathcal{E} = 0$ the bodies fail to return periodically. Bodies on parabolic orbits leave and return from infinity with vanishing kinetic energy: they are in a state of rest at infinity.

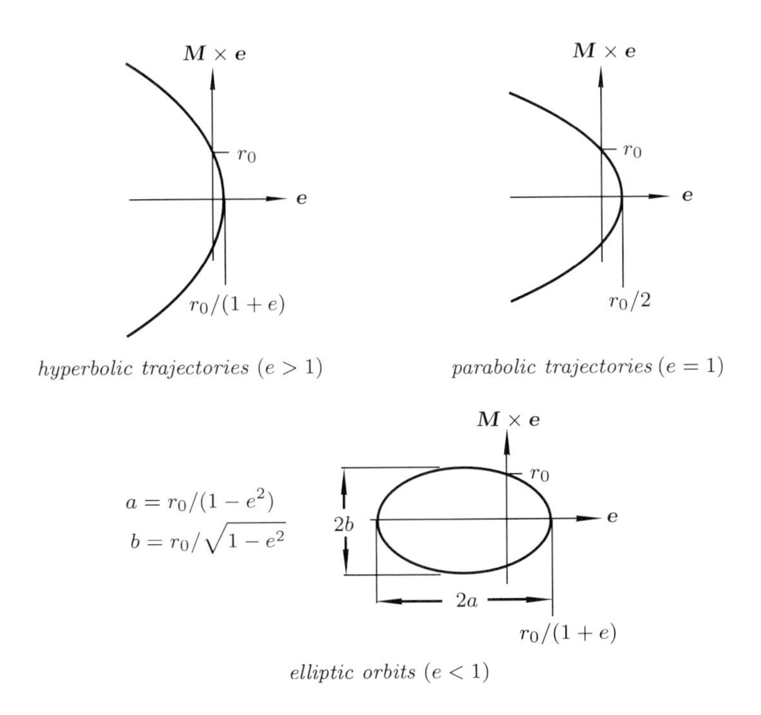

Figure 4-13. The Trajectories of Two-Body, Inverse-Square Force Motion. Free trajectories corresponding to $e > 1$ ($\mathcal{E} > 0$) are hyperbolas. They are separated from bound orbits by the parabolic trajectory for which $e = 1$ ($\mathcal{E} = 0$). Bound orbits are ellipses characterized by $e < 1$ ($\mathcal{E} < 0$).

The known action of the Kepler problem is contained in Eqs. (4.34)–(4.37) for the vector representation and in Eqs. (4.46) for the spinor representation. The trajectories in configuration space are obtained as a canonical transformation from the invariant set of the elementary flow for which the action is the generating function. For the vector representation, the total action is $S(r, \phi, \theta, I, t) = -\mathcal{E}t + S_r(r) + S_\phi(\phi) + S_\theta(\theta)$ and the invariant set is $I = (\mathcal{E}, M, M_3)$. The transformation to the primitive coordinates (r, θ, ϕ) follows the canonical transformation of Eqs. (3.40).

With the coordinates oriented so that $M_3 = M$, the θ equation simply yields $p_\theta = 0$ and $\theta = const = \pi/2$. The trajectory in space $\phi(r)$ is contained in

$$\partial S/\partial M = \phi + \partial S_r/\partial M = const,$$

while that in time $r(t)$ is given by

$$\partial S/\partial \mathcal{E} = -t + \partial S_r/\partial \mathcal{E} = const.$$

With the action $S_\phi(\phi) = M\phi$ and $S_r(r)$ taken from Eq. (4.37), the trajectory equations are

$$\phi = \int \frac{M\,dr}{r^2\sqrt{2\mu\,(\mathcal{E} + k/r - M^2/2\mu r^2)}} \tag{4.50}$$

and

$$t = \int \frac{\mu\,dr}{\sqrt{2\mu\,(\mathcal{E} + k/r - M^2/2\mu r^2)}}. \tag{4.51}$$

Equation (4.50) yields the orbit in space showing how ϕ and r are related while Eq. (4.51) gives the orbit in time relating r to t. Each of these orbit integrals has an interesting story to tell not only about the motion of two bodies but about the mathematical nature of orbits generally.

Take first the orbit in space, the integral (4.50). The invariants $\mathcal{E}$ and M may be expressed in terms of e and r_0 according to Eq. (4.48) and the equation may be integrated directly with the result

$$r(\phi) = r_0/(1 + e\cos\phi). \tag{4.52}$$

Equation (4.52) describes the family of conic sections with the closest focus at the origin, eccentricity e, and latus rectum r_0. The latus rectum of the complete set of orbits turns out to be the same as the radius of the ground state orbit: $r_0 = M^2/k\mu$. It does not depend on the energy as does the eccentricity. (The properties of conic sections are reviewed in **Note 4**.) The three ranges for the energy correspond to three ranges for the eccentricity,

$$\begin{aligned} &\text{elliptic orbits}: && \mathcal{E} < 0, && 0 \leqslant e < 1,\\ &\text{parabolic orbits}: && \mathcal{E} = 0, && e = 1,\\ &\text{hyperbolic orbits}: && \mathcal{E} > 0, && e > 1. \end{aligned}$$

For $e < 1$, the orbit is an ellipse with one focus at the origin. (The "ellipse" for $e = 0$ is a circle with center at the origin.) For $e = 1$, the trajectory is a parabola. For $e > 1$, the trajectory is a hyperbola with focus at the origin. All three trajectories are illustrated in Fig. 4-13.

There is a royal road to the trajectory in space bypassing Eq. (4.50). It is provided by the hidden invariant of Kepler motion, the eccentricity vector. Project the eccentricity vector onto the position vector and obtain

$$\mathbf{r} \cdot \mathbf{e} = re \cos\phi = \mathbf{r} \cdot (\mathbf{p} \times \mathbf{M})/k\mu - r. \tag{4.53}$$

Since $\mathbf{r} \cdot (\mathbf{p} \times \mathbf{M}) = (\mathbf{r} \times \mathbf{p}) \cdot \mathbf{M} = M^2$, the orbit in space (4.52) also follows from Eq. (4.53) and the fact that $M^2/k\mu = r_0$. Equation (4.52) which follows from the invariance of $\mathbf{e}$ describes a closed trajectory. It therefore manifests the direct connection between this additional invariant and the closed-orbit degeneracy.

Let us examine the bound orbits which are ellipses corresponding to $\mathcal{E} < 0$ and $e < 1$ in Fig. 4-13 *(a)*. The major axis of the ellipse is aligned with the eccentricity vector which is drawn from the origin of the ellipse at the focus closest to pericentron to its center. As shown in **Note 4**, the semi-major axis a and semi-minor axis b of the ellipse are given by

$$a = r_0/(1 - e^2), \qquad b = r_0/\sqrt{1 - e^2}, \tag{4.54}$$

which, for the conditions (4.48) of Kepler motion, are

$$a = k/2|\mathcal{E}|, \qquad b = M/\sqrt{2\mu|\mathcal{E}|}.$$

The semi-major axis a depends only upon the total energy of the bodies. On the other hand, the semi-minor axis depends upon both the angular momentum and the energy. The tightest orbit is the circle corresponding to the ground state for which $a = r_0 = k/2|\mathcal{E}_0|$. The orbits widen for $|\mathcal{E}| \to 0$ approaching the limit of the parabolic orbit at $\mathcal{E} = 0$.

Children learn to draw the ellipse by using two invariants: the distance between two pegs driven into a surface and the length of a loop of string slipped over the pegs. A pencil traces out the orbit as it draws the string taut. The length of celestial string required for two heavenly bodies is related to the mechanical invariants $\mathcal{E}$ and M as $(1+e)k/|\mathcal{E}|$ and the heavenly pegs must be separated by the distance $ek/|\mathcal{E}|$.

The parabolic trajectory corresponding to $e = 1$ approaches from infinity, crosses the x_2 axis at a distance r_0 from the origin, and crosses the eccentricity vector at exactly half this distance from the origin as shown in Fig. 4-13 *(b)*.

The free trajectories are hyperbolic and correspond to $e > 1$. They pass from infinity and cross the eccentricity vector a distance $r_{\min} = a/(e + 1)$ from the origin where the parameter a takes the value $a = r_0/(e^2 - 1)$ for the hyperbolic case. Like all the other trajectories, they also cross the x_2 axis in Fig. 4-13 *(c)* at a distance r_0 from the origin. Bodies on free trajectories are in motion at infinity ($\mathcal{E} > 0$). For very energetic bodies $\mathcal{E} \gg |\mathcal{E}_0|$, the trajectory is a straight line which passes a distance r_0/e from the origin.

The elliptical orbits of the planets are shown in Fig. 4-14. The orbits of the giant planets are nearly circular. Neptune has the most circular orbit of all the planets with an eccentricity of 0.009. Pluto (the most distant planet from the sun) and Mercury (the closest) possess the most eccentric orbits ($e = 0.25$ and $e = 0.21$, respectively). The earth's orbit is one of small eccentricity $e = 0.017$. The asteroid Icarus, whose orbit lies within the inner planets, possesses an eccentricity $e = 0.87$. Ceres and its associated asteroids which lie between Mars and Jupiter have eccentricity $e = 0.076$.

How long does it take the bodies to traverse their trajectories? The typical time over which the bodies significantly interact may be directly found from the mechanical quantities which describe the motion: $\mathcal{E}$, M, k, and μ. There are two time scales which one can form from these quantities, one based upon $\mathcal{E}$ and one based upon M. One can see that there are two different time scales by imagining first that the angular momentum M vanishes. The time scale is then established by the rate at which the inverse-square force draws the bodies together along a straight line and is determined by the total energy $\mathcal{E}$. On the other hand, imagine the energy vanishes. In this case the angular momentum dominates the motion and it determines the time scale of the interaction.

The set of quantities containing $\mathcal{E}$ (but not M) may be combined to form a quantity solely possessing the dimensions of time: the Kepler frequency,

$$\omega \equiv \sqrt{8(-\mathcal{E})^3/k^2\mu} = \sqrt{k/\mu}/a^{3/2}. \tag{4.55}$$

The Kepler frequency has the dimensions of a reciprocal time and it will be shown that its reciprocal indeed gives the time it takes the bodies to orbit one another for the elliptic case or the time over which the bodies significantly deflect one another for the hyperbolic case. The Kepler frequency is appropriate for both elliptic and hyperbolic orbits for which $\mathcal{E} \neq 0$; but it vanishes for the parabolic case for which $\mathcal{E} = 0$. It is here that the time scale formed from M, k, and μ rather than $\mathcal{E}$ becomes appropriate. A parabolic frequency appropriate for this case may be formed as

$$\omega_0 = k^2\mu/M^3 = \sqrt{k/\mu}/r_0^{3/2}.$$

While the Kepler frequency ω sets the time scales for elliptic and hyperbolic orbits, the time parameter which describes parabolic orbits is ω_0. The two time scales are related by the eccentricity as

$$\omega/\omega_0 = (1 - e^2)^{3/2}.$$

To find the exact expression for the time along the trajectory one must integrate the time equation (4.51). This equation may be made more transparent for the elliptic and hyperbolic cases by introducing the Kepler frequency as well as the orbit parameters a and e in favor of $\mathcal{E}$ and M:

$$\omega t = \int \frac{r\,dr}{a\sqrt{a^2e^2 - (r-a)^2}}. \tag{4.56}$$

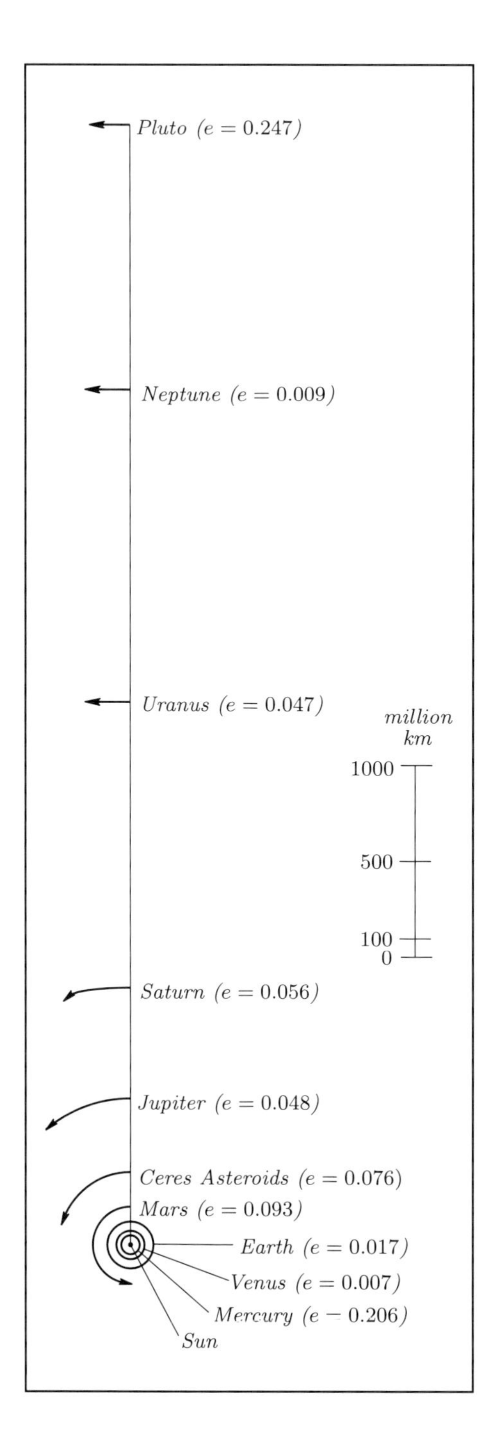

Figure 4-14. Orbits of the Planets. The planets are shown in scaled distances from the sun. Orbit tracks for each planet are shown for one earth-year. Mercury (the nearest to the sun) and Pluto (the most distant) have the most eccentric orbits.

The parabolic case is special. The integral (4.51) for parabolic orbits for which $\mathcal{E} = 0$ simplifies to

$$\omega_0 t = \int \frac{r\,dr}{r_0^2\sqrt{(2r/r_0 - 1)}}. \tag{4.57}$$

Unlike the space relationship between r and ϕ which is integrable in elementary functions—the conic sections (4.52)—the time relationship (4.56) is not directly integrable in elementary functions. Mathematicians of the late seventeenth and eighteenth century found that an integrable result in elementary functions could be obtained through a transformation from the angular variable ϕ to an new angular variable ψ such that

$$\begin{aligned} r &= a(1 - e\cos\psi), && e < 1, \\ r &= \frac{r_0}{2}\sec^2(\psi/2), && e = 1, \\ r &= a(1 - e\cosh\psi). && e > 1. \end{aligned} \tag{4.58}$$

Astronomers call the angle ψ the eccentric anomaly. The relationship between the eccentric anomaly ψ and the true anomaly ϕ may be obtained from Eqs. (4.52) and (4.58) so that one finds the angles are related as

$$\begin{aligned} \tan(\phi/2) &= \sqrt{\frac{1+e}{1-e}}\tan(\psi/2), && e < 1, \\ \tan(\phi/2) &= \tan(\psi/2), && e = 1, \\ \tan(\phi/2) &= \sqrt{\frac{e+1}{e-1}}\tanh(\psi/2), && e > 1. \end{aligned} \tag{4.59}$$

For the elliptic case ($\mathcal{E} < 0$) the angle ϕ revolves continuously with period 2π as ψ sweeps through its range of values from $-\infty$ to $+\infty$; but for the hyperbolic case ($\mathcal{E} > 0$), the sweep of ψ over its range sweeps the angle ϕ only over the half-turn from $-\pi/2$ to $+\pi/2$ as required by the third of Eqs. (4.59) in which the tangent is made equal to the hyperbolic tangent. Thus, the continuous parameter ψ naturally generates both the infinitely periodic, repeating, bound orbits of the elliptic case and the single-pass, free orbit of the hyperbolic case. The parabolic case ($\mathcal{E} = 0$) again demonstrates its special character: it is integrable in elementary functions as signaled by the equality between ϕ and ψ in the second of Eqs. (4.59).

Introduction of the eccentric anomaly through Eqs. (4.58) into the time equations (4.56) and (4.57) releases simple integrals for elliptic, parabolic, and hyperbolic trajectories:

$$\begin{aligned} \omega t &= \psi - e\sin\psi, && e < 1, \\ \omega_0 t &= \tan(\phi/2)[1 + \tan^2(\phi/2)/3], && e = 1, \\ \omega t &= e\sinh\psi - \psi, && e > 1, \end{aligned} \tag{4.60}$$

where the constants of integration have been chosen so that $\psi = \phi = 0$ at $t = 0$.

The orbit variables r and t are parametrically represented in terms of ψ through Eqs. (4.58) and (4.60). As ψ traces out its excursion from $-\infty$ to $+\infty$, r traces excursions out and back over its range between $r_{\min}$ and $r_{\max}$, ϕ completes periodic revolutions of 2π for the elliptic case or a half-turn from $-\pi/2$ to $+\pi/2$ for the parabolic and hyperbolic cases, and t increases with ψ monotonically save for a periodic fluctuation proportional to the eccentricity e.

While the elliptic and hyperbolic trajectories must be represented in terms of the eccentric anomaly ψ, the parabolic orbit is directly expressible in terms of the true anomaly ϕ by the second of Eqs. (4.60).*

For elliptic orbits, ψ generates periodic orbits in ϕ according to the first of Eqs. (4.59). To find the time T for one complete orbit on the ellipse, set $t = T$ and $\psi = 2\pi$ in the first of Eqs. (4.60) and find

$$T = \frac{2\pi}{\omega} = \pi k \sqrt{\mu/2(-\mathcal{E})^3}. \tag{4.61}$$

The period of the orbit is revealed to depend only upon the square root of the reduced mass of the bodies and the inverse $\frac{3}{2}$ power of their total energy. Massive bodies have long periods. Energetic bodies have short periods. It is also possible to express the period in terms of the linear dimension of the orbit by eliminating $\mathcal{E}$ in favor of a from the first of Eqs. (4.54):

$$T = 2\pi \sqrt{\mu/k}\, a^{3/2}. \tag{4.62}$$

The time T to complete a revolution around an elliptical orbit may also be conveniently determined from Kepler's second law by separating the variables:

$$dA = (M/2\mu)\, dt,$$

from which

$$T = \int_{A_{\text{ellipse}}} (2\mu/M)\, dA = 2\mu A_{\text{ellipse}}/M.$$

Since the area of an ellipse is πab, the period is $T = 2\pi\sqrt{\mu/k}\, a^{3/2}$ when the relations (4.54) are used to eliminate M and b. Kepler's third law is contained in Eq. (4.62): *The periods of the planets are proportional to the linear dimension of the orbit to the* $\frac{3}{2}$ *power.*

The time dependence of the hyperbolic orbits is the same as that for the elliptic orbits with hyperbolic functions replacing circular functions. (The

* This elementary integral is obtained by substituting the parabolic expression for r from Eq. (4.58) into the time integral (4.57) with the result $\omega_0 t = \int \sec^4(\phi/2)\, d\phi$. The integral is then developed by breaking the integrand into two terms through the identity $\sec^4(\phi/2) = \sec^2(\phi/2)\left[1 + \tan^2(\phi/2)\right]$ which leads to the second of Eqs. (4.60).

times for the hyperbolic trajectories can be obtained from those of the elliptic trajectories by making the angle and Kepler frequency imaginary: $\psi \to i\psi$, $\omega \to i\omega$. This is because of the square-root dependence of ω on $\mathcal{E}$ which switches sign for the two cases. Only a finite amount of time is required for the body to repeat its trajectory on an elliptic orbit since the path length of the orbit is finite. Since the path lengths of the parabolic and hyperbolic trajectories are infinite and the body moves at finite velocity, the time required to complete a transit over the entire trajectory is likewise infinite.

The Kepler Equation

Equations of disarming simplicity often provoke epochal developments in mathematics. One such equation is the time equation of two-body orbits,

$$\omega t = \psi - e \sin \psi .$$

It is known as *Kepler's equation.* This modest equation spurred the work of many of mathematics' most fruitful contributors including Newton, Lagrange, Euler, and Augustin-Louis Cauchy. Equations (4.60), of which this is the elliptic case, are more useful to astronomers when the eccentric anomaly is expressed as a function of time. Inversion of the Kepler equation to $\psi = \psi(t)$ served as a major challenge to mathematicians of the seventeenth and eighteenth century, giving birth to major developments in analysis as well as the theory of functions of complex variables (**Note 5**).

Before the seventeenth century algebraic expressions dominated analysis. [An algebraic expression is a polynomial of the form $\psi(t) = \alpha_0 + \alpha_1 t + \alpha_2 t^2 + \cdots + \alpha_n t^n$ with a finite number of terms.] It was natural to wonder if the inversion of Kepler's equation could be developed as an algebraic expression. In a masterful proof in the *Principia*, Newton showed this is impossible; moreover, his proof has even wider implications. It shows that the area of any smooth oval cannot be expressed algebraically. This connection which relates algebraic area integrals generally and the time integral for Kepler motion springs from Kepler's second law: the times for orbits are area integrals over these orbits.*

Since the time is an area integral of the ellipse—a smooth oval—it increases periodically by the same amount every time the true anomaly ϕ completes a cycle of 2π. As the eccentric anomaly ψ sweeps over its range, infinitely many cycles of ϕ result. The resulting equation $\psi = \psi(t)$ must therefore have infinitely many roots. (They are, in fact, $t = nT$; $n = 0, \pm 1, \pm 2, \ldots$.) Since an algebraic expression has only a finite number

---

* A discussion of Newton's proof may be found in V. I. Arnol'd, *Huygens and Barrow, Newton and Hooke*, Birkhauser, Basel (1990).

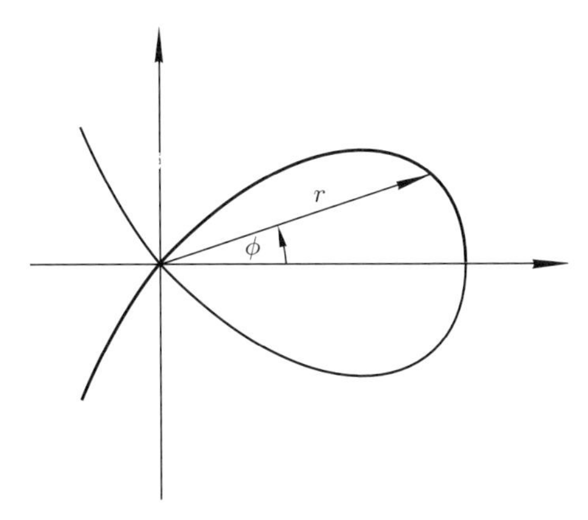

Figure 4-15. An Algebraically Integrable Oval. The oval $r(\phi) = (1-\tau^2)\sqrt{1+\tau^2}$ where $\tau = \tan\phi$ has an algebraic area integral $\int (r^2/2)\, d\phi = \tau^3\left(\tau^2/5 - 1/3\right)$.

of roots equal in number to the order n of the polynomial, it cannot be the time integral of Kepler motion.

This result, framed in the language of time and the Kepler ellipse, is more general. Newton showed the area integral of any smooth oval cannot be algebraic! This does not mean the area integral of any closed curve cannot be algebraic. Orbits which are *not* smooth can possess algebraic area integrals. An example is the oval with kink of Fig. 4-15,

$$r(\phi) = (1-\tau^2)\sqrt{1+\tau^2},$$

where $\tau = \tan\phi$. It has the area integral

$$\int (r^2/2)\, d\phi = \tau^3\left(\tau^2/5 - 1/3\right).$$

Functions which are not algebraic are said to be *transcendental* and have expression as infinite series. The exponential and trigonometric functions are the most familiar transcendental functions. They are the real and imaginary parts of the exponential function of $z = x + iy$ whose infinite series expansion is

$$e^z = e^x(\cos y + i\sin y) = \sum_{m=0}^{\infty} z^m/m! = 1 + z + z^2/2! + z^3/3! + \cdots.$$

The infinite series representing the transcendental solution of Kepler's equation (which is not so familiar) is

$$\psi(t) = \omega t + 2\sum_{m=1}^{\infty} \frac{1}{m} J_m(me)\sin m\omega t, \tag{4.63}$$

where J_m is the Bessel function. The detailed development of this solution may be found in **Note 5**.

The Bessel function $J_m(me)$ appearing in Eq. (4.63) vanishes for $e = 0$ and oscillates with a diminishing amplitude for large values of its argument me. This is just the required behavior to allow the angle for circular orbits ($e = 0$) to be directly proportional to time: $\psi = \omega t$. But for $e > 0$, the Bessel function has the property of speeding up and slowing down the sweep of the angle giving the orbit angle a nonlinear dependence on time. As a result, the angle sweeps rapidly around the orbit near pericentron and more slowly around the region near the apocentron. The devices of deferents and epicycles which Ptolemaic astronomers introduced to force this effect in the motion of the planets is here naturally portrayed in the motion of Eq. (4.63) through the oscillatory property of the Bessel function.

Collisions

It is impossible for two point particles in one another's attractive field to collide if the angular momentum remains finite. The separation between the bodies is given by the relative position vector r. Two bodies may approach no closer than the distance $r_{\min}$ which is proportional to the square of the angular momentum in Eqs. (4.48) and (4.49). This distance is measured from the center of mass of the bodies. A small caveat is therefore worth noting. Although two point particles with finite angular momentum may not collide, finite bodies may collide if their collisional cross-section radius (which is the sum of their body radii, $r_1 + r_2$,) is greater than $r_{\min}$.

When $M \to 0$, a two-body collision becomes possible. A collisional trajectory has eccentricity $e = 1$ even though the energy need not vanish. This is easily seen from Eq. (4.48) since $\mathcal{E}_0 \to \infty$ when $M \to 0$. In the collisional limit the Kepler ellipse becomes more and more elongated and degenerates to a straight line oscillation along the eccentricity vector of Fig. 4-16 with $r_{\min} = 0$ and $r_{\max} = 2a$ from Eqs. (4.48) and (4.49). The origin representing the center of mass and the pericentron merge, the distant focus and the apocentron merge, and the moving point representing the bodies oscillates between these points over the distance $2a$. Infinitely many collisions may occur with period T, a collision occurring each time the moving point returns to the origin.

In the parabolic and hyperbolic cases ($\mathcal{E} \geqslant 0$) the collisional trajectory is also a straight line passing through the origin. In contrast to the elliptic case, the straight-line parabolic and hyperbolic trajectories pass to infinity. Only one collision occurs in the parabolic and hyperbolic cases since there is only one nonperiodic contact with the origin.

Collisions introduce a delicate feature into the classical mechanics of point particles: the existence of singularities. A singularity is a point where one or more quantities become infinite. The inverse-square force is infinite

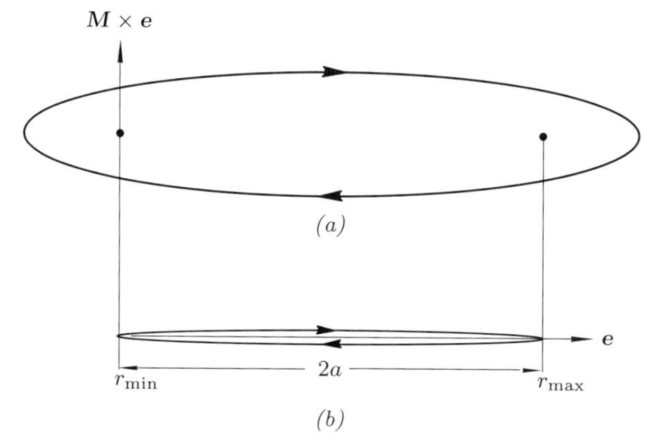

Figure 4-16. Collisional Trajectories. The ellipse *(a)* of an incipient collisional orbit degenerates to straight-line oscillations along the eccentricity vector when $e \to 1$ *(b)*. The pericentron and origin merge at $r_{\min} = 0$ and the distant focus and apocentron merge at $r_{\max} = 2a$.

at the point of collision. Why then does it not lead to an infinite speed of the particles? The answer is that the impulse integral $\int(k/r^2)\,dt$ containing the point $r = 0$ is finite.

For two-body motion with gravitational and electrical forces, infinite speeds of particles can never be established in finite times. Remarkably, in the case of several bodies attracted by inverse-square forces it is possible to arrange them in a way so as to accelerate one of the bodies to infinite speed in finite time—the many-bodies act like "sling-shots" on one of the bodies. One can also devise peculiar force laws that accelerate two bodies to infinite speeds in finite time.*

Heavenly Tori

When the orbits are bound, Kepler motion (like all bound, integrable motion) may be reduced to a form of ultimate simplicity: the motion lies upon an invariant torus in phase space most naturally described in angle-action coordinates. Since the relative motion is three-dimensional, this torus is a three-torus built up as a Cartesian product of three circles, each of which is described by one of the three angle coordinates. The three-torus is a toroidal body foliated with nested two-tori. Since the motion is fully degenerate, the flow winds *once* around these two-tori on closed streamlines.

Let us exhibit the invariant tori of Kepler motion. The invariant set I of the motion consists of the single-valued action coordinates J and the

---

* See Florin N. Diacu, "Painlevé's Conjecture," *The Mathematical Intelligencer*, **15**, No. 2 (1993).

trajectory flow is a composition of three periodic oscillations, each characterized by an invariant amplitude and an angle which rotates at an invariant frequency. These angle-action coordinates of integrable motion arise in a canonical transformation from either the vector or spinor representation of the separated coordinates and momenta.

Because of the periodic nature of bound motion, the angle coordinates are of two kinds: rotations and librations. A rotation coordinate repeatedly passes through periods of 2π. A libration coordinate oscillates back and forth between two extremes. In Kepler motion the angular coordinate ϕ is a rotation coordinate successively passing through periods of 2π and the radial coordinate r is a libration coordinate oscillating between r_{min} and r_{max}.

The action coordinates of the vector representation are obtained from Eqs. (4.34) through (4.37):

$$\begin{aligned} J_r &= (2\pi)^{-1} \oint \sqrt{2\mu \left[\mathcal{H} + k/r - M^2/2\mu r^2\right]}\, dr, \\ J_\phi &= (2\pi)^{-1} \oint M_3\, d\phi, \\ J_\theta &= (2\pi)^{-1} \oint \sqrt{M^2 - {M_3}^2/\sin^2\theta}\, d\theta. \end{aligned} \tag{4.64}$$

The complicated appearance of these integrals is misleading; for they are quite simple. This is because they represent the contour integral of a complete orbit in phase space. (They would not be so simple if the integral were not over the complete orbit.) Since they are invariants, one knows the actions must be expressible in terms of the Casimir set $(\mathcal{H}, M, M_3)$, or since the energy may equally well be expressed in terms of the Kepler constant, in the form in which all quantities have the dimensions of action, (h, M, M_3).

The polar angle action from Eq. (4.64) is immediately integrable as

$$J_\phi = (2\pi)^{-1} \int_0^{2\pi} M_3\, d\phi = M_3. \tag{4.65}$$

The two invariants M_3 and M define a cone on whose surface the angular momentum vector must lie. The half-angle χ of this cone is given by $\sin\chi = (M_3/M)$. As the motion of the orbit traces out a complete period, the latitudinal angle θ librates between $-\chi$ and χ. The latitudinal action variable from Eq. (4.64) thus covers the range $\oint d\theta = 2\int_{-\chi}^{\chi} d\theta$:

$$J_\theta = (\pi)^{-1} M \int_{-\chi}^{\chi} \sqrt{1 - \sin^2\chi/\sin^2\theta}\, d\theta$$

and has the integral

$$J_\theta = M(1 - \sin\chi) = M - M_3. \tag{4.66}$$

The radial coordinate r librates between $r_{\rm min}$ and $r_{\rm max}$. The range for the integral J_r is therefore $\oint dr = 2\int_{r_{\rm min}}^{r_{\rm max}} dr$. This integral is most naturally evaluated with the theory of residues, a method of evaluating closed contour integrals in the complex plane.* It turns out to be simply

$$J_r = h - M. \tag{4.67}$$

It follows from (4.66) and (4.67) that the sum of the actions is the Kepler constant,

$$h = J_r + J_\phi + J_\theta,$$

The action coordinates are invariants. The more familiar forms of the invariants for separated Kepler motion in spherical-polar coordinates—the invariant set $(\mathcal{H}, M, M_3)$—may be expressed in terms of them through Eqs. (4.65)–(4.67) and the condition $\mathcal{H} = -k^2\mu/2h^2$:

$$\begin{aligned} M_3 &= J_\phi, \\ M &= J_\phi + J_\theta, \\ \mathcal{H} &= -\frac{k^2\mu/2}{(J_r + J_\phi + J_\theta)^2}. \end{aligned} \tag{4.68}$$

The phase velocities in angle-action coordinates are $\omega_i = \partial\mathcal{H}/\partial J_i$. It is apparent from the last of Eqs. (4.68) that all three phase velocities ω_r, ω_ϕ, ω_θ will be identical for the Kepler Hamiltonian since the action coordinates appear in the sum $(J_r + J_\phi + J_\theta)$. This single, common phase velocity is found from Eq. (4.68) to be the Kepler frequency (4.55):

$$\omega = \frac{k^2\mu}{(J_r + J_\phi + J_\theta)^3}. \tag{4.69}$$

The equality of all three frequencies manifests the complete degeneracy of Kepler motion. Equality of ω_ϕ and ω_θ corresponding to the appearance of J_ϕ and J_θ in the sum $(J_\phi + J_\theta)$ is the signature of the central force degeneracy. Equality of ω_r with ω_ϕ and ω_θ corresponding to the appearance of all three action coordinates in the sum $(J_r + J_\phi + J_\theta)$ is the signature of the additional inverse-square force degeneracy.

The simplicity achievable with the angle-action representation is further evidenced by the Hamiltonian. From Eq. (4.69), it may be expressed in action coordinates in the simple form

$$\mathcal{H} = -J_i\omega_i/2$$

---

* Arnold Sommerfeld, *Mechanics*, Academic Press, New York (1956); H. Goldstein, *Classical Mechanics* (2nd Ed.), Addison Wesley, Reading Mass., pp. 472–478 (1980).

and may be compared with the far more involved form of Eq. (4.29). The final invariant of Kepler motion, the eccentricity e, is also expressible in action invariants. Since $e^2 = 1 - M^2/h^2$, the eccentricity is

$$e^2 = 1 - \left(\frac{J_\phi + J_\theta}{J_r + J_\phi + J_\theta}\right)^2. \tag{4.70}$$

Reduction to angle-action coordinates is also possible for the spinor representation of Kepler motion. In this case the action coordinates are J_ξ, J_η, and J_ϕ and they are expressible in the spinor Casimir set $(\mathcal{H}, E_3, M_3)$. The action variable J_ϕ is identical to that for spherical-polar coordinates. The remaining action coordinates may be constructed from Eqs. (4.46) as

$$\begin{aligned} J_\xi &= (2\pi)^{-1}\frac{1}{2}\oint \sqrt{2\mu\left[\mathcal{H} + (1 - e_3)k/\xi - {M_3}^2/2\mu\xi^2\right]}\, d\xi, \\ J_\eta &= (2\pi)^{-1}\frac{1}{2}\oint \sqrt{2\mu\left[\mathcal{H} + (1 + e_3)k/\eta - {M_3}^2/2\mu\eta^2\right]}\, d\eta, \end{aligned} \tag{4.71}$$

(note the factor 1/2 in front of these integrals, a sign that these are spinor representations). These integrals may be evaluated by again using the theory of residues with the result

$$\begin{aligned} J_\phi &= M_3, \\ J_\xi &= h/2 - (M_3 + E_3)/2, \\ J_\eta &= h/2 - (M_3 - E_3)/2. \end{aligned} \tag{4.72}$$

Since the parabolic coordinate form of Kepler motion is a spinor representation, one expects the parabolic actions to be spinors. This is indeed the case; the parabolic actions J_ξ and J_η are found from Eqs. (4.22) and (4.72) to be none other than the spinor invariants of the vectors $\boldsymbol{S}$ and $\boldsymbol{D}$ that belong to the pair of two-sphere families that cover the three-sphere of four-dimensional rotational symmetry:

$$J_\xi = S^\pm, \qquad J_\eta = D^\pm.$$

The spinor Casimir set (h, E_3, M_3) is expressible in action coordinates as

$$M_3 = J_\phi, \qquad E_3 = J_\eta - J_\xi, \qquad h = J_\eta + J_\xi + J_\phi. \tag{4.73}$$

The corresponding Hamiltonian has a form similar to that for the vector representation,

$$\mathcal{H} = -\frac{k^2\mu/2}{(J_\phi + J_\xi + J_\eta)^2}.$$

In passing, one notes that the Poincaré invariant $\mathcal{S} = \oint p_i\, dq_i$ of integrable motions is a sum of the action invariants, $\mathcal{S} = 2\pi \sum_i J_i$. The Poincaré invariant of Kepler motion is independent of the representation; for either the vector or spinor representation it is (aside from a factor of 2π) the Kepler constant:

$$\mathcal{S} = 2\pi h.$$

The Heavenly Spheres

As a bound, integrable motion, Kepler motion lies upon a three-torus. From the viewpoint of its symmetries, however, Kepler motion lies upon a three-sphere which is the basis of its four-dimensional rotational symmetry (the Kepler three-sphere is explicitly exhibited in **Note 6**). One therefore has two different topological representations of Kepler motion: a three-torus filled with two-tori upon which streamlines close after one turn and a three-sphere filled with a pair of families of two-spheres upon which the trajectories also wind.

How can the motion both lie upon a torus and a sphere? This unique occurrence is due to the exceptional properties of four-dimensional space and the three-sphere supporting its rotational symmetry. Although the sphere and torus are in general topologically distinct objects, in the remarkable case of the three-sphere, they come into coincidence. Kepler motion simultaneously lies upon a sphere and a torus! Let us see how the three-torus and three-sphere are united in Kepler motion.

An instructive way of seeing the topological equivalence of the three-sphere and the three-torus is to turn one into the other with operations that preserve topology. One begins with the three-sphere as represented by a pair of three-dimensional spherical bodies joined along a great two-sphere in Fig. 4-17 *(b)*. First, cut the bodies apart along the great two-sphere (the common "skin") which joins them, separating them as in Fig. 4-17 *(c)*. (The bodies will subsequently be glued back together on this surface.) Next, open a toroidal cavity with surface C in *one* of the bodies, say A_1, as shown in Fig. 4-17 *(d)*. Next slice through the body A_1 to the inner edge of the toroidal cavity thereby exposing the surfaces B_1', B_1'' as in Fig. 4-17 *(e)*. Now draw out these exposed surfaces so that the surface C of the toroidal cavity becomes the outer surface of a growing toroid in Fig. 4-17 *(f)*. The formerly outer spherical surfaces A_1', A_1'' along with B_1', B_1'' now become the cross-sectional surface of the emerging toroid. (It is important to note that although the sliced surfaces B_1', B_1'' can be joined back together to complete the torus as in *(g)*, the surfaces A_1', A_1'' cannot be joined in this way because they were severed along their edges, not their faces. A single two-spherical body *cannot* be turned into a torus. It is the second spherical body joined with the first that makes completion of the torus possible.) The surfaces A_2', A_2'' of the second spherical body can now be joined to the surfaces A_1', A_1'' of the first as they were in the original *(a)*–*(b)*. The surfaces B_1', B_1'' can

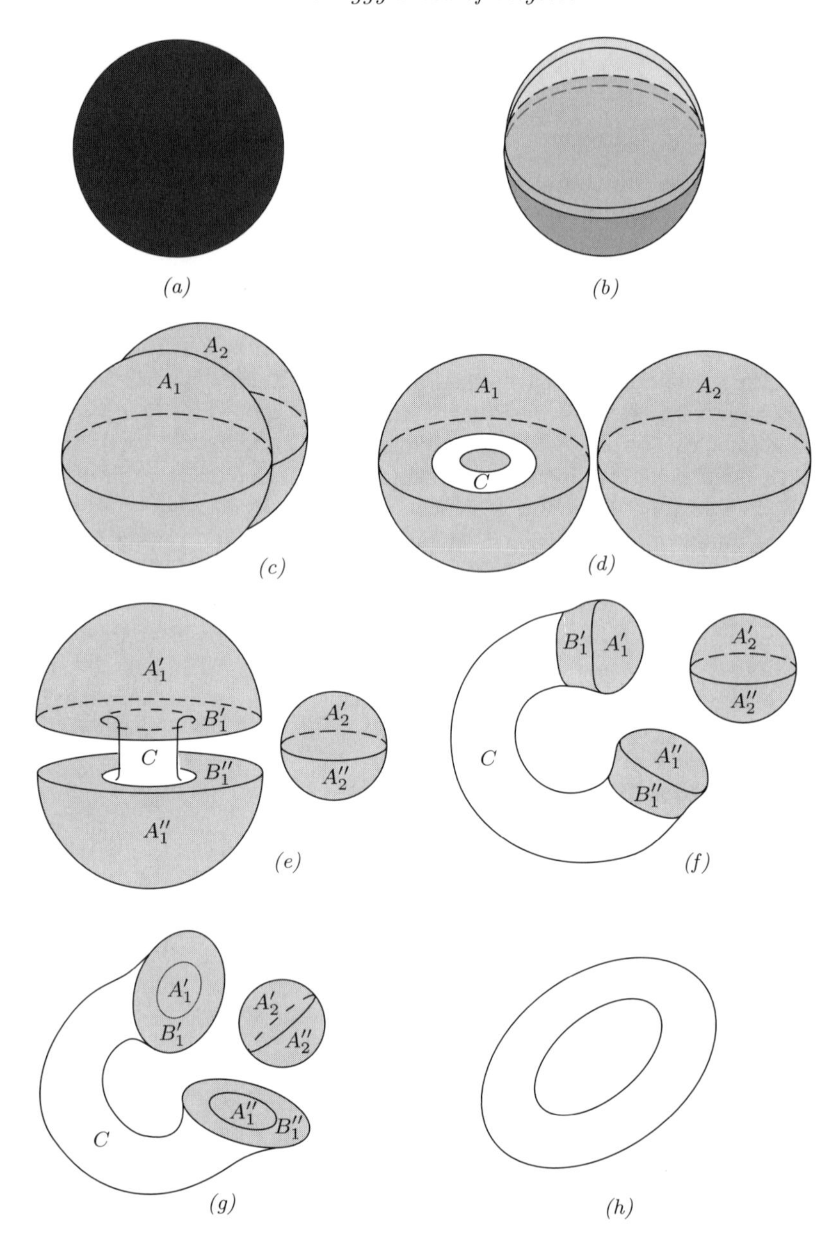

Figure 4-17. Topological Equivalence of the Three-Sphere and Three-Torus of Kepler Motion. The Kepler three-sphere *(a)* is topologically equivalent to two interpenetrating spherical bodies *(b)* and may be turned into a three-torus *(h)* by unfastening the bodies along their surface skin and separating them *(c)*, opening up a toroidal cavity C in one of the bodies *(d)*, slicing into the cavity and drawing out a growing proto-torus *(e)*, *(f)*, and closing it by rejoining the sliced surfaces and refastening the spherical bodies by their skins *(g)* to yield the integrable three-torus upon which the motion also lies *(h)*.

also be joined together as in Fig 4-17 *(g)* thereby creating the three-torus foliated with two-tori *(h)*.

The complete motions of the two-body problem of celestial mechanics, "The Kepler Problem," have been revealed; and we have rediscovered the perfection of the heavenly spheres in new raiment. The motion of two bodies in its most penetrating and elegant mathematical form consists of perfect circles which build invariant tori. Heavenly bodies lie upon heavenly tori; and these heavenly tori are coincident with heavenly spheres, a correspondence unique to Kepler motion.

The underlying symmetries of the solar system are indeed the symmetries of perfect spheres—the rotational symmetries of spheres in four dimensions. We do not directly see this underlying spherical symmetry in the configuration space of our experience—we see ellipses. But mathematics provides us the ability to see into phase space. And in the full world of motion that is phase space, spherical symmetry shines forth. In so doing it reveals that the heavenly spheres have a richer structure than the ancients imagined. The heavens are an outstanding foliation of spheres within spheres which, by the magic of four-dimensional space, are at the same time a foliation of tori within tori.

Reprise

The solution of the Kepler problem epitomizes the unity and harmony of mathematics and the material world. One begins with the symmetries of space and time and the simple law of least action. It offers up the equations of motion, the universal Galilean invariants, and a new invariant which exists only for the inverse-square force, the force of gravity and electricity. One asks for the bound orbits by restricting the energy to the range $\mathcal{E} < 0$. An ellipse results. The geometric properties of the ellipse—its axes and the periodic time to complete a circuit around the orbit (a, b, T)—though pure mathematical quantities are, through the laws of motion, endowed with mass, energy, and angular momentum—the stuff of materiality:

$$a = k/2\mathcal{E}, \qquad b = M/\sqrt{2\mu|\mathcal{E}|}, \qquad T = \pi k\sqrt{\mu/2|\mathcal{E}|^3}.$$

The inverse-square force of gravity within the space and time of rotational and translational symmetry is one of the great unifying themes of the physical world and a key to understanding all heavenly motions. Yet this force is the unifying feature not only for the planetary motions of the heavens but also for the atom and the elements which are the stuff of heavenly bodies. The unity is even more manifold than Newton, Galileo, and Copernicus might have imagined.

One stands in amazement before the far-reaching presence of the four-dimensional rotational symmetry of Kepler motion in the world. The ultimate building block of this symmetry is the spinor, the most elementary object possessing rotational symmetry. A single spinor builds a three-vector.

A pair of spinors also builds a pair of three-vectors which describe a four-dimensional rotation. A pair of spinors turns out to be the mathematical signature of the electron and its antiparticle the positron. The same rotational symmetry generalized to complex quantities and higher dimensions explains much of the structure of the most elementary particles known today.

CHAPTER 5

Quantum Mechanics: The Elements

Consider the solar system: an ensemble of isolated point-like planets orbiting the sun through the vast tenuousness of interplanetary space. Consider the elements: rich textures and colors, a diversity of gaseous, liquid, and solid phases, abundant forms of chemical reactivity. These two aspects of the natural world appear beyond compare. Yet the heavens and the elements are possessed of a striking unity.

The heavens and the elements are both the manifestation of particles in motion bound by the inverse-square force: sun and planets which constitute the solar system; electrons and charged nuclei which constitute the atom. Both are governed by the principle of least action and shaped by the same Hamiltonian structure. Both partake of the same symmetries and possess the same invariants. The prototypical planetary orbit and the prototypical element both blossom from four-dimensional rotations.

How, then, explain the differences? To be sure, the heavens and the elements are sustained by different forces, gravity on the one hand and electricity on the other. Though both are inverse-square forces, the strength of the electrical force between two electrons is forty two orders of magnitude greater than that of their gravitational force. So there is a decoupling that takes place. The large-scale world bound by gravity goes its way with only a loose connection to the microscopic world bound by electricity. Yet this is not the whole story.

The decisive difference between the heavens and the elements, between solar system and atom, turns on the fact that quite independently of the kinds of forces acting, all particles are ultimately wave-like. Pressed to its limit, a particle is not a little marble; it is a bundle of waves with a definite wavelength that specifies the distance between the alternate crests and valleys of its wave structure. The gravitational force binding the solar system does not push the sun and planets to this limit; but the electrical force binding the electrons to the nucleus of an atom does push atomic scale particles to this limit. The mechanics of the bodies of the solar system is a particle-like classical motion. The mechanics of the microscopic particles which constitute the atom is a wave-like or quantum motion.

The movement from classical mechanics to quantum mechanics is revolutionary. Yet the central concepts of mechanics are preserved in integrity. The law of motion has a universal character. Least action, invariants, symmetries, and Hamiltonian structure pervade both the classical and the quantum world. Indeed, the quantum enhances and deepens their significance as it introduces new features unique to the quantum realm.

The quantum scale is extraordinarily small by human standards, roughly a billion times smaller than the human hand. This is why the discovery of quantum phenomena is a late development in human history. The world of the quantum is a microscopic world of particles—the atom, electrons, protons, neutrons, and a host of others. Yet the effects of the quantum are not confined to the microscopic. The quantum signs its presence in the large-scale world in the unique structure of the elements out of which the chemistry of nature and the diversity of matter arise. The diversity of the elements rests upon the quantum interaction of charged particles. We begin our journey into quantum mechanics with a survey of these particles.

Atom: Source of the Elements

The atom consists of a nucleus bound by the electrical force to a constellation of orbiting electrons. The nucleus of the hydrogen atom consists of a single proton. But all other atomic nuclei consisting of more than one proton also contain neutrons which bind the protons together with the strong nuclear force (the force must indeed be "strong" since it binds protons into a nucleus with a greater strength than their repulsive electrical force which would otherwise fling them apart). The nucleus extends over an exceedingly small region of space whose breadth is ten thousand times smaller than the spatial extent of the electronic orbits. If the nucleus were represented by the sun, the electronic orbits would be located in the vicinity of and beyond Pluto, the most distant planet.

The total number of negatively charged electrons in an atom is equal to the number of positively charged protons in its nucleus. The atom as a whole is therefore electrically neutral. The prototypical atom is the hydrogen atom. It consists of a nucleus formed of one proton to which a single electron is bound by the inverse-square electrical force.

Each element is characterized by the total number of electrons (or protons) which constitute its atoms. The weight of an element (or the atomic weight) has a one-to-one correspondence with the number of orbiting electrons. The periods of the Periodic Table which are based upon atomic weight are therefore uniquely correlated with the number of electrons (or protons) of the atom of a given element. Periods of **2**, **8**, and **18** which describe the chemical and physical properties of the elements are the inescapable pattern in the Periodic Table as shown in Fig 2-4. Atoms with 2 electrons behave similarly to atoms with 8 and 18 electrons. Atoms with 3 electrons behave similarly to atoms with 9 and 19 electrons, and so on.

The elliptical patterns and transit times in the motions of the heavenly bodies are a manifestation of the law of motion for an inverse-square gravitational force. It is no small marvel that the rhythmic patterns of the Periodic Table are a manifestation of the same law of motion with an inverse-square force identical to that of gravity.

The action principle for gravitational potential energy governs the universe in the large. If the action principle with an electrical potential energy identical in form to that of gravity governs the atom, then it should lead to geometrical structures for the atom identical to those of the solar system. Two attractive charges may form bound elliptical orbits. They may also form unbound parabolic and hyperbolic trajectories.

So there it is: a miniature solar system with a charged nucleus playing the role of the sun, orbiting electrons playing the role of the planets, and the electrical force playing the role of gravity. It is remarkable that this model proved to be correct. It possesses devastating contradictions.

First, an atomic scale solar system has an amorphous structure. An electron orbiting the nucleus according to the mechanics of Newton, Lagrange, and Hamilton would indeed move in an elliptical orbit; but an ellipse in itself has no obvious connection to the repeating properties of the Periodic Table. Changes in the energy and angular momentum of the elliptical orbit are not reflected in any obvious way with changes in the chemical properties of the elements. The energy and the angular momentum change continuously. But the chemical shifts are in discrete jumps from element to element. Shifts in the number of electrons elliptically orbiting a nucleus also do not in themselves explain the dramatic shifts of chemical properties and repeating cycles which the chemical elements exhibit. Structured as the ellipse might be, it does not possess a microscopic richness that flowers with the diversity of the elements.

Second, such a miniature solar system is unstable. Electromagnetic theory predicts that an orbiting electron should radiate energy. As the electron loses energy by radiation, its orbit should spiral inward to the nucleus in an ultimate collapse of the atom. The Newtonian mechanics of a charged nucleus with an electrically bound electron yields an inherently unstable atom.

Numerous experiments of the early twentieth century substantiated the existence of the nucleus and the electron and verified that the binding force between these two particles was indeed the electrical inverse-square force. Unfortunately, when these two particles are allowed to interact according to the laws of mechanics revealed by Newton, Lagrange, and Hamilton, they yield the family of conic sections—dynamical structures that cannot explain the behavior of the elements.

Niels Bohr resolved this contradiction by proposing that at the microscopic level of the atom one must regard the motion of a point particle orbiting another particle as a wave. A wave arrayed around a closed orbit cannot close on itself unless it consists of an integer number of wavelengths.

The length of the orbit must therefore be an integer multiple of a fundamental length (fixed by a new constant of nature first discerned by Max Planck). The orbit is *quantized*: the states of motion of the electron are discrete waves denumerated by integers.

What Kepler did for classical mechanics with the ellipse, Bohr did for quantum mechanics with the quantized orbit. As Newton would subsequently produce the Kepler ellipse from a more fundamental law of motion, Wolfgang Pauli would subsequently produce the Bohr orbit from the more fundamental laws of quantum mechanics established by P. A. M. Dirac and Werner Heisenberg. Overshadowing these developments in the law of motion are the symmetries of two-body motion with gravitational and electrical forces. The eccentricity invariant of Jacob Hermann persists from the classical realm into the quantum realm where Pauli would reveal its quantum form. The symmetries of four-dimensional rotations play a central role in both the classical and quantum motion of two bodies.

Particles and Waves: Dichotomy and Unity

At first blush, particles and waves seem contradictory manifestations of objects such as electrons and protons. It does not seem possible that an electron can be both a point particle and a wave. The resolution of this dichotomy hinges on a proper recognition of the importance of *scale* in determining which kind of properties the electron can be expected to exhibit. The length scale of a particle such as an electron is its wavelength. When it interacts with objects whose length scale is much longer than its wavelength, it acts like a point particle. When it interacts with objects whose length scale is comparable to or shorter than its wavelength, it acts like a wave.

An elementary wave consisting of a single frequency ω and traveling in the direction of the wave vector $\boldsymbol{k}$ may be represented as a wave function,

$$\Psi(\mathbf{x}, t) = Ae^{i(\boldsymbol{k}\cdot\mathbf{x}-\omega t)}, \tag{5.1}$$

where A is a complex amplitude which specifies both the strength and phase of the wave. The intensity of the wave which is measured (and must therefore be a real quantity) is given by the square amplitude $|\Psi|^2 = \Psi\Psi^*$, where Ψ^* is the complex conjugate of Ψ. Particles display a spectrum of such elementary waves.

Under classical conditions one says a particle is located at the point $\mathbf{x}$. Under quantum conditions, one says the particle is located somewhere within the region where $|\Psi(\mathbf{x})|^2$ has significant value. The point $\mathbf{x}$ of a classical particle is replaced by the "smear" $|\Psi(\mathbf{x})|^2$ of a quantum particle as shown in Fig. 5-1. The $|\Psi(\mathbf{x})|^2$ smear represents the probability of finding the particle at the point $\mathbf{x}$. The particle has no likelihood of being

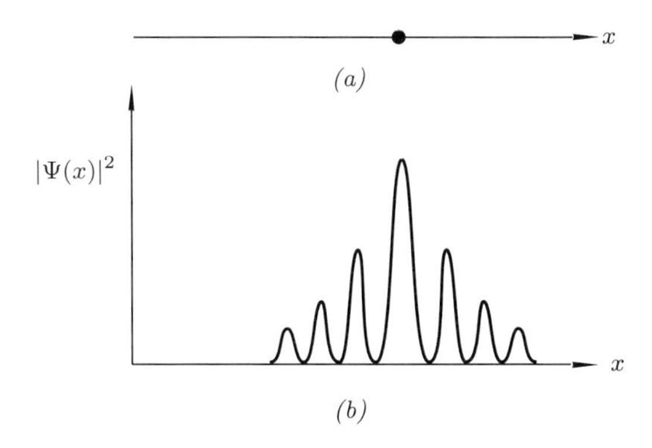

Figure 5-1. Classical Point Particle and Quantum Wave. The classical point $\mathbf{x}$ of a particle's position *(a)* becomes the quantum smear $|\Psi(\mathbf{x})|^2$ *(b).*

in regions where $|\Psi(\mathbf{x})|^2$ vanishes and a very great likelihood of being in regions where $|\Psi(\mathbf{x})|^2$ has large values.

This idea is more precisely illustrated in Fig. 5-2. For particle wavelengths short compared to the scale of the objects with which it interacts, the point properties of a quantum particle are pre-eminent. The wave-like oscillation of the particle cannot be detected when it passes through a slit long compared to its wavelength. The particle acts as if it were flying freely through an open window and strikes the target with the precision of its original aim. The particle, even though it be a wave, acts in this situation just as if it were a point.

But for particles whose wavelength becomes comparable to or longer than the length scales of the objects with which it interacts, the wave nature becomes pre-eminent. Even if it is aimed precisely at the center of the slit, the particle does not fly freely through it. The particle waves collide with the slit walls causing it to be diffracted and its location is made uncertain. It no longer lies on the straight-line trajectory through the center of the slit, but is distributed over the target.

If a particle is to be identified with a wave represented by the wave function Ψ, the wave function must be equivalent to that mechanical quantity embodying its full dynamics, its action S. The action of a motion is equal to the sum of the actions of its independent parts. The wave function of a motion, however, represents a probability; and the probability of a sum of independent parts is the product of the probabilities of the parts. The identity between the wave function Ψ and its corresponding action S is an identity between functions that are multiplicative on the one hand and additive on the other.

The wave function must therefore be an exponential of the action

$$\Psi = Ae^{iS/\hbar}, \tag{5.2}$$

where $\hbar$ is a constant of nature required to render the wave function di-

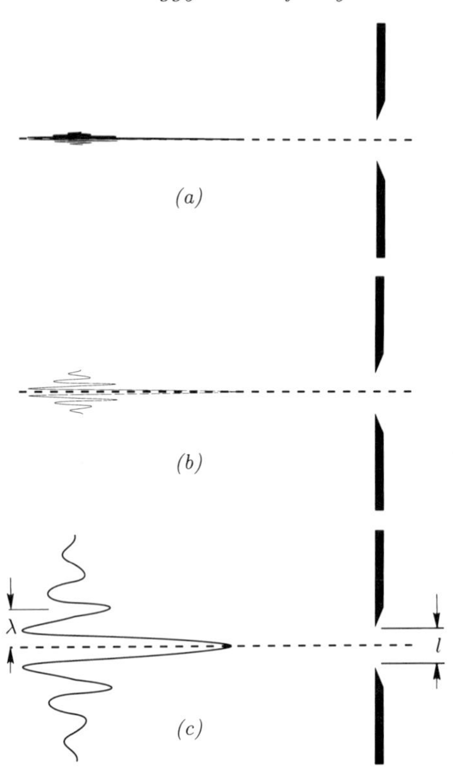

Figure 5-2. Particle-Wave Duality. At wavelengths short compared to the slit with which it interacts *(a)*, the particle appears as a point flying freely through the slit. But at wavelengths λ comparable to or longer than the slit width l the particle appears as a wave and is diffracted by the slit *(b)*, *(c)*.

mensionless. The action has dimensions $length^2 \times time$ and the dimensions of the constant $\hbar$, which arises in the identity between particles and waves, must likewise have the dimensions of $length^2 \times time$, the dimensions of action.

The identification of Eq. (5.2) with Eq. (5.1) reveals that the action is represented in terms of wave vector and frequency as $S = \hbar(\boldsymbol{k} \cdot \mathbf{x} - \omega t)$ and has the derivatives

$$\partial S/\partial \mathbf{x} = \hbar \boldsymbol{k}, \qquad \partial S/\partial t = -\hbar\omega. \tag{5.3}$$

But the space and time derivatives of the action are the momentum and energy as given by Eqs. (3.26):

$$\partial S/\partial \mathbf{x} = \mathbf{p}, \qquad \partial S/\partial t = -H.$$

The wave vector and frequency of a wave are therefore given in terms of the momentum and energy of a particle by

$$\boldsymbol{k} = \mathbf{p}/\hbar, \qquad \omega = H/\hbar. \tag{5.4}$$

Equation (5.4) embodies the duality between particles and waves. Dynamical quantities such as $\boldsymbol{p}$ and H are equivalent to the wave vector $\boldsymbol{k}$ and frequency ω of a wave.

The quantum revolution can be regarded as a transfiguration of the classical in which the phase flow, rather than consisting of particle streamlines, consists of waves which both constructively and destructively interfere. Mechanical quantities which are properties of the flow such as momentum, angular momentum, energy, eccentricity, and so on are no longer regarded as ordinary algebraic functions of space–time points. Rather, they are regarded instead as more creative mathematical objects which will manifest themselves in waves. Such wave-like mechanical objects are known as *operators.*

An operator is a supple mathematical creation. It has both the capacity to generate wave-like objects as well as those which compact down to a point. An operator is not the measurable value of the physical quantity it represents in the way classical mechanical variables take the measurable values of quantities they represent. Instead, operators generate a whole spectrum of values unique to a particular mechanical quantity which oscillate, decay, or grow like interacting waves of light or sound. A measurable value of a physical quantity turns out to be one of the values in the spectrum.

The nature of an operator may be established by noting that the wave function (5.1) reproduces itself upon differentiation with coefficients which are its wave vector and frequency:

$$\partial\Psi/\partial\mathbf{x} = i\boldsymbol{k}\Psi, \qquad \partial\Psi/\partial t = -i\omega\Psi. \tag{5.5}$$

On making the identification of $(\boldsymbol{k}, \omega)$ with $(\boldsymbol{p}, H)$ in Eqs. (5.4), the forms in Eq. (5.5) can be expressed as

$$(-i\hbar\partial/\partial\mathbf{x})\Psi = \boldsymbol{p}\Psi, \qquad (i\hbar\partial/\partial t)\Psi = H\Psi. \tag{5.6}$$

These identities introduce *dynamical operators*:

$$\hat{\mathbf{p}} \equiv -i\hbar\partial/\partial\mathbf{x}, \qquad \hat{H} \equiv i\hbar\partial/\partial t, \tag{5.7}$$

in terms of which they become

$$\hat{\boldsymbol{p}}\Psi = \boldsymbol{p}\Psi, \qquad \hat{H}\Psi = H\Psi. \tag{5.8}$$

The wave-particle duality leads us to the remarkable identities (5.8). These austere mathematical relationships—uncanny in their simplicity and mesmerizing in their iconography—tell us that an operator such as $\hat{\boldsymbol{p}}$ applied to a function Ψ yields precisely that same function with a coefficient which is the observed value $\boldsymbol{p}$ of that dynamical quantity.

The Quantum Rendering of Mechanics

Particles beget mechanical quantities which behave as algebraic variables. Waves beget the same mechanical quantities but they behave as operators. When particle wavelengths become short compared to the scale of motion, quantum operators become classical mechanical quantities. Quantum waves settle down into streamlines and become the classical phase flow. One may therefore regard quantum mechanics as the fundamental theory of motion which contains classical mechanics as its limit at large quantum wavelengths. The quantum world of the atom therefore appropriates the classical mechanics of Chaps. 3 and 4 with the simple, if epochal, change that mechanical quantities become operators.

Thus, to every mechanical quantity A (which may represent momentum, angular momentum, energy, eccentricity, and so on) there corresponds an operator $\hat{A}$. A dynamical quantity and its operator are linked by the fundamental identity connecting particles and waves for which Eqs. (5.8) are the prototypes:

$$\hat{A}\Psi = A\Psi. \tag{5.9}$$

Let us illustrate the operators for the most important mechanical quantities. The operators of space and time are postulated to be identical to their observables:

$$\hat{\mathbf{x}} = \mathbf{x}, \qquad \hat{t} = t.$$

On the other hand, Eqs. (5.7) indicate that the operators corresponding to the momentum and energy, $\hat{\mathbf{p}} = -i\hbar\partial/\partial\mathbf{x}$ and $\hat{H} = i\hbar\partial/\partial t$, as well as the angular momentum,

$$\hat{\boldsymbol{M}} = -i\hbar\,\mathbf{x}\times\partial/\partial\mathbf{x},$$

are differential operators. More complex quantities which involve position, momentum, and energy may be built up from the momentum and energy operators.

The operator appropriate to the kinetic energy involves the squared magnitude of the momentum. For a single particle of mass m the kinetic energy operator is $\hat{T} = \hat{p}^2/2m$. The operator $\hat{p}^2$ is

$$\hat{p}^2 = -\hbar^2\left(\partial^2/\partial x_1^2 + \partial^2/\partial x_2^2 + \partial^2/\partial x_3^2\right).$$

The quantum kinetic energy operator, aside from constant factors, is the Laplacian $\hat{p}^2 = -\hbar^2\nabla^2 = -\hbar^2\partial^2/\partial x_i^2$.

The action S is the fundamental mechanical quantity describing classical motion and the Hamilton–Jacobi equation which governs it can be taken as the master equation governing classical motion. The action may be extended into the quantum realm through identity with the wave function, $\Psi \sim e^{iS/\hbar}$. What is the corresponding equation for the action in the quantum realm? It is thc Schroedinger equation—the quantum identity for the energy. This equation results when one writes $\hat{H} = i\hbar\partial/\partial t$ on the one hand, substitutes the Laplacian $\hat{p}_\alpha^2 = -\hbar^2\nabla_\alpha^2$ in $\hat{H} = \hat{T} + \hat{V} = \sum_\alpha \hat{p}_\alpha^2/2m_\alpha + V(\mathbf{x}_\alpha)$ on the other, and equates the results:

$$i\hbar\frac{\partial\Psi}{\partial t} = -\sum_\alpha(\hbar^2/2m_\alpha)\nabla_\alpha^2\Psi + V(\mathbf{x}_\alpha)\Psi. \tag{5.10}$$

The Schroedinger equation is the master equation of quantum motions, the quantum transfiguration of the classical Hamilton–Jacobi equation. The Hamilton–Jacobi equation is the limiting form of the Schroedinger equation in the classical limit. To see this connection, express the wave function in terms of the action through $\Psi \sim e^{iS/\hbar}$. Then one has $\partial\Psi/\partial t = (i\Psi/\hbar)\partial S/\partial t$ with $\boldsymbol{\nabla}_\alpha\Psi = (i\Psi/\hbar)\boldsymbol{\nabla}_\alpha S$. The Laplacian of the wave function expands to two terms in the action:

$$\nabla_\alpha^2\Psi = (i\Psi/\hbar)\nabla_\alpha^2 S - (\Psi/\hbar^2)\boldsymbol{\nabla}_\alpha S\cdot\boldsymbol{\nabla}_\alpha S.$$

The Schroedinger equation expressed in terms of the action is then

$$\partial S/\partial t + i\hbar\sum_\alpha(1/2m_\alpha)\nabla_\alpha^2 S + \sum_\alpha(1/2m_\alpha)\boldsymbol{\nabla}_\alpha S\cdot\boldsymbol{\nabla}_\alpha S + V(\mathbf{x}_\alpha) = 0$$

and should be compared with the classical Hamilton–Jacobi equation (3.38). The Schroedinger equation is a second-order wave equation whereas the Hamilton–Jacobi equation is first-order. It is the second derivative terms in the Schroedinger equation proportional to $\hbar$ that are the uniquely quantum terms. These terms vanish in the limit $\hbar \to 0$; and in this limit the Schroedinger equation becomes the classical Hamilton–Jacobi equation:

$$\partial S/\partial t + \sum_\alpha(1/2m_\alpha)\boldsymbol{\nabla}_\alpha S\cdot\boldsymbol{\nabla}_\alpha S + V(\mathbf{x}_\alpha) = 0.$$

Another way of looking at the correspondence is to recognize that the Hamilton–Jacobi equation also happens to be the equation of geometrical optics—a motion for which wavelengths of rays are short compared to the lengths of the gradients with which they interact (Hamilton and Jacobi primarily thought of their equation in the optical rather than mechanical context; but it properly describes both phenomena). The quantum wave equation—the Schroedinger equation—is a second-order wave equation; but when the wavelengths of the particles are short compared to the lengths of the gradients with which they interact, the second derivative terms may be neglected compared to the first derivative terms and the Schroedinger equation, like the radiation equations, reduces to the classical Hamilton–Jacobi equation.

The observed values A of quantum-mechanical operators $\hat{A}$ are real quantities. The requirement that the observables of an operator always be real imposes an important condition on the operator. An operator whose observables are always real satisfies the *Hermitean condition.* Let Ψ and Φ be any two wave functions operated upon by the same operator. The operator is Hermitean if it satisfies

$$\int \Phi^* \hat{A}\Psi \, d^3x = \int \Psi \hat{A}^* \Phi^* \, d^3x,$$

where the integral is over all space and $()^*$ denotes the complex conjugate. Hermitean operators give the same results when the functions upon which they operate are interchanged and the complex conjugate is taken.

The necessity of the Hermitean requirement for operators representing real mechanical quantities may be seen by multiplying the first of two operator equations,

$$\hat{A}\Psi = A\Psi, \qquad \hat{A}\Phi = A\Phi,$$

by Φ^*, taking the complex conjugate of the second and multiplying it by Ψ, and integrating over all space to produce

$$\int \Phi^* \hat{A}\Psi \, d^3x = A \int \Phi^* \Psi \, d^3x \tag{5.11a}$$

and

$$\int \Psi \hat{A}^* \Phi^* \, d^3x = A^* \int \Psi \Phi^* \, d^3x. \tag{5.11b}$$

It is required that the observable A be a real mechanical quantity and hence it must be true that $A = A^*$. The left-hand sides of Eqs. (5.11) show that this will only be the case if $\hat{A}$ satisfies the Hermitean condition.

All quantum-mechanical operators must be Hermitean. It may be directly shown that the fundamental dynamical operators $\hat{\mathbf{x}}$, $\hat{t}$, $\hat{\mathbf{p}}$, $\hat{H}$, and $\hat{\boldsymbol{M}}$ are all Hermitean. But operators that must be built up from them are not necessarily Hermitean when the mechanical variables are simply transcribed into operator notation according to the rules $H \to i\hbar\partial/\partial t$ and $\mathbf{p} \to -i\hbar\partial/\partial\mathbf{x}$. A noteworthy example is the eccentricity

$$\mathbf{e} = \mathbf{p} \times \boldsymbol{M}/km - \mathbf{r}/r.$$

The quantum-mechanical operator for $\mathbf{e}$ *cannot* be

$$\hat{\mathbf{e}} = \hat{\mathbf{p}} \times \hat{\boldsymbol{M}}/km - \hat{\mathbf{r}}/r$$

because (surprise) $\hat{\mathbf{p}} \times \hat{\boldsymbol{M}}$ is not Hermitean.

What is the axiomatic prescription for constructing operators which are built up as functions of the elementary dynamical variables? Remarkably, save for the requirement they be Hermitean, there is none. There is an art involved in the formation of operators which are built up from the fundamental dynamical quantities $\mathbf{x}$, $\mathbf{p}$, t, and H. The artistic openness of quantum mechanics is another manifestation of an incompleteness in the theory. It is possible that the artistic freedom which requires us to frame anew each unique description of the microscopic world in quantum form is inherent in our interaction with nature.

The classical description of motion involves only one thing: a mechanical quantity A which is observed. The quantum description of motion involves three things: the wave function Ψ, a mechanical quantity A, and its operator $\hat{A}$. The wave function Ψ represents a state of nature. The operator $\hat{A}$ acts upon Ψ and represents the *process of observation* of nature. The observable A is the *result* of that process of observation. These three quantities—the state of nature Ψ under observation, the operator $\hat{A}$ representing the observation, and the observable A representing the result of that observation—are intimately linked by the fundamental identity of quantum mechanics (5.9).

The essence of the *quantum* is contained in the remarkable mathematical properties of the quantum identity $\hat{A}\Psi = A\Psi$. This is an unusual kind of equation; for Ψ appears linearly on both sides but $\hat{A}$ and A are distinctly different mathematical objects. The identity cannot be satisfied by any function Ψ (other than the trivial one $\Psi = 0$) *unless* A has rather particular values. Although the quantum identity appears formally to be an equation for Ψ, it is an equation for *both* Ψ *and* A. This mathematical property is precisely what is required physically. For it is the value of the observable A which one set out to determine in postulating the identity between particles and waves.

The quantum identity of an operator is generally satisfied by not one but many values of Ψ and A (often infinitely many). This is certainly required physically; for these are the observable values and one knows that mechanical quantities generally take a multitude of values. Each of the values A is called an *eigenvalue* of $\hat{A}$ (after the German word *eigen*, meaning *special*) and the corresponding function Ψ that goes with that eigenvalue is called an *eigenstate* or *eigenfunction* of $\hat{A}$.

Let A_ν denote these eigenvalues and Ψ_ν the corresponding eigenfunctions with the index ν running over all the permissible states. These eigenfunctions satisfy the normalization and orthogonality condition*

$$\int \Psi_\nu \Psi_\mu^* \, d^3x = \delta_{\nu\mu}.$$

---

* The orthonormality condition may be verified by setting $A{=}A_\nu$ in Eq. (5.11*a*), $A = A_\mu$ in Eq. (5.11*b*), and $\Psi = \Psi_\nu$, $\Phi = \Psi_\mu$ in both. Take the difference of the resulting equations and find

$$\int \Psi_\mu^* \hat{A} \Psi_\nu \, d^3x - \int \Psi_\nu \hat{A}^* \Psi_\mu^* \, d^3x = (A_\nu - A_\mu) \int \Psi_\nu \Psi_\mu^* \, d^3x.$$

Since $\hat{A}$ is Hermitean, the left-hand side vanishes and orthonormality follows.

The orthonormality condition distinguishes the various eigenstates guaranteeing that they are independent; for if $\int \Psi_\nu \Psi_\mu \, d^3x \neq 0$, then Ψ_ν and Ψ_μ are not distinct states.

The fact that the quantum identity is satisfied only for special values of an observable shows that nature need not be present to us on the "full real number line" of classical physics. Whereas a mechanical quantity may take any value on the real number line in classical mechanics, it may only take the eigenvalues of its operator in quantum mechanics. These eigenvalues are not generally dense on the real line. The angular momentum operator, as a leading example, possesses discrete rather than continuous eigenvalues.

Discrete spectra signal a qualitative reduction in the permissible values for mechanical quantities—a countable number of eigenvalues denumerated by the integers rather than the uncountable number of values on the real number line permitted in classical motions. The existence of discrete spectra for quantum observables is the origin of the term "quantum." Classically, we have the illusion of the full real number line because discrete quanta blur together giving the illusion of a continuum on classical length scales.

A variety of mechanical operators are appropriate to every physical situation and each operator has a multiplicity of eigenstates Ψ_ν. The state function describing that situation generally contains all these eigenstates; for nature is rarely in a *pure state*, that is, one in which $\Psi = \Psi_\nu$ for one particular ν of one particular mechanical quantity. Rather, the state function of a generic motion is a composite of all eigenfunctions,

$$\Psi = \sum_\nu a_\nu \Psi_\nu,$$

where ν now runs over all the eigenfunctions and the a_ν are weights that fix the amounts of each eigenstate to be superposed in the ensemble so as to describe the particular physical circumstance in which nature finds itself.

The Algebra of Operators

In the classical description of motion one uses ordinary algebraic relations between mechanical quantities. For example, the component M_3 of the angular momentum is equal to $(x_1p_2 - x_2p_1)$. It makes no difference whether one writes the products of $\mathbf{x}$ and $\boldsymbol{p}$ as x_1p_2 or p_2x_1. This property of ordinary algebraic variables is that of *commutativity*. In ordinary algebra the variables in a product commute: $AB = BA$. In quantum-mechanical algebra, dynamical operators do not necessarily commute: $\hat{A}\hat{B} \neq \hat{B}\hat{A}$.

The failure of quantum operators to commute is rooted in the wave nature of the mechanics which they describe. The operator $\hat{A}\hat{B}$ corresponds to an observation of B followed by an observation of A. The operator $\hat{B}\hat{A}$

corresponds to an observation of A followed by an observation of B. The *commutator* $(\hat{A}\hat{B} - \hat{B}\hat{A})$ describes the interference of one observation with another. If $(\hat{A}\hat{B} - \hat{B}\hat{A}) = 0$, nature will be in identically the same state if the observation of A follows that of B or that of B follows that of A. On the other hand, if $(\hat{A}\hat{B} - \hat{B}\hat{A}) \neq 0$, then A and B are mechanically so intertwined that the act of observing one of them interferes with the observation of the other.

The duality between particles and waves brings a nonvanishing commutator into mechanics—a quantity that never arises in the classical regime revealed by Newton, Lagrange, and Hamilton. Nonetheless, this quantum-mechanical quantity is deeply rooted in classical mechanics. In a brilliant proposal (**Note 7**) which utilizes only the properties of Poisson brackets, P. A. M. Dirac suggested that the commutator of quantum-mechanical operators is (aside from the multiplicative constant $i\hbar$) the classical Poisson bracket of these operators:

$$(\hat{A}\hat{B} - \hat{B}\hat{A}) = i\hbar[\hat{A}, \hat{B}]. \tag{5.12}$$

The Dirac postulate connecting the Poisson bracket with the quantum commutator is the central bridge that carries classical mechanics into quantum mechanics. It directly reveals those dynamical variables for which wave-like interference is inherent. Dirac's postulate (5.12) is the touchstone between the classical and quantum realms—and perhaps the most beautiful mathematical statement in quantum mechanics.

If two quantities are to be simultaneously observed when the ensemble is in a given eigenstate, it is obvious that the operators corresponding to these quantities must have this eigenstate in common. Conversely, if two operators have the same eigenstate in common, the quantities which they represent may be simultaneously observed. Simultaneity of observation and commonality of eigenstates imply and are implied by one another.

The operators of simultaneously observable quantities also commute. The operator equation for an observation of $\hat{A}$ followed by an observation of $\hat{B}$ for an ensemble in an eigenstate Ψ of $\hat{A}$ is

$$\hat{B}\hat{A}\Psi = \hat{B}A\Psi.$$

If $\hat{A}$ and $\hat{B}$ commute, then the identity can be written

$$\hat{A}(\hat{B}\Psi) = A(\hat{B}\Psi).$$

But this requires that $\hat{B}\Psi$ is Ψ itself save for a multiplicative factor, or $\hat{B}\Psi = B\Psi$. The multiplicative factor is therefore an eigenvalue of $\hat{B}$ and Ψ must be an eigenstate of $\hat{B}$ as well as $\hat{A}$.

The nature of observation in the quantum realm may be summed up in the following principle: *simultaneous observation of mechanical quantities requires that their operators commute and that these operators possess eigenstates in common.*

In the classical limit in which $\hbar \to 0$ the commutator of mechanical quantities vanishes no matter what the value of the Poisson bracket. This is consistent with the classical situation: operators become the observables themselves and the algebra becomes ordinary. Observation of one quantity has no effect on the observation of any other.

The most elementary nonvanishing quantum commutator is that of the canonical coordinates: position and momentum. Since the classical Poisson bracket was shown in Chap. 3 to have the value $[q_i, p_j] = \delta_{ij}$, the fundamental quantum commutator is

$$(\hat{q}_i \hat{p}_j - \hat{p}_j \hat{q}_i) = i\hbar\, \delta_{ij}. \tag{5.13}$$

The two primal vectors of motion—position and momentum—can never be simultaneously observed on the quantum level. Every canonical pair of coordinates possess a Poisson bracket which is unity. Canonical coordinates are so mechanically entwined by their symplectic union that they necessarily interfere with the observation of one another. Equation (5.13) presents us with one of the most important revelations of quantum mechanics: *the uncertainty principle.* It tells us that only half the coordinates of motion may be exactly observed.

Probability: The Unique Quantum Invariant

Quantum mechanics preserves the classical invariants of motion. But quantum mechanics also opens us to new invariants unique to the quantum realm. The premier invariant of quantum motion has no classical counterpart. It springs from that uniquely quantum creation, the wave function Ψ itself. It is the probability that nature will be found in a state with precisely this wave function. Let us see how it comes about that the *probability of a state* is the central invariant of quantum motion.

At the quantum scale in which particles behave as waves, a particle is not said to be at a given *point* in space and time. Rather, it is specified as being in a given *state.* This state is described by the wave function $\Psi(\mathbf{x}, t)$. This mathematical object has asserted itself into the description of nature through the identity between particles and waves. It has a natural and beautiful role to play in that description

Because of the uncertainty principle, the full details of particle motions are not open to us. But statistical information about these inaccessible motions is available. The state function is the bearer of this statistical information. The square amplitude $|\Psi|^2 = \Psi\Psi^*$ naturally represents the probability density that the particle in the state $\Psi(\mathbf{x}, t)$ is at the point $(\mathbf{x}, t)$.

This connection was first discerned by Max Born in 1924; and it is the feature of quantum mechanics Einstein accepted only tentatively and whose overthrow he awaited to the end of his life. Perhaps Einstein will in the end be vindicated by a new mechanics which captures the wave nature of matter with the spirit and continuity of classical mechanics. But it is more likely that the next mechanical revolution will flow from new, more refined ideas about uncertainty and determinism on both the classical and quantum level. Einstein, whose reverence for the truth was constant, might rejoice in such a theory. We shall return to this theme in Chap. 7.

The probabilistic nature of $|\Psi|^2|$ may be drawn out of the Schroedinger equation (5.10). The time evolution of the probability density is

$$\frac{\partial |\Psi|^2}{\partial t} = \Psi \frac{\partial \Psi^*}{\partial t} + \frac{\partial \Psi}{\partial t} \Psi^*. \tag{5.14}$$

The time derivatives are taken from the Schroedinger equation and its complex conjugate. The product of Ψ^* with $\nabla^2 \Psi = \partial^2 \Psi / \partial x_i^2$ may be expressed as

$$\Psi^* \nabla^2 \Psi = \frac{\partial}{\partial x_i} \left(\Psi^* \frac{\partial \Psi}{\partial x_i} \right) - \frac{\partial \Psi^*}{\partial x_i} \frac{\partial \Psi}{\partial x_i}$$

with a corresponding complex conjugate. One then finds the products $(\partial \Psi^* / \partial x_i)(\partial \Psi / \partial x_i)$ and $\Psi^* V \Psi$ vanish and Eq. (5.14) yields the probability continuity equation

$$\frac{\partial |\Psi|^2}{\partial t} + \frac{\partial J_i}{\partial x_i} = 0, \tag{5.15}$$

where $\boldsymbol{J}$ is the *probability current*

$$\boldsymbol{J} = \sum_\alpha \frac{i\hbar}{2m_\alpha} \left(\Psi \frac{\partial \Psi^*}{\partial \mathbf{x}_\alpha} - \frac{\partial \Psi}{\partial \mathbf{x}_\alpha} \Psi^* \right). \tag{5.16}$$

The probability current represents the flux of probability through a surface of unit area normal to $\boldsymbol{J}$. One may regard the current of any physical quantity as the product of the density of that quantity and its transport velocity. The probability current is similarly expressible in terms of a transport velocity—the particle velocity operator $\hat{\dot{\mathbf{x}}} = \hat{\mathbf{p}}/m = (-i\hbar/m)\partial/\partial \mathbf{x}$. It has the form

$$\boldsymbol{J} = \sum_\alpha (\Psi \, \hat{\dot{\mathbf{x}}}_\alpha \Psi^* + \Psi^* \hat{\dot{\mathbf{x}}}_\alpha^* \Psi)/2.$$

The transport velocity for the probability is the Hermitean average of the particle velocity operator and the density $\Psi\Psi^*$.

The total probability contained in some region of space is $\int |\Psi|^2 \, d^3x$ where the integral extends over the region of interest. The continuity equation for the probability density implies that this quantity is an invariant for a bound system of particles. This can be seen by direct calculation using the continuity equation:

$$\frac{d}{dt} \int |\Psi|^2 \, d^3x = \int \frac{\partial |\Psi|^2}{\partial t} \, d^3x = - \int \frac{\partial J_i}{\partial x_i} \, d^3x. \tag{5.17}$$

The volume integral (5.17) may be transformed into an integral over the surface enclosing the particles,

$$\frac{d}{dt}\int |\Psi|^2 \, d^3x = -\int \boldsymbol{J} \cdot \boldsymbol{n} \, d^2x, \tag{5.18}$$

where $\boldsymbol{n}$ is the outward directed normal to the surface. The probability current $\boldsymbol{J}$ and accordingly the right-hand side of Eq. (5.18) vanish outside the region of the particles. The total integrated probability is therefore an invariant:

$$\frac{d}{dt}\int |\Psi|^2 \, d^3x = 0. \tag{5.19}$$

A system of bound particles can be located with certainty within the entire space of their motion. If a probability value of unity represents certainty, then the probability density for bound particles must satisfy the condition

$$\int |\Psi|^2 \, d^3x = 1, \tag{5.20}$$

where the integral is over all space.*

An invariant is the manifestation of an underlying symmetry. What is the symmetry underlying the invariant $\int |\Psi|^2 \, d^3x$? It is the openness of nature to quantum waves of all phases. Just as there is no unique origin for classical point particles, there is no unique phase for a quantum wave. The state Ψ differs from the state

$$\Psi' = e^{i\lambda}\Psi$$

by the phase λ, the symmetry parameter of the transformation. These two states cannot be distinguished from one another since $|\Psi'|^2 = |\Psi|^2$ and it is this quantity (not Ψ itself) which is observable.

The probability density may be used to build up statistical information about the inaccessible motions. The most fundamental source of information on statistical processes is the *expectation value.* It represents the average of a run of observations one makes on a variable which is observed imperfectly. The *expectation value* of the position $\mathbf{x}$ of a particle in a state Ψ is

$$\langle \mathbf{x} \rangle = \int \Psi \mathbf{x} \Psi^* \, d^3x.$$

---

* In the case of particles free to move off to infinity, there does not exist a well-defined total integrated probability with which the state functions can be normalized such as that given by Eq. (5.20). This is because the probability current and the surface integral in Eq. (5.18) do not vanish at infinity where particles are present. Although an invariant like $\int |\Psi|^2 \, d^3x$ for the total probability over all space does not always exist, the probability density always satisfies the continuity equation (5.15); and it is the continuity equation which is the more fundamental expression of the conservation of probability. A local increase in the density $\partial|\Psi|^2/\partial t$ comes at the expense of the current in the divergence $\partial J_i/\partial x_i$.

The general rule for finding the expectation value of any mechanical quantity A is postulated to be

$$\langle A \rangle = \int \Psi \hat{A} \Psi^* d^3x. \tag{5.21}$$

These definitions of expectation values imply that the eigenstates have been normalized by the condition $\langle \Psi\Psi^* \rangle = 1$.

The appropriateness of the expectation postulate (5.21) can be seen as follows. If a particle happens to find itself in a pure eigenstate of $\hat{A}$, say Ψ_ν with eigenvalue A_ν, then A is a fully accessible variable since $\hat{A}\Psi_\nu = A_\nu \Psi_\nu$ and Eq. (5.21) yields $\langle A \rangle = A_\nu$.

On the other hand, if (as is more often the case) the particle is in a state that is not a pure eigenstate of $\hat{A}$, then its state Ψ consists of a superposition of all the eigenstates Ψ_ν of $\hat{A}$: $\Psi = \sum_\nu a_\nu \Psi_\nu$. The expectation value will then not be a single eigenvalue of $\hat{A}$ but an average over many of them:

$$\langle A \rangle = \sum_\nu |a_\nu|^2 A_\nu.$$

From this it follows that the squares of the weights, $|a_\nu|^2$, represent the probabilities that the particle is in the state ν. A mechanical quantity which approaches complete accessibility will have a sharply peaked probability density about that accessible eigenvalue (the $|a_\nu|^2$ of the accessible eigenvalue approaches unity and all the other a_ν vanish). On the other hand, a quantity which is not fully accessible will have probability density functions with appropriately less degrees of sharpness.

The probabilistic nature of quantum mechanics is both its *sine qua non* and its Achilles heel. The state function is governed by the Schroedinger equation and the probability embodied in $|\Psi|^2$ propagates according to the continuity equation (5.15) derived from it. These equations are fully deterministic! How then does probability enter?

To answer this question one must recall that the probabilistic structure of quantum mechanics stems from the indeterminacy of quantum motions: the impossibility of observing half the physical quantities of nature exactly. The state function and the Schroedinger equation are ready-made quantum vehicles for propagating indeterminacy through space and time given an appropriate initial condition (these equations are first order differential equations in time); but they in themselves cannot set up the initial condition that describes a particular mix of indeterminacy. There are no equations in quantum mechanics which can do this.

The most useful way found thus far to complete the probabilistic structure of quantum mechanics was advanced by Bohr and has come to be known as the "Copenhagen interpretation." In it, probability is introduced through the act of observation—an activity which lies outside the formal mathematics of the theory. Probability in the present theory of quantum mechanics enters at the boundary of its mathematical structure as the initial conditions of the equations of motion. This happens every time one makes an observation. The evolution of nature governed by the Schroedinger and continuity equations is interrupted whenever an observation is made. (The state function is said to "collapse" at the instant of observation.) New initial conditions are simultaneously established by the act of observation. This new state function possesses a mix of indeterminacy different from that of the old. Some information is therefore lost; and it is through this loss that probability enters the theory. Nature then evolves from these new initial conditions until the next observation.

Inadequate as this procedure may be from the viewpoint of the integrity of the theory, it has proved extraordinarily successful in dealing with the quantum mechanics of atoms, molecules, light, and nuclear particles. As long as one remains in the quantum realm, the Copenhagen interpretation works successfully. It is only when one attempts to leap from the quantum world into the classical world that paradox presents itself.*

Symmetry and Invariants in Quantum Mechanics

Symmetry, invariants, and degeneracy are profound aspects of mechanics. These features of motion are universal for both the classical mechanics of heavenly bodies and the quantum mechanics of the elements. While quantum mechanics introduces a new invariant unique to the quantum world, the probability $\int |\Psi|^2 \, d^3x$, it also preserves the invariants and symmetries of the classical world. Indeed, quantum mechanics reveals the even more vital role symmetry plays in our interaction with the world. Classically we believe we observe the world quite independently of whatever symmetries are present. Quantum mechanics tells us that the very act of observation rests upon the symmetries with which nature presents us.

How is observation dependent upon symmetry? The simultaneous observation of any set of physical quantities requires that their operators commute or, classically, that they have vanishing Poisson brackets. But these are the Casimir sets of mechanics. Aspects of nature which can be simultaneously observed must possess an invariance with respect to a particular symmetry of nature. The question of which quantities are observable without quantum uncertainty has therefore already been answered classically: they are the Casimir sets.

The two Casimir sets of translational symmetry consist of all the elements of the translation group itself, the components of $\boldsymbol{P}$ (or $\boldsymbol{N}$). All the components of $\boldsymbol{P}$ (or $\boldsymbol{N}$) are therefore simultaneously observable. The

---

* Interested readers may wish to examine these paradoxes in J. S. Bell, *Speakable and Unspeakable in Quantum Mechanics*, Cambridge University Press (1987) and J. A. Wheeler and W. H. Zurek (eds.), *Quantum Theory and Measurement*, Princeton University Press, Princeton, N. J. (1983).

Casimir set of rotational symmetry consists of the pair (M, M_3). It tells us that different angular momentum components may not be simultaneously observed. On other hand, any single component, say M_3, and the total magnitude M commute and may be simultaneously observed. The simultaneous observation of these two quantities clearly fixes the total magnitude of the remaining two components. But the manner in which that magnitude is split between these two components is inaccessible.

One knows the Casimir sets are the quantum observables, but one explicitly knows only the Casimir sets of the fundamental symmetries: translations in space and time and three-dimensional rotations. There remain unknown invariants for general motions since the additional symmetries beyond the fundamental ones are unknown. There is one motion for which all the Casimir sets may be explicitly exhibited. This is the motion of two bodies; and it will shortly become the center of our attention.

The manner in which invariants and degeneracy manifest themselves in quantum mechanics is different from their manifestation in classical mechanics. In classical mechanics motion consists of the trajectory flow of particles in phase space. Each isolating invariant introduces a degeneracy into the motion in which the full phase space available to the trajectories degenerates into a submanifold. Increasing numbers of isolating invariants correspond to increasing degeneracies of the motion which restrict the trajectories to increasingly restricted submanifolds of phase space. For the maximum possible number of isolating invariants, the motion is completely degenerate resulting in a closed orbit.

In quantum mechanics it is not possible to observe trajectories. One observes instead the eigenvalues of the operators representing mechanical quantities. The phase space of the classical world is replaced by two spaces in the quantum world: the state space spanned by the eigenfunctions Ψ_ν and its corresponding eigenvalue space spanned by the eigenvalues A_ν. It is these spaces that particles are free to explore in quantum motion. An invariant on the classical level leads to a degeneracy of the phase space. An invariant on the quantum level leads to a degeneracy in the state and eigenvalue spaces. Whereas a degeneracy on the classical level is manifested in a collapse of the general manifold in which the particle trajectories may lie to a submanifold, a degeneracy on the quantum level is manifested in a collapse of the general manifold of eigenvalues in which the ensemble may reside to a submanifold. As a result, different eigenfunctions, rather than corresponding to uniquely different eigenvalues, will correspond to the same eigenvalue. Roughly, one may say that a manifold of classical trajectories isolated by a degeneracy to the same orbit become quantum states isolated by that degeneracy to the same eigenvalue.

Any invariant and the total energy necessarily have a vanishing Poisson bracket and, by Dirac's postulate, a vanishing quantum commutator. A vanishing quantum commutator with the energy is therefore connected with each invariant and its corresponding degeneracy. On the basis of the quantum identity (5.9) and Dirac's postulate (5.12), it follows that any two invariants (e.g., the angular momentum components M_1, M_2) which each commute with the energy $\hat{H}$ but *do not* commute with one another possess distinctly *different* eigenstates which correspond to the *same* eigenvalue of the energy. Let Ψ_1 and Ψ_2 be the eigenstates of $\hat{M}_1$ and $\hat{M}_2$. Since $\hat{M}_1$ and $\hat{M}_2$ commute with the energy $\hat{H}$ they are also eigenstates of $\hat{H}$ and satisfy

$$\hat{H}\Psi_1 = H\Psi_1, \qquad \hat{M}_1\Psi_1 = M_1\Psi_1,$$
$$\hat{H}\Psi_2 = H\Psi_2, \qquad \hat{M}_2\Psi_2 = M_2\Psi_2.$$

But since $\hat{M}_1$ and $\hat{M}_2$ do not commute, their eigenstates must be distinct:

$$\hat{M}_1\Psi_2 \neq M_1\Psi_2, \qquad \hat{M}_2\Psi_1 \neq M_2\Psi_1.$$

Further, each additional invariant that does not commute with any other invariant introduces a new level of degeneracy in which the additional eigenstates of these invariants again correspond to that *same* eigenvalue of the energy. The number of additional eigenstates for each eigenvalue depends upon the degree of degeneracy or, equivalently, the number of additional isolating invariants. These additional eigenstates are necessary to describe the additional invariants. But since any invariant commutes with the energy, these eigenstates all correspond to the same eigenvalue of the energy.

Our paradigmatic example is the motion of two bodies bound by the inverse-square force. The relative motion of these two bodies possesses a full set of isolating invariants: it is completely degenerate. In the motion of two heavenly bodies complete degeneracy is manifested in isolating invariants consisting of the energy, angular momentum, and eccentricity. These invariants generate a closed orbit lying completely within a single plane.

In the quantum motion of the electron about a nucleus complete degeneracy is also manifested in a complete set of identical invariants. But rather than generate a closed orbit, the operators corresponding to these isolating invariants generate multiple eigenstates of angular momentum and eccentricity which correspond to the same eigenvalue of the energy.

In the quantum regime the dynamics of particles are no longer expressed in trajectories. Instead, particle dynamics are expressed in the eigenstates and eigenvalues which the ensemble may assume. As the determination of the trajectories of particles is the essence of classical mechanics, the determination of the eigenstates and eigenvalues of the particle ensemble is the essence of quantum mechanics.

How are the quantum states that a system of particles may assume to be determined? They are the solutions—the eigenvalues and eigenstates—of the quantum identities (5.9) for any dynamical quantity of interest. For most quantities other than $\mathbf{x}$ and t these identities are differential equations. Quantum states—eigenvalues and eigenstates—are the solutions of these differential operator equations.

The states of the translational invariants whose Poisson brackets and quantum commutators vanish are completely accessible and the eigenvalues of their operators form a continuous spectrum. The Poisson bracket and quantum commutator of the linear momentum vanishes: $[P_i, P_j] = 0$. The eigenvalues of each component P_i form a continuous spectrum $-\infty \leq P_i \leq +\infty$. This is also true for the mass-center invariant $\boldsymbol{N}$. The components of $\boldsymbol{N}$ are all simultaneously observable and their eigenvalues also form a continuous spectrum.

Quantum Rotational Symmetry

The states of the angular momentum $\boldsymbol{M}$ are different from those of $\boldsymbol{P}$ and $\boldsymbol{N}$ since different angular momentum components do not commute. Angular momentum commutes according to Dirac's postulate (5.12) as

$$(\hat{M}_i \hat{M}_j - \hat{M}_j \hat{M}_i) = i\hbar \epsilon_{ijk} \hat{M}_k. \tag{5.22}$$

One of the generous aspects of quantum operator equations is that their eigenvalues may be determined by algebraic rather than differential equation methods. The quantum commutator is the key to the algebraic determination of the eigenvalue spectrum of mechanical quantities. The commutator establishes the multiplicative rule of this algebra and operators called *creation* and *annihilation operators* are its central elements. The creation and annihilation operators have the property of generating new quantum states from old ones with incrementally more or less quanta. In so doing they generate, step-by-step, the complete lattice of states. Here is how they unfold the eigenvalues of vectors possessing rotational symmetry for which the angular momentum is the prototype.

As described in Chap. 4, the quantities

$$\hat{M}_+ = \hat{M}_1 + i\hat{M}_2, \quad \hat{M}_- = \hat{M}_1 - i\hat{M}_2$$

are equivalent to $\hat{M}_1$ and $\hat{M}_2$; however, their Poisson bracket with $\hat{M}_3$ results in self- rather than cyclic-reproduction. Using Eq. (5.22), it is an easy matter to show

$$(\hat{M}_3 \hat{M}_\pm - \hat{M}_\pm \hat{M}_3) = \pm\hbar \hat{M}_\pm. \tag{5.23}$$

It is this self-reproducing aspect of $\hat{M}_+$ and $\hat{M}_-$ that gives them the property of creating or annihilating quanta of angular momentum when they act upon $\hat{M}_3$. That is why they are the creation and annihilation operators of rotational symmetry. Further, although $\hat{M}_+$ and $\hat{M}_-$ do not represent quantities simultaneously observable with M and M_3, their product *does* represent an observable since it is completely expressible in terms of the observables $\hat{M}$ and $\hat{M}_3$:

$$\hat{M}_\pm \hat{M}_\mp = \hat{M}_1^2 + \hat{M}_2^2 \pm \hbar \hat{M}_3 = \hat{M}^2 - \hat{M}_3^2 \pm \hbar \hat{M}_3. \tag{5.24}$$

These creation and annihilation operators may now be used to develop the eigenvalue spectrum of any vector possessed of the rotational symmetry (5.22) using the angular momentum as the illustration. Since angular momentum has the same dimensions as $\hbar$, the eigenvalues of $\hat{M}_3$ and $\hat{M}^2$ may be expressed as

$$M_3 = m\hbar, \qquad M^2 = \beta\hbar^2,$$

where m and β are pure numbers. The operator identities for the angular momentum are then

$$\hat{M}_3\Psi_{\beta m} = m\hbar\Psi_{\beta m}, \qquad \hat{M}^2\Psi_{\beta m} = \beta\hbar^2\Psi_{\beta m}. \tag{5.25}$$

Notice that $\Psi_{\beta m}$ is the same eigenstate for both indentities since the eigenstates of any two commuting operators are common to both. (The momentum component operators $\hat{M}_1$ and $\hat{M}_2$ *do not* possess $\Psi_{\beta m}$ as an eigenstate since they do not commute with $\hat{M}_3$.) The determination of the eigenvalues of the angular momentum reduces to the task of finding the permissible values of m and β.

The eigenvalues m and β may be determined by first showing that the states between which an angular momentum component such as M_3 may change are integral multiples of $\hbar$. To see this, apply $\hat{M}_+$ to the first of Eqs. (5.25):

$$\hat{M}_+\hat{M}_3\Psi_{\beta m} = m\hbar\hat{M}_+\Psi_{\beta m}. \tag{5.26}$$

The operator product $\hat{M}_+\hat{M}_3$ may be eliminated in favor of $\hat{M}_3\hat{M}_+$ by use of the commutation condition (5.23). Equation (5.26) then becomes

$$\hat{M}_3\hat{M}_+\Psi_{\beta m} = (m+1)\hbar\hat{M}_+\Psi_{\beta m}. \tag{5.27}$$

This equation may be recognized as just the operator equation for $\hat{M}_3$, the first of Eqs. (5.25), with m advanced to $m+1$,

$$\hat{M}_3\Psi_{\beta m+1} = (m+1)\hbar\Psi_{\beta m+1}, \tag{5.28}$$

where $\hat{M}_+\Psi_{\beta m}$ is identified with $\Psi_{\beta m+1}$. The characteristic role of $\hat{M}_+$ as a creation operator is therefore revealed:

$$\hat{M}_+\Psi_{\beta m} = const \times \Psi_{\beta m+1}. \tag{5.29}$$

The operator $\hat{M}_+$ acting on the state $\Psi_{\beta m}$ transforms it into the state $\Psi_{\beta m+1}$ (aside from a multiplicative constant) thereby demonstrating its property of creating a state with one more quantum of angular momentum than that upon which it operates. The eigenvalue of the state $\Psi_{\beta m+1}$ differs from that of the state $\Psi_{\beta m}$ by one quantum of angular momentum of magnitude $\hbar$.

A symmetrical analysis beginning with the application of the annihilation operator $\hat{M}_-$ to the operator equation for $\hat{M}_3$ reveals that $\hat{M}_-$ removes one quantum of angular momentum from the state $\Psi_{\beta m}$ upon which it acts thereby creating the state $\Psi_{\beta m-1}$:

$$\hat{M}_- \Psi_{\beta m} = const \times \Psi_{\beta m-1}. \tag{5.30}$$

The angular momentum states of a given component M_3 are therefore

$$M_3 = m\hbar$$

with the values of m differing by integers.

Now turn to the states of the magnitude M^2. Its quantum number β may be determined by noting that a component $M_3 = m\hbar$ of a vector of magnitude $M^2 = \beta\hbar^2$ can possess only a certain maximum magnitude. Let $|m| = j$ be the maximum value of this component corresponding to a fixed magnitude β. The quantum number m therefore lies in the range

$$-j \leqslant m \leqslant j. \tag{5.31}$$

Set $m = j$ in Eq. (5.29), and find $\hat{M}_+ \Psi_{\beta j} = const \times \Psi_{\beta j+1}$. But j is the maximum value of the permitted states for m, so the state $\Psi_{\beta j+1}$ must vanish identically thereby requiring

$$\hat{M}_+ \Psi_{\beta j} \equiv 0. \tag{5.32}$$

The state $\Psi_{\beta j}$ is not an eigenstate of $\hat{M}_+$, but an operator can be created for which $\Psi_{\beta j}$ is an eigenstate by applying $\hat{M}_-$ to Eq. (5.32) thereby creating the operator $\hat{M}_- \hat{M}_+$:

$$\hat{M}_- \hat{M}_+ \Psi_{\beta j} = 0. \tag{5.33}$$

The operator $\hat{M}_- \hat{M}_+$ from Eq. (5.24) involves only $\hat{M}_3$ and $\hat{M}^2$, both of which possess $\Psi_{\beta j}$ as an eigenstate (recall $\hat{M}_- \hat{M}_+$ represents a quantity simultaneously observable with M^2 and M_3). Equations (5.24) and (5.33) therefore yield

$$\hat{M}_- \hat{M}_+ \Psi_{\beta j} = [\beta - j(j+1)]\, \hbar^2 \Psi_{\beta j} = 0, \tag{5.34}$$

and the eigenvalue β is given in terms of j by

$$\beta = j(j+1). \tag{5.35}$$

Since β is given in terms of the more fundamental quantum number j, it is customary to use j to label the eigenstate as Ψ_{jm}. The angular momentum eigenvalues are

$$M_3 = m\hbar, \qquad M^2 = j(j+1)\hbar^2. \tag{5.36}$$

The permissible values of j may be determined by noting that for j=0, it is necessary that $m = 0$ and the angular momentum vanishes identically. The first value of j for which two values of m exist differing by an integer (corresponding to one quantum of angular momentum) is $j = 1/2$ for which $m = 1/2$ and $m = -1/2$. Continuing with increasing j so that each succeeding set of m states differ by an integer, one finds the permitted values of j are $j = 0,\ 1/2,\ 1,\ 3/2,\ \ldots$.

It may seem that the determination of the quantum states of the angular momentum is a rather specialized exercise. But the result we have just obtained is a key that will unlock many doors; for the quantum states given by Eqs. (5.36) are ubiquitous in the physical world. They are the states of *any* rotationally symmetric vector. We have developed them using the angular momentum as the illustration; but they are the quantum states of any vector $\boldsymbol{S}$ which satisfies commutation conditions of rotational symmetry:

$$\hat{S}_i\hat{S}_j - \hat{S}_j\hat{S}_i = i\hbar\epsilon_{ijk}\hat{S}_k. \tag{5.37}$$

Such a rotationally symmetric vector has a single Casimir invariant—its magnitude S—and is characterized by the Casimir set consisting of the pair (S, S_3). The quantum states of a rotationally symmetric vector are

$$S_3 = m\hbar, \qquad S^2 = j(j+1)\hbar^2. \tag{5.38}$$

The number j, which specifies the magnitude, takes integer or half-integer values. The number m, which specifies one of the components, takes either the integer or half-integer values in the range $-j \leqslant m \leqslant j$. For each j there are $2j+1$ values of m permitted. For a given vector of magnitude $S = \sqrt{j(j+1)}\hbar$ there are $2j+1$ possible orientations, each one having a component S_3 of magnitude $m\hbar$ where m takes either the integers or half-integers lying inclusively between $-j$ and $+j$.

Reflection Symmetry

The gravitational and electrical force laws show that they possess an additional symmetry of motion beyond those of translations and rotations. These symmetries are *space and time reflections* in which the space and time coordinates are reflected through the origin:

$$\mathbf{x}' = -\mathbf{x},$$
$$t' = -t.$$

Reflection symmetry is a fundamental symmetry of nature common to both the classical and quantum mechanics of gravitation and electricity; but it plays no role in the classical motion of particles. On the other hand, it is an important symmetry in quantum mechanics. Reflections in quantum mechanics are observable and they generate new conservation laws. This difference between the classical and the quantum arises because classical particles are structureless points; they possess no mechanism for responding to reflections. A quantum particle, by contrast, is a superposition of waves. These waves are eigenstates $\Psi(\mathbf{x}, t)$ with a rich mathematical structure which responds to space and time reflections

The potential V of Galilean relativity given by Eq. (3.4) is unchanged by reflection of the space coordinates, $\mathbf{x} \to -\mathbf{x}$. It is also unchanged by reflection of the time coordinate $t \to -t$ because it is independent of time. Reflections of the space and time coordinates also do not change the kinetic energy because space and time appear quadratically through the momenta. Since the Hamiltonian is a sum of the kinetic and potential energies it is also an even function of the space and time coordinates. The Hamiltonian of gravity and electricity in Galilean relativity possesses reflection symmetry: it is both space- and time-reversible.

Reflections are discrete rather than continuous symmetries. The symmetry parameters that describe them do not take a continuous range of values; rather, they are binary variables taking one of two mutually exclusive values. Reflections may be described by two binary quantities: the parity $\mathcal{P}$ which describes the response of mechanical quantities to reflections of the space coordinates and the time-reversibility $\mathcal{T}$ which describes their response to reflections of the time coordinate.

An eigenstate possesses definite values of $\mathcal{P}$ and $\mathcal{T}$ depending upon whether it is an even or odd function of space and time. An even function of the space coordinates, $\Psi(\mathbf{x}, t) = \Psi(-\mathbf{x}, t)$, is said to possess *even* parity signified by the value $\mathcal{P} = 1$ whereas an odd function, $\Psi(\mathbf{x}, t) = -\Psi(-\mathbf{x}, t)$, possesses *odd* parity signified by $\mathcal{P} = -1$. In like fashion an even function of the time coordinate, $\Psi(\mathbf{x}, t) = \Psi(\mathbf{x}, -t)$, is time-reversible with $\mathcal{T} = 1$ and an odd function of time, $\Psi(\mathbf{x}, t) = -\Psi(\mathbf{x}, -t)$, is anti-time-reversible with the value $\mathcal{T} = -1$.

The operator $\hat{\mathcal{P}}$ is the parity operator. When acting upon a state $\Psi(\mathbf{x}, t)$ it inverts the space coordinates:

$$\hat{\mathcal{P}}\Psi(\mathbf{x}, t) = \Psi(-\mathbf{x}, t).$$

The operator $\hat{\mathcal{T}}$ is the time-reversibility operator. It inverts the time coordinate:

$$\hat{\mathcal{T}}\Psi(\mathbf{x}, t) = \Psi(\mathbf{x}, -t).$$

Parity and time-reversibility are temporal invariants. Once established by initial conditions, they never change. This is because they commute with the Hamiltonian,

$$\hat{\mathcal{P}}\hat{H} - \hat{H}\hat{\mathcal{P}} = 0, \qquad \hat{\mathcal{T}}\hat{H} - \hat{H}\hat{\mathcal{T}} = 0,$$

which is an even function of space and time. Notice, however, that although parity is a rotational invariant (it commutes with $\hat{\boldsymbol{M}}$) time-reversibility is not and neither of them is a translational invariant (they do not commute with $\hat{\boldsymbol{P}}$ and $\hat{\boldsymbol{N}}$). But the combined reflection of both space and time represented by the operator $\hat{\mathcal{P}}\hat{\mathcal{T}}$ is both a translational and rotational invariant.

The eigenvalues of parity and time-reversibility are easily determined. First take the parity by examining the quantum identity for the operator $\hat{\mathcal{P}}^2$:

$$\hat{\mathcal{P}}^2\Psi = \mathcal{P}^2\Psi.$$

Now apply the parity operator twice to a state noting that this results in no change in the state:

$$\hat{\mathcal{P}}^2\Psi = \Psi.$$

Comparison of these two expressions shows that the eigenvalues of $\hat{\mathcal{P}}$ are

$$\mathcal{P} = \pm 1.$$

A given state function has a definite value of parity $\mathcal{P} = 1$ or $\mathcal{P} = -1$. This value of parity is invariant: once fixed by initial conditions it cannot change. Identical reasoning for time-reversibility shows that the eigenvalues of $\mathcal{T}$ are also

$$\mathcal{T} = \pm 1$$

and they are likewise invariants.

It is easy to see that the Lagrangian $L = T - V$, like the Hamiltonian, is an even function of the space and time coordinates for the potential of gravitation and electricity. This means that trajectories with both even and odd parity and time-reversibility satisfy the same law of motion. Nature apparently has no preference for one over the other. The time-reversed trajectory of any motion in which "all runs backward" is as permissible as the forward running trajectory. The law of motion of gravitation and electricity is time-reversible. This is not so in the nucleus; the potential of the weak nuclear force, unlike that for gravity and electricity, is not time-reversible. It is the only known aspect of nature that violates time-reversibility at a fundamental level.

Our experience of the world is one of *time-irreversibility*: time has a unique and vivid direction and the world seems to move inexorably in that direction. How can this be? There is no compelling evidence that the time-irreversibility of the weak nuclear force in and of itself is the explanation for the time-irreversibility of the large-scale world. The time-reversibility of gravitation and electricity which govern the large-scale world when the world is manifestly not time-reversible is one of the great unsolved problems of physics.

Hydrogen: Prototypical Atom, Primal Element

We now come to the quantum rendering of the two-body problem: the motion of an electron and a positively charged nucleus bound by the electrical inverse-square force. The classical mechanics of the heavens describes the trajectories of two heavenly bodies. The quantum mechanics of the atom describes the eigenvalues and eigenstates of two microscopic particles. The bound motions of these two particles constitute hydrogen, the prototypical atom and the first element of the Periodic Table.

The quantum motion, like the celestial motion, may be decomposed into a center of mass component and a relative component. The center of mass translates uniformly through space at a constant velocity. This motion is one of a free particle whose mass is equal to the total mass of the electron and the nucleus. It possesses the translational invariants of the motion: the linear momentum $\boldsymbol{P}$ and the mass-center invariant $\boldsymbol{N}$.

It is in the relative motion that the structure of the hydrogen atom is to be found just as the motion of the planets was found in the relative celestial motion. The invariants of the relative motion are the energy $\mathcal{H}$, angular momentum $\boldsymbol{M}$, and the eccentricity—either in the dimensionless form $\boldsymbol{e}$ or in the form of the eccentrum $\boldsymbol{E} = h\boldsymbol{e}$ where $h = \sqrt{k^2\mu/(-2\mathcal{H})}$ is the Kepler constant.

The leap into the quantum world is made by establishing the appropriate operators for the hydrogen atom invariants $\mathcal{H}$, $\boldsymbol{M}$, $\boldsymbol{E}$. The operators $\hat{\mathcal{H}}$ and $\hat{\boldsymbol{M}}$ are elementary Hermitean operators. But the eccentricity and eccentrum do not directly provide a prescription for their operator representations. They are quantities built up from the canonical coordinates $\boldsymbol{r}$ and $\boldsymbol{p}$ for which the direct transcription of the classical expression into operator form turns out not to be Hermitean.

Another way of looking at this difficulty is that the transcription of the classical expression for the eccentricity is not unique. One possibility is

$$\hat{\mathbf{e}} = \quad \hat{\mathbf{p}} \times \hat{\boldsymbol{M}}/k\mu - \hat{\mathbf{r}}/r. \tag{5.39}$$

But an equally possible form is

$$\hat{\mathbf{e}} = -\hat{\boldsymbol{M}} \times \hat{\mathbf{p}}/k\mu - \hat{\mathbf{r}}/r. \tag{5.40}$$

Both forms are identical in the limit in which $\hat{\mathbf{p}}$ and $\hat{\boldsymbol{M}}$ become classical quantities. However, as operator expressions they can only be identical if they are simultaneously observable. The two operators must therefore commute. But $\hat{\mathbf{p}}$ and $\hat{\boldsymbol{M}}$ do not commute since the Poisson bracket of any vector with the angular momentum cyclically reproduces the vector yielding the commutator

$$(\hat{\mathbf{p}} \times \hat{\boldsymbol{M}} + \hat{\boldsymbol{M}} \times \hat{\mathbf{p}}) = i2\hbar\hat{\mathbf{p}}. \tag{5.41}$$

The two potential operators (5.39) and (5.40) are not the same. They differ by the factor $i2\hbar\hat{\mathbf{p}}$.

The artistic choice of creating an Hermitean operator using both orderings of $\hat{\mathbf{p}}$ and $\hat{\boldsymbol{M}}$ by taking the average of the two possibilities has proved to give results in exact accord with the atomic world:

$$\hat{\mathbf{e}} = (\hat{\mathbf{p}} \times \hat{\boldsymbol{M}} - \hat{\boldsymbol{M}} \times \hat{\mathbf{p}})/2k\mu - \hat{\mathbf{r}}/r. \tag{5.42}$$

Using the commutator (5.41), the eccentricity operator (5.42) may be expressed as

$$\hat{\mathbf{e}} = (\hat{\mathbf{p}} \times \hat{\boldsymbol{M}} - i\hbar\hat{\mathbf{p}})/k\mu - \hat{\mathbf{r}}/r. \tag{5.43}$$

The eccentricity operator (5.43) is the image of the classical eccentricity (4.7) but with the additional quantum term involving $i\hbar\hat{\mathbf{p}}$.

Only four of the six components of $\boldsymbol{M}$ and $\mathbf{e}$ are independent since they satisfy the subsidiary conditions (4.8) which can be expressed as

$$e^2 = 1 - (\mathcal{E}_0\hbar^2)^{-1}\mathcal{H}M^2, \qquad \boldsymbol{M} \cdot \mathbf{e} = 0, \tag{5.44}$$

where $\mathcal{E}_0 \equiv -k^2\mu/2\hbar^2$ will be shown to be the ground state or minimum energy level of the hydrogen atom. These classical subsidiary equations become quantum operator equations for $\hat{e}^2$:

$$\hat{e}^2 = 1 - (\mathcal{E}_0\hbar^2)^{-1}\hat{\mathcal{H}}(\hat{M}^2 + \hbar^2), \qquad \hat{\boldsymbol{M}} \cdot \hat{\mathbf{e}} = 0. \tag{5.45}$$

It is a notable fact that M^2 in the first of Eqs. (5.44) appears in the form $\hat{M}^2 + \hbar^2$ in the first of Eqs. (5.45), its operator counterpart. Readers may wish to verify that when the operator $\hat{\mathbf{e}}$ of Eq. (5.43) is squared, additional quantum terms appear which are not present in the square of the classical eccentricity $\mathbf{e}$.* The net effect of these quantum terms is to evolve M^2 in the classical expression (5.44) into $\hat{M}^2 + \hbar^2$ in the quantum expression (5.45).

The form of the eccentricity invariant which is most useful in revealing the structure of the hydrogen atom is the eccentrum. The subsidiary equations (5.45) in $\hat{\mathbf{e}}$ may be transformed into operator equations in $\hat{\boldsymbol{E}}$. Since $\hat{\mathcal{H}}$ commutes with $\hat{\mathbf{e}}$ and $\hat{\boldsymbol{E}}$, the squared operators are related as $\mathcal{E}_0\hbar^2\hat{E}^2 = \hat{\mathcal{H}}\hat{e}^2$. Equations (5.45) accordingly become the operator counterparts of Eqs. (4.11),

$$(\hat{M}^2 + \hat{E}^2 + \hbar^2)\hat{\mathcal{H}} = \mathcal{E}_0\hbar^2, \quad \hat{\boldsymbol{M}} \cdot \hat{\boldsymbol{E}} = 0. \tag{5.46}$$

The hydrogen atom is described by the invariant operators $\hat{\boldsymbol{M}}$, $\hat{\boldsymbol{E}}$, and $\hat{\mathcal{H}}$. The mutual interaction of these invariants gives the atom its structure which in turn gives hydrogen its peculiar chemical and physical properties. This structure originates in the unique way in which symmetry orchestrates the invariants through the commutation or Poisson bracket algebra. These symmetries are summarized by the three Casimir sets for two-body motion with inverse-square forces:

$$(\hat{\mathcal{H}}, \hat{M}, \hat{M}_3), \qquad (\hat{\mathcal{H}}, \hat{M}, \hat{M}_1), \qquad (\hat{\mathcal{H}}, \hat{E}_3, \hat{M}_3).$$

---

* In the square of the quantum eccentricity, $\hat{\mathbf{e}} = (\hat{\mathbf{p}} \times \hat{\boldsymbol{M}} - i\hbar\hat{\mathbf{p}})/k\mu - \hat{\mathbf{r}}/r$, noncommuting products generate the commutators $i\hbar[(\boldsymbol{p} \times \boldsymbol{M})_i, x_i/r]$ and $\hbar^2[p_i, x_i/r]$. The first is $2\hat{\mathbf{p}} \cdot \hat{\mathbf{r}}/r$ and cancels an identically occurring product. The second involves the identity $[p_i, x_i/r] \equiv 2/r$ and combines with $\hbar^2\hat{p}^2/k^2\mu^2$ to create $(-2\hbar^2/k^2\mu)(\hat{p}^2/2\mu - k/r) = \hat{\mathcal{H}}/\mathcal{E}_0$.

The magnitude $\hat{M}$ is the Casimir invariant of three-dimensional rotations in configuration space, a symmetry which persists in the presence of the wider four-dimensional rotational symmetry in phase space. It therefore commutes with all the components of $\hat{\boldsymbol{M}}$ as well as with their magnitudes [see Eqs. (4.10) for the full symmetry algebra]; but it does not commute with the components of $\hat{\boldsymbol{E}}$ because $\hat{\boldsymbol{E}}$ is not a Casimir invariant of three-dimensional rotations.

The observability of the eccentrum magnitude $\hat{E}^2$ (or eccentricity $\hat{e}^2$) turns on the question of the observability of $\hat{\mathcal{H}}$ and $\hat{M}$ since it depends only upon these two quantities according to the first of either Eqs. (5.45) or (5.46). It can be seen that $\hat{E}^2$ is actually a member of the first Casimir set though it does not appear explicitly in the triplet $(\hat{\mathcal{H}}, \hat{M}, \hat{M}_3)$ because the set is identified with only three independent invariants. Its presence is indicated by the presence of $\hat{\mathcal{H}}$ and $\hat{M}$. The magnitude $\hat{E}^2$ (or $\hat{e}^2$) is observable along with $(\hat{\mathcal{H}}, \hat{M}, \hat{M}_3)$.

In classical two-body motion a state of relative motion with three degrees of freedom is specified by six invariants (β, I) where β and I are each Casimir sets. In quantum two-body motion the invariants specifying a state must all be simultaneously observable. The invariant operators must therefore all commute. But the operators corresponding to the six invariants (β, I) of the relative motion *do not* all commute since $[\beta_i, I_j] = \delta_{ij}$. It is not possible to specify six invariants in quantum motion. We have come face to face with the uncertainty principle: half the invariants of quantum motion cannot be observed when the other half are specified.

What then is the quantum situation? The uncertainty principle requires that the quantum motion of two particles be cast in a radically different way from the classical motion of two heavenly bodies. Here lies the rift in the otherwise intimate correspondence between the quantum and the classical worlds. A classical state of relative motion is specified by six invariants. A quantum state of the hydrogen atom is specified by only three of these six invariants. There are three possibilities which correspond to the three Casimir sets $(\hat{\mathcal{H}}, \hat{M}, \hat{M}_3)$, $(\hat{\mathcal{H}}, \hat{M}, \hat{M}_1)$, $(\hat{\mathcal{H}}, \hat{E}_3, \hat{M}_3)$; however the first two are essentially the same motion with different labels. They are both vector representations of two-body motion. The uniquely different states are specified by either the triplet $(\hat{\mathcal{H}}, \hat{M}, \hat{M}_3)$ or the triplet $(\hat{\mathcal{H}}, \hat{E}_3, \hat{M}_3)$, the first the vector representation, the second the spinor representation.

In the vector case $(\hat{\mathcal{H}}, \hat{M}, \hat{M}_3)$ the eccentrum components are not observable (though the magnitude of the eccentrum is observable because $\hat{E}^2$ commutes with $\hat{M}$ and $\hat{M}_3$). In the spinor case $(\hat{\mathcal{H}}, \hat{E}_3, \hat{M}_3)$ the eccentrum component parallel to an observable angular momentum component is observablc; but the total magnitudes of the angular momentum and eccentrum are not.

The sets of state specifications $(\hat{\mathcal{H}}, \hat{M}, \hat{M}_3)$, or $(\hat{\mathcal{H}}, \hat{E}_3, \hat{M}_3)$ are direct images of the separation constants of the Hamilton–Jacobi equation in the two symmetric coordinate systems (spherical-polar and parabolic) of the classical motion. For each of the state specifications one wishes to know the actual structure of the hydrogen atom. This information is contained in the observable values of the invariants—the eigenvalues of the operators $(\hat{\mathcal{H}}, \hat{M}, \hat{M}_3)$ on the one hand and $(\hat{\mathcal{H}}, \hat{E}_3, \hat{M}_3)$ on the other. Let us now draw out these eigenvalues.

The States of Hydrogen

The ground state energy level is the fundamental eigenvalue of the hydrogen atom. This is the minimum energy available to the relative motion of the electron about the nucleus. The ground state energy of the motion of one heavenly body about another found in Chap. 4 is

$$\mathcal{E}_0 = -k^2\mu/2M^2.$$

In classical motions this energy is unbounded from below since the angular momentum can take arbitrarily small values for which the trajectory becomes progressively a highly elongated ellipse passing closer and closer to the center of mass. The distance of closest approach of the two bodies is, from Eqs. (4.48) and (4.49), $r_{\text{min}} = M^2/2k\mu$ and is proportional to M^2. In the classical motion of heavenly bodies there is no restriction on the magnitude of the angular momentum. Hence, as $M \to 0$, the distance of closest approach has the limit $r_{\text{min}} \to 0$ and the two bodies are free to collapse into one another.

In quantum two-body motion a new constant $\hbar$ has appeared bearing the dimensions of angular momentum. It establishes an irreducible minimum energy level which does not exist in the classical motion of two heavenly bodies; and this minimum energy barrier prevents the electron from collapsing into the nucleus. The ground state energy eigenvalue of the hydrogen atom will be shown to be

$$\mathcal{E}_0 = -k^2\mu/2\hbar^2. \tag{5.47}$$

In the ground state energy of the hydrogen atom, Planck's constant, the fundamental unit of angular momentum, replaces the angular momentum M which appears in the classical ground state energy.

The ground state energy of two heavenly bodies depends upon the initial conditions of the motion. However the ground state energy of the hydrogen atom is completely expressible in terms of fundamental physical constants, the electron and proton mass and charge and Planck's constant. It is therefore a universal constant independent of particular initial conditions.

The four-dimensional rotational symmetry of the hydrogen atom is the Rosetta stone for deciphering its eigenvalues. The ground state energy eigenvalue $\mathcal{E}_0$ and the full set of eigenvalues of the invariants flow from the algebra of their commutation relationships. It is not necessary to involve the complexity of the coordinate system and the expression of the invariants as coordinate-dependent differential operators. (This independence from the details of the coordinate system was foreshadowed by Bohr's simple model which used only the properties of the invariants in fixing the ground state orbit.) The eigenvalues of the hydrogen atom were first deduced from the full set of invariants by Wolfgang Pauli in the mid-1920s.*

The eigenvalues of the energy $\hat{\mathcal{H}}$, the angular momentum $\hat{M}^2$, the eccentrum component $\hat{E}_3$, and the angular momentum component $\hat{M}_3$ may be represented in terms of quantum numbers n, l, q, and m:

$$\mathcal{E} = \mathcal{E}_0/n^2, \qquad M^2 = l(l+1)\hbar^2, \qquad E_3 = q\hbar, \qquad M_3 = m\hbar. \tag{5.48}$$

The determination of the eigenvalues of the invariants then reduces to the determination of the quantum numbers (n, q, m) and (n, l, m). We already know the quantum numbers m and l for the rotationally symmetric vector $\boldsymbol{M}$. They are given by Eqs. (5.36). But we do not yet know the quantum numbers n and q nor do we know the relationships they bear to m and l.

The eigenvalues of the hydrogen atom flow from its four-dimensional rotational symmetry. This symmetry may be made transparent in quantum motion just as it was in classical motion by utilizing the rotational operators $\hat{\boldsymbol{S}}$ and $\hat{\boldsymbol{D}}$ of the two familes of two-spheres which are the projections of the four-dimensional rotational symmetry:

$$\hat{\boldsymbol{S}} = (\hat{\boldsymbol{M}} + \hat{\boldsymbol{E}})/2, \qquad \hat{\boldsymbol{D}} = (\hat{\boldsymbol{M}} - \hat{\boldsymbol{E}})/2, \tag{5.49}$$

in terms of which $\hat{\boldsymbol{M}}$ and $\hat{\boldsymbol{E}}$ are

$$\hat{\boldsymbol{M}} = \hat{\boldsymbol{S}} + \hat{\boldsymbol{D}}, \qquad \hat{\boldsymbol{E}} = \hat{\boldsymbol{S}} - \hat{\boldsymbol{D}}. \tag{5.50}$$

The description in terms of $\hat{\boldsymbol{S}}$ and $\hat{\boldsymbol{D}}$ is equivalent to that in terms of $\hat{\boldsymbol{M}}$ and $\hat{\boldsymbol{E}}$. A state of the hydrogen atom specified by $(\mathcal{H}, E_3, M_3)$ is the same as that specified by $(\mathcal{H}, S_3, D_3)$.

The subsidiary equations (5.46) may also be transformed into operator equations for $\hat{\boldsymbol{S}}$ and $\hat{\boldsymbol{D}}$ and follow those for the classical motion (4.13) with the addition of a uniquely quantum term which appeared in squaring the eccentricity operator:

$$(\hat{S}^2 + \hat{D}^2 + \hbar^2/2)\hat{\mathcal{H}} = \mathcal{E}_0\hbar^2/2, \qquad \hat{S}^2 - \hat{D}^2 = 0. \tag{5.51}$$

As in the classical motion, the magnitudes $\hat{S}^2$ and $\hat{D}^2$ are identical and one of them, say $\hat{D}^2$, may be eliminated. Since $\hat{\mathcal{H}}$ commutes with $\hat{S}^2$ and $\hat{D}^2$

* W. Pauli, "Über das Wasserstoffspektrum vom Standpunkt der neuen Quantenmechanik," *Z. Phys.* ***36*** (1926), 336-363; English translation in *Sources of Quantum Mechanics*, B. L. Van der Waerden (ed.), Dover Books (1968), 387–415.

in the first of Eqs. (5.51), it may be represented in terms of its eigenvalues $\mathcal{E} = \mathcal{E}_0/n^2$. The first subsidiary condition of Eq. (5.51) them becomes solely an equation in $\hat{S}^2$ and n:

$$\hat{S}^2 = (n^2 - 1)\hbar^2/4. \tag{5.52}$$

The quantum number n is now known. It is directly related to the quantum numbers of the rotationally symmetric vector $\boldsymbol{S}$ given in Eqs. (5.53). The vector $\boldsymbol{D}$ has identical structure. Since the magnitude S^2 has eigenvalues $s(s+1)\hbar^2$, the quantum number n may be expressed in terms of s from Eq. (5.52) as $S^2 = s(s+1)\hbar^2 = (n^2-1)\hbar^2/4$ or

$$n^2 = 1 + 4s(s+1). \tag{5.53}$$

The angular momentum and eccentrum quantum numbers m and q follow from Eqs. (5.50). The $\hat{S}_3$ and $\hat{D}_3$ eigenvalues can be represented as

$$S_3 = \sigma\hbar, \qquad D_3 = \delta\hbar; \qquad -s \leqslant \sigma, \delta \leqslant s.$$

The quantum numbers m and q are then

$$m = \sigma + \delta, \qquad q = \sigma - \delta, \tag{5.54}$$

where σ and δ now take all the integer and half-integer values lying between $-s$ and s. One thus passes through all values of s taken from the sequence $s = 0, \frac{1}{2}, 1, \frac{3}{2}, \ldots$. In so doing the corresponding values of σ and δ arise. In this manner one builds up all the quantum numbers n, q, and m according to Eqs. (5.53) and (5.54). In particular, one finds immediately from Eq. (5.53) that the quantum number n takes all the positive integers excluding zero: $n = 1, 2, \ldots$.

For the energy level $n = 1$ it is required that $s = 0$. Only one state exists, the ground state, for which q and m take the values

$$q = 0, \quad m = 0.$$

For $n = 2$ one finds $s = \frac{1}{2}$ from Eq. (5.53) and

$$\sigma = \pm\tfrac{1}{2}, \qquad \delta = \pm\tfrac{1}{2}.$$

Four states of q and m correspond to the four possible combinations of these values allowed by Eq. (5.54):

$$q = 0, \;\; m = \pm 1; \quad q = \pm 1, \;\; m = 0.$$

For $n = 3$, the conditions (5.53) and (5.54) generate nine different states of q and m:

$$\begin{gathered} q = \;\; 0, \;\; m = \;\; 0; \\ q = 1, \;\; m = -1; \quad q = -1, \;\; m = 1; \\ q = 0, \;\; m = \pm 2; \quad q = \pm 1, \;\; m = \pm 1; \quad q = \pm 2, \;\; m = 0. \end{gathered}$$

For $n = 4$, there are sixteen states of q and m:

$$\begin{array}{llll} q = 0, & m = \pm 1; & q = \pm 1, & m = 0; \\ q = \pm 1, & m = \mp 2; & q = \pm 2, & m = \mp 1; \\ q = 0, & m = \pm 3; & q = \pm 3, & m = 0; \\ q = \pm 1, & m = \pm 2; & q = \pm 2, & m = \pm 1. \end{array}$$

Each energy level n possesses n^2 distinct states corresponding to the n^2 degeneracies brought about by the concerted action of the central and inverse-square natures of the electrical force.

Although the quantum numbers s, σ, and δ take both integer and half-integer values, the half-integer values generate quantum numbers n, q, and m which are always integer. Half-integer states of angular momentum and eccentrum therefore do not appear in the hydrogen atom.

The eigenvalues of the total energy $\mathcal{E} = \mathcal{E}_0/n^2$ can now be used to show the intimate connection between the Kepler constant h and the Planck constant $\hbar$. Use $\mathcal{E} = \mathcal{E}_0/n^2 = -k^2\mu/2h^2$ with the definition of the quantum ground state energy $\mathcal{E}_0 = -k^2\mu/2\hbar^2$ and find

$$h = n\hbar. \tag{5.55}$$

The Kepler constant is quantized in units of $\hbar$. For low values of the energy quantum number n the discrete character of the Kepler constant is inescapable. Only in the limit $n \gg 1$ does it acquire the semblance of a continuous mechanical quantity.

The eigenvalues of the eccentricity may also now be determined. The eccentricity $\boldsymbol{e}$ is related to the eccentrum $\boldsymbol{E}$ by $\boldsymbol{e} = \boldsymbol{E}/h$. Its eigenvalues therefore follow from those of $E_3 = q\hbar$ and $h = n\hbar$:

$$e_3 = q/n. \tag{5.56}$$

The eigenvalues of the hydrogen atom in the spherical-polar state given by the vector specification $(\mathcal{H}, M, M_3)$ are now also known. The eigenvalues of $\hat{\mathcal{H}}$ and $\hat{M}_3$ are the same as those in the spinor state specification $(\mathcal{H}, E_3, M_3)$. The angular momentum magnitude M replaces the eccentrum E_3 in this specification. The eigenvalues of the angular momentum magnitude $\hat{M}$ have already been exhibited in Eqs. (5.36). However only the integer states of angular momentum appear in the hydrogen atom according to Eq. (5.53). The eigenvalues of the angular momentum are $M^2 = l(l+1)\hbar^2$ and $M_3 = m\hbar$ as given in Eqs. (5.48) where the l are the positive integers including zero and m has values $-l \leqslant m \leqslant l$. The eigenvalues of the energy are $\mathcal{H} = \mathcal{E}_0/n^2$; however the angular momentum eigenvalues permitted depend upon the energy eigenvalues. This dependence may be found from the eccentricity magnitude e^2.

The squared eccentricity operator is given by the first of Eqs. (5.45). The eigenvalue equation $\hat{e}^2\Psi = e^2\Psi$ is therefore

$$\left[1 - (\mathcal{E}_0\hbar^2)^{-1}\hat{\mathcal{H}}(\hat{M}^2 + \hbar^2)\right]\Psi_{nlm} = e^2\Psi_{nlm}. \tag{5.57}$$

The eigenstates Ψ_{nlm} of $\hat{e}^2$ are also eigenstates of the operators $\hat{\mathcal{H}}$ and $\hat{M}^2$. The eigenvalues of $\hat{e}^2$ can therefore be directly read off Eq. (5.57) by replacing $\hat{\mathcal{H}}$ and $\hat{M}^2$ with their eigenvalues:

$$e^2 = 1 - \frac{1 + l(l+1)}{n^2}. \tag{5.58}$$

Since $e^2 \geqslant 0$, Eq. (5.58) shows that the angular momentum quantum number l and the total energy quantum number n must bear the relationship

$$l \leqslant n - 1.$$

For a given quantum number n, the quantum number l takes all the values $0 \leqslant l \leqslant n-1$ and m takes all the values $|m| \leqslant l$. For $n = 1$ the single state of l and m is

$$l = 0,\ m = 0.$$

For $n = 2$ there are four states of l and m:

$$\begin{gathered} l = 0,\ m = 0; \\ l = 1,\ m = 0; \quad l = 1,\ m = \pm 1. \end{gathered}$$

For $n = 3$ the nine states of l and m are

$$\begin{gathered} l = 0,\ m = 0; \\ l = 1,\ m = 0; \quad l = 1,\ m = \pm 1; \\ l = 2,\ m = 0; \quad l = 2,\ m = \pm 1; \quad l = 2,\ m = \pm 2. \end{gathered}$$

For $n = 4$ there are sixteen states of l and m:

$$\begin{gathered} l = 0,\ m = 0; \\ l = 1,\ m = 0; \quad l = 1,\ m = \pm 1; \\ l = 2,\ m = 0; \quad l = 2,\ m = \pm 1; \quad l = 2,\ m = \pm 2; \\ l = 3,\ m = 0; \quad l = 3,\ m = \pm 1; \quad l = 3,\ m = \pm 2; \quad l = 3,\ m = \pm 3. \end{gathered}$$

As was the case for the parabolic states, the spherical-polar states of the hydrogen atom advance with energy level n in the sequence $1–4–9–16–\cdots$, in accord with n^2 degeneracies. Both state specifications of the hydrogen atom possess this fundamental property.

The semi-major axis of the ellipse of classical motion is given by the first of Eqs. (4.54) as $a = k/2|\mathcal{E}|$ and is a function only of the energy $\mathcal{E}$. The semi-minor axis is given in terms of the semi-major axis and the eccentricity as $b = a\sqrt{1-e^2}$. A correspondence with classical mechanical quantities may be obtained by substituting the eigenvalues of $\hat{\mathcal{H}}$ and those for $\hat{e}^2$ given by Eq. (5.58) into these expressions:

$$a = n^2 a_0, \qquad b = \sqrt{1 + l(l+1)}\, n a_0, \tag{5.59}$$

where $a_0 = k/2|\mathcal{E}_0| = \hbar^2/k\mu = 0.529 \times 10^{-8} cm$ is the Bohr radius. This atomic radius was first proposed by Niels Bohr in 1913 in his simple quantum model of the atom in which the electron moves in a circular orbit about the nucleus.

In the ground state $n = 1$, $l = 0$, $m = 0$, the mechanical quantities a, b, and e have the values $a = a_0$, $b = a_0$, $e = 0$. To the extent that it can be described in the language of classical trajectories, the ground state orbit has circular symmetry, consistent with Bohr's original insight. Unlike the motion of a heavenly body (and unlike Bohr's model), the electron possesses no observable orbital angular momentum in the ground state. The ground state electron is not a point particle moving on a circular trajectory. The electron is not confined to an orbit but is a wave distributed over the entire space surrounding the nucleus. In the ground state this distribution is completely symmetrical in angle. The most likely radial position of the electron (described by the probability density distribution $|\Psi_{n00}|^2$) corresponds to $r = a = n^2 a_0$; but there is always a non-negligible probability that an electron in a given state may be found anywhere in the space surrounding the nucleus.

In the higher energy levels $n > 1$, no "circular" states exist in which the eccentricity vanishes except in the limit $n \to \infty$. The minimum eccentricity corresponds to the maximum value of the angular momentum quantum number $l_{\max} = n - 1$ and is given by

$$e^2_{\min} = \frac{(n-1)}{n^2}.$$

These are states for which $e_3 = q\hbar = 0$ and the eccentricity vector lies completely in the plane perpendicular to the polar axis.

Note that if the energy and momentum eigenvalues are simply substituted into the classical expression for the squared eccentricity, the first of Eqs. (4.8), one obtains

$$e^2 = 1 - \frac{l(l+1)}{n^2},$$

an expression which differs from Eq. (5.58), the quantum result, by the term $1/n^2$. Only in the limit of large quantum number n do the two expressions become identical.

The State Functions of Hydrogen

The invariant operators of the hydrogen atom generate the eigenvalues which are observable and measurable. But quantum operators beget not only eigenvalues; they also beget eigenstates which define the probability density distributions of the electron about the nucleus. These probability density distributions are the shapes of the electron waves corresponding to the different states.

The shapes of the hydrogen atom electron waves are inherently three-dimensional. The classical relative motion of two heavenly bodies is essentially two-dimensional (the motion lying in a plane perpendicular to the angular momentum vector), but the quantum relative motion of two bodies is always three-dimensional. There is always a finite probability that the motion may lie outside the plane perpendicular to any component of the angular momentum which one may specify, such as M_3. This is because one cannot precisely specify the angular momentum; hence a plane perpendicular to the angular momentum vector in which the relative motion of the atom must take place cannot be precisely defined.

The symmetries of Kepler motion are described by the vector Casimir set $(\mathcal{H}, M, M_3)$ and the spinor set $(\mathcal{H}, E_3, M_3)$. If the atom is a vector state $(\mathcal{H}, M, M_3)$, the coordinates that reflect this symmetry are spherical-polar coordinates r, ϕ, θ shown in Fig. 4-9. The coordinate lines in this system are the intersections of the surfaces of spheres $r = const$, cones $\theta = const$, and planes $\phi = const$.

If a state of the atom is specified by the spinor Casimir set $(\mathcal{H}, E_3, M_3)$, the corresponding symmetric coordinates are the parabolic coordinates ξ, η, ϕ shown in Fig. 4-10. These coordinates are formed by the intersections of surfaces of paraboloids of revolution $\xi = const$, $\eta = const$ about the x_3 axis whose foci are the origin and by the planes $\phi = const$ where ϕ is the polar angle as in spherical-polar coordinates.

The eigenstates of the hydrogen atom are the wave functions Ψ_{nlm} for the state specified by $(\mathcal{H}, M, M_3)$ and the quantum numbers (n, l, m) or Ψ_{nqm} for the state specified by $(\mathcal{H}, E_3, M_3)$ and the quantum numbers (n, q, m). These wave functions generate the electron probability density distribution functions $|\Psi_{nlm}|^2$ and $|\Psi_{nqm}|^2$.

Separability of the Hamilton–Jacobi equation on the classical level is transformed into separability of operator equations on the quantum level. To see this connection, recall that the wave function is given in terms of the action by $\Psi = e^{iS/\hbar}$. The separable action decomposes into a sum $S(r, \theta, \phi) = S_r(r) + S_\theta(\theta) + S_\phi(\phi)$ for the spherical-polar states or $S(\xi, \eta, \phi) = S_\xi(\xi) + S_\eta(\eta) + S_\phi(\phi)$ for the parabolic states. On the quantum level the wave function for the spherical-polar states is also decomposable as

$$\Psi_{nlm} = e^{i[S_{nl}(r)+S_{lm}(\theta)+S_m(\phi)]/\hbar},$$

with a similar decomposition for the parabolic states,

$$\Psi_{nqm} = e^{i[S_{nqm}(\xi)+S_{nqm}(\eta)+S_m(\phi)]/\hbar}.$$

The three functions $S_{nl}(r)$, $S_{lm}(\theta)$, $S_m(\phi)$ on the one hand or $S_{nqm}(\xi)$, $S_{nqm}(\eta)$, $S_m(\phi)$ on the other are determined from the three operator identities for $(\mathcal{H}, M, M_3)$ or $(\mathcal{H}, E_3, M_3)$. A correspondence exists between the three separated Hamilton–Jacobi equations for the three partial actions on the classical level and the three operator equations for the actions on the quantum level. The Casimir sets contain the separation constants. The quantum numbers (n, l, m) or (n, q, m) appear as the separation constants on the quantum level. Both sets are ordinary differential equations which are integrable. The details of integration of the quantum operator equations may be found in **Note 9** for the spherical-polar states and in **Note 10** for the parabolic states. The eigenstates must be normalized by the condition $\int \Psi^*_{nlm}\Psi_{nlm}d^3x = 1$. Since the partial actions are only determined up to a constant, the normalization condition fixes the constants of integration.

The polar eigenstates $e^{iS_m(\phi)/\hbar}$ are common to both the spherical-polar and parabolic eigenstates and are quite simple. The operator equation

$$\hat{M}_3\Psi_{nlm} = m\hbar\Psi_{nlm}$$

reduces to

$$-i\hbar\frac{\partial}{\partial\phi}e^{iS_m(\phi)/\hbar} = m\hbar e^{iS_m(\phi)/\hbar}$$

with the solution

$$S_m(\phi) = m\phi\hbar + const.$$

Normalization determines the constant of integration resulting in

$$e^{iS_m(\phi)/\hbar} = \frac{1}{\sqrt{2\pi}}e^{im\phi}.$$

It is interesting to observe that the hydrogen atom probability density for both the spherical-polar and parabolic states is independent of the polar angle ϕ. Since $S_m(\phi) = m\phi\hbar + const$ is purely real, the probability density is given by

$$|\Psi_{nlm}|^2 = e^{i[S_{nl}(r)-S^*_{nl}(r)+S_{lm}(\theta)-S^*_{lm}(\theta)]/\hbar}$$

for the spherical-polar states and

$$|\Psi_{nqm}|^2 = e^{i[S_{nqm}(\xi)-S^*_{nqm}(\xi)+S_{nqm}(\eta)-S^*_{nqm}(\eta)]/\hbar}$$

for the parabolic states. On the other hand, the probability current $\boldsymbol{J}$ possesses only a polar component,

$$J_\phi = \frac{i\hbar}{2\mu r}\left(\Psi\frac{\partial\Psi^*}{\partial\phi} - \frac{\partial\Psi}{\partial\phi}\Psi^*\right) = \frac{m\hbar}{2\mu r}|\Psi|^2,$$

and is also independent of polar angle (though it does depend upon r). There is therefore no flux of probability across spherical surfaces $r = const$, nor is there a flux of probability across the conical surfaces $\theta = const$. Probability flux only swirls across the planes $\phi = const$ of Figs. 4-9 and 4-10 and this flux J_ϕ is itself uniform in polar angle. The probability continuity equation (5.15) therefore yields $\partial J_\phi / \partial \phi = 0$ and a time-independent probability density $|\Psi|^2$ consistent with the steady-state of the atom.

The partial actions S_r and S_θ are not elementary functions. They are particular kinds of polynomials in r which are forms of the confluent hypergeometric function called Laguerre polynomials and another family of polynomials in θ called Legendre polynomials. The spherical-polar eigenstates are determined in detail in **Note 9** and a large collection of the lowest quantum number states are explicitly exhibited there.

The probability density of the hydrogen atom for both the spherical-polar and parabolic states is always azimuthally symmetric: there is no distinguished azimuthal angular position ϕ. This is not so for the latitudinal angle θ. The probability density does depend upon this angle because different values of the quantum numbers l and m reflect preferred locations in the latitude of the electron which yield the angular momenta specified by these values. On the other hand, there is no way to distinguish a "top" and "bottom" of the atom. The probability density is mirror-symmetric across the plane $x_3 = 0$, reflected in the condition $|\Psi(\theta)|^2 = |\Psi(\pi/2 - \theta)|^2$ which one may verify for all states.

The ground state,

$$\Psi_{100} = \frac{1}{\sqrt{\pi}} a_0^{-3/2} e^{-r/a_0},$$

depends only upon the radial coordinate r and is spherically symmetric. All eigenstates fall off exponentially from the origin over a length scale equal to the Bohr radius a_0. The higher quantum number states fall off more slowly with the factor e^{-r/na_0}. All the eigenstates Ψ_{n00} in which the angular quantum numbers l and m vanish are spherically symmetric. For example, the state Ψ_{300} shown in Fig. 5-3 is

$$\Psi_{300} = \frac{1}{81\sqrt{3\pi}} a_0^{-3/2} \left[27 - 18 \left(\frac{r}{a_0} \right) + 2 \left(\frac{r}{a_0} \right)^2 \right] e^{-r/3a_0}.$$

In addition to falling off exponentially in r the higher states also exhibit wave-structure in r. The number of nodes (points at which Ψ_{nlm} vanishes) increases with increasing quantum number n. The number of radial nodes for an eigenstate Ψ_{nlm} is equal to $n - 1$. While the ground state has no nodes—only a solitary wave which falls off in r—the higher states are radial waves that oscillate in space with increasing numbers of nodes as n increases. The wave functions become more intricate as n increases;

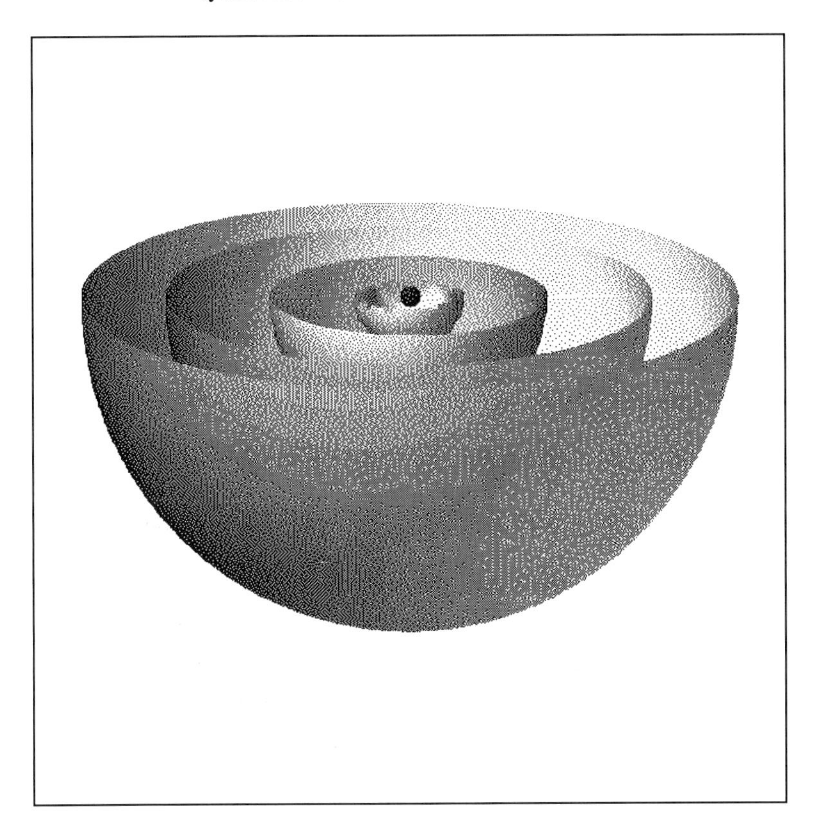

Figure 5-3. Spherical-Polar Eigenstates of the Hydrogen Atom. The state is represented as surfaces upon which probability densities $|\Psi_{300}|^2$ are uniform. In addition to falling off exponentially in r, these spherically symmetric states also exhibit wave structure indicated by the multiple surfaces. The probability density $|\Psi|^2$ vanishes on two nodal surfaces which are located between the inner two and outer two surfaces. *(Courtesy George D. Purvis III)*

but the intricate parts of these wave functions are of progressively smaller amplitude relative to the eigenstates of lower quantum number.

The effects on the state function induced by the total angular momentum quantum number l are symmetric about the polar axis. The effects upon Ψ_{nlm} wrought by increasing angular momentum quantum number m are oriented about the polar axis and are mirror-symmetric across the plane $x_3 = 0$ (the plane in which the corresponding classical two-body motion would lie). The angular momentum effects can best be seen for the case of maximum angular momentum for a given energy level, that is, states for which $l = n - 1$. Each of these states is distinguished by the polar angular momentum quantum number $m \leqslant l$. A sequence of such states is exhibited in Figs. 5-4 for the case $l = 2$ corresponding to the energy level $n = 3$. When $|m| = l$, the angular momentum vector is aligned with the polar axis and the electron is most likely to be located near a plane perpendicular to the angular momentum vector.

Figure 5-4 (a). Spherical-Polar Eigenstates of the Hydrogen Atom. A probability density surface for the state $\Psi_{32\pm2}$ illustrates the case $m = l$. In this case the angular momentum vector is aligned with the polar axis (perpendicular to plane of the torus-like surface) and the electron is most likely to be located near this plane. This state corresponds most closely to the classical motion. *(Courtesy George D. Purvis III)*

This condition corresponds most closely to the classical motion and the effect can be seen in the state $\Psi_{32\pm2}$,

$$\Psi_{32\pm2} = \frac{1}{162\sqrt{\pi}} a_0^{-3/2} \left(\frac{r}{a_0}\right)^2 e^{-r/3a_0} \sin^2\theta e^{\pm 2i\phi},$$

which is illustrated in Fig. 5-4 *(a)*. The electron probability distribution is concentrated in the plane perpendicular to the polar or 3-axis.

In Fig. 5-4 *(b)*, the same states are shown, but with the polar component diminished to $m = 1$:

$$\Psi_{32\pm1} = \frac{1}{81\sqrt{\pi}} a_0^{-3/2} \left(\frac{r}{a_0}\right)^2 e^{-r/3a_0} \sin\theta \cos\theta e^{\pm i\phi}.$$

The electron now is distributed over two torus-like surfaces and its net contribution to the polar angular momentum is correspondingly diminished.

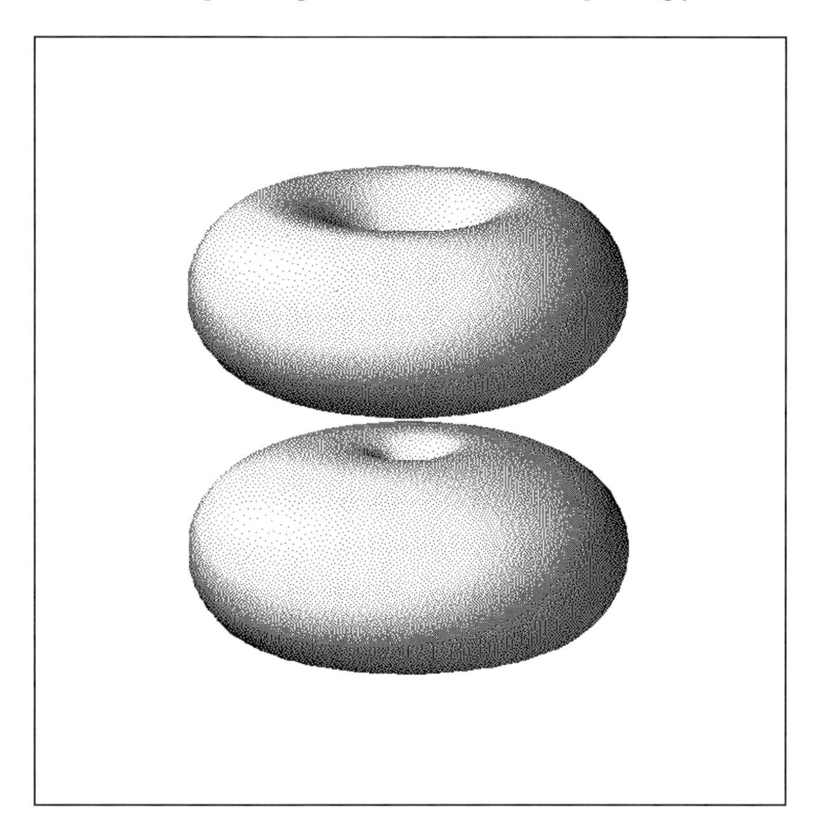

Figure 5-4 (b). Spherical-Polar Eigenstates of the Hydrogen Atom. A probability density surface for the state $\Psi_{32\pm1}$ is shown. The electron is distributed over two torus-like surfaces. The angular momentum is maximal but no longer totally polar. *(Courtesy George D. Purvis III)*

When the polar angular momentum is reduced to $m = 0$, the states are most unlike the classical motion. The electron must execute its motion in such way that it does not generate an observable polar component of angular momentum. An example of such a state is

$$\Psi_{320} = \frac{1}{81\sqrt{6\pi}} a_0^{-3/2} \left(\frac{r}{a_0}\right)^2 e^{-r/3a_0} \left(3\cos^2\theta - 1\right)$$

and shown in Fig. 5-4 *(c)*. One sees that the electron wave function posseses upper and lower lobes which cancel the angular momentum on the central torus-like zone around the polar axis so that it possesses significant angular momentum components M_1 and M_2 but vanishing M_3. The interested

Figure 5-4 (c). Spherical-Polar Eigenstates of the Hydrogen Atom. A probability density surface for the state Ψ_{320} illustrates a state of maximal total angular momentum quantum number $l = n - 1$ but a vanishing polar component $m = 0$. The electron is distributed over upper and lower lobes and the central torus-like zone in such a way that the polar angular momentum vanishes. This state differs most from the classical motion. *(Courtesy George D. Purvis III)*

reader may examine the variety of structure in the full set of states up through $n = 3$ found in **Note 9**.

The parity of the spherical-polar states of hydrogen turns on the quantum number l (see **Note 9** for the details). The parity of a given state is given by

$$\mathcal{P} = (-1)^l.$$

The spherical-polar states illustrated thus far have all been even parity states corresponding to even values of l. An example of an odd parity state is

$$\Psi_{310} = \frac{2}{27\sqrt{\pi}} a_0^{-3/2} \left(\frac{r}{a_0}\right) \left[1 - \frac{1}{6}\left(\frac{r}{a_0}\right)\right] e^{-r/3a_0} \cos\theta,$$

shown in Fig. 5-5.

Figure 5-5. Spherical-Polar Eigenstates of the Hydrogen Atom. A probability density surface for an odd parity state: the state Ψ_{310}. *(Courtesy George D. Purvis III)*

Consider now the set of eigenstates for the hydrogen atom in a spinor state specified by $(\mathcal{H}, E_3, M_3)$ with corresponding parabolic quantum numbers (n, q, m). The detailed determination of these eigenstates may be found in **Note 10**. The angle coordinate ϕ and its corresponding rotational invariant M_3 are common to both the spherical-polar and parabolic representations. The polar states $e^{iS_m(\phi)/\hbar} = e^{im\phi}/\sqrt{2\pi}$ in parabolic coordinates are precisely the same as the polar states in spherical-polar coordinates. The remaining parabolic states $e^{iS_{nqm}(\xi)/\hbar}$ and $e^{iS_{nqm}(\xi)/\hbar}$ are not elementary functions. These two functions are, however, formally identical, differing only in the sign of the quantum number q as shown in **Note 10**. The parabolic functions have the property

$$S_{nqm}(\xi) = S_{n-qm}(\eta). \tag{5.60}$$

The parabolic state functions are confluent hypergeometric functions similar to the Laguerre polynomial. They are determined in **Note 10**.

The parabolic ground state is

$$\Psi_{100} = \frac{1}{\sqrt{\pi}} a_0^{-3/2} e^{-(\xi+\eta)/2a_0}.$$

This state is identical to the spherical-polar ground state. It must indeed be so since for $n = 1$ it is required that $q = m = l = 0$. The exponential fall-off terms all appear with the common factor $(\xi+\eta)/2na_0$. In parabolic coordinates the factor $(\xi+\eta)/2$ is the magnitude of the position vector r. The exponential fall-off terms are therefore of the form

$$e^{-(\xi+\eta)/2na_0} = e^{-r/na_0}$$

and are identical to the exponential fall-off factors of the state functions Ψ_{nlm} in spherical-polar coordinates. Since these factors depend only upon the energy level quantum number n and this quantum number is common to both state specifications, the two factors are identical. Moreover, since the ground state is completely symmetrical in both state specifications, the ground state functions Ψ_{100} are identical for both spherical-polar and parabolic coordinates.

The spherical-polar and parabolic states actually overlap in an even more general way: they are identical whenever $q = 0$ for the parabolic states and $l = n - 1$ for the spherical-polar states. Using the relationships $\xi = r(1+\cos\theta)$ and $\eta = r(1-\cos\theta)$ it is an easy matter to show that the parabolic state Ψ_{n0m} is identical to the spherical-polar state Ψ_{nn-1m}. The states for which $q = 0$ and $l = n-1$ are also the states most closely related to the classical motion; for in this case the eccentricity vector lies completely in the plane of motion perpendicular to the M_3 component of the angular momentum.

There are no parabolic states Ψ_{nqm} of the form Ψ_{n00} when n is even. It is not possible for the angular momentum and eccentrum to simultaneously vanish in such states. All parabolic states with even energy level quantum numbers n are spherically *asymmetric*. On the other hand, states with odd quantum numbers n do possess spherically symmetric states.

In contrast to the spherical-polar states, the parabolic states are not generally symmetric with respect to the plane $x_3 = 0$. They are only so for states Ψ_{nqm} with $q = 0$ as illustrated by

$$\Psi_{300} = \frac{1}{9\sqrt{\pi}} a_0^{-3/2} \left[1 - \frac{1}{3}\left(\frac{\xi}{a_0}\right)\right]\left[1 - \frac{1}{3}\left(\frac{\eta}{a_0}\right)\right] e^{-(\xi+\eta)/6a_0}$$

which is shown in Fig. 5-6 *(a)*.

Parabolic states with nonvanishing eccentricity q are asymmetric and are illustrated by the state

$$\Psi_{3-20} = \frac{1}{9\sqrt{\pi}} a_0^{-3/2} \left[1 - \frac{2}{3}\left(\frac{\eta}{a_0}\right) + \frac{1}{18}\left(\frac{\eta}{a_0}\right)^2\right] e^{-(\xi+\eta)/6a_0}.$$

Figure 5-6 (a). Parabolic Eigenstates of the Hydrogen Atom. A probability density surface for the state $\Psi_{nqm} = \Psi_{300}$ illustrates a parabolic state symmetric with respect to the plane $x_3 = 0$. The x_3 axis is perpendicular to the toroidal plane and the atom has been rotated toward the reader for better viewing. *(Courtesy George D. Purvis III)*

This state is shown in Fig. 5-6 *(b)*. The full set of parabolic states up to $n = 3$ may be found in **Note 10**.

Parity manifests itself in parabolic states as an interchange symmetry as described in the discussion following Eq. (4.45). Inversion of the space coordinates interchanges ξ and η. Since the eccentricity is a function of odd parity, the sign of its quantum number q is also changed on coordinate inversion.

The hydrogen atom—the two-body motion with inverse-square force rendered by quantum mechanics—possesses eigenstates which form a rich collection of three-dimensional structures. Quantum states stand in marked contrast to classical trajectories for which the orbit lies wholly in the plane perpendicular to the angular momentum vector. The eigenstates are intricate geometric shapes formed of exponential, trigonometric, and polynomial functions. They contrast dramatically against the classical motion for which the geometric structures are the circle and ellipse. This collec-

Figure 5-6 (b). Parabolic Eigenstates of the Hydrogen Atom. A probability density surface for an asymmetric parabolic state: the state $\Psi_{nqm} = \Psi_{3-20}$. The surfaces encircle the x_3 axis which is perpendicular to the planes of the flattened spheroid structures. *(Courtesy George D. Purvis III)*

tion of eigenstates of quantum two-body motion provides the basis for the manifold structure of the elements—structure not contained in the ellipse of classical motions.

Mechanics of the Periodic Table

The total number of eigenstates that correspond to a given value of n (and hence to a given energy level $\mathcal{E}_n$) increases in the pattern $1+4+9+\cdots$. This pattern closely corresponds to the pattern $2+8+18+\cdots$ of the Periodic Table. But not quite. Aside from the fact that the quantum pattern is off by a factor of two from the Periodic Table, there is no requirement that a nucleus with a given number of electrons would have its electrons occupying all these states. In fact, the likely condition would be one of the lowest total energy. All the electrons would seek to occupy the ground state $n = 1, l = m = q = 0$.

The rule which governs the permissible states that electrons may occupy thereby removing the arbitrariness described above (and hence provides the basis for the atomic description of the Periodic Table) lies beyond the inverse-square electrical force of the two-body problem. Although the inverse-square force law (3.4) is adequate for the basic description of the interaction between an electron and a positively charged nucleus, it is quite incomplete for a description of the quantum properties of the electron itself.

This is the first glimmer of approaching crisis; for to inquire into the quantum properties of the electron is to plunge into the breakdown of the Galilean symmetries of the two-body problem. We need not yet face this crisis because the implications of the electron's quantum properties for the hydrogen atom turn out to be quite simple. We can summarize them as follows. The eigenstates of the electron must be regarded not as a scalar but as a four-dimensional quantity,

$$\Psi = (\Psi_0, \Psi_1, \Psi_2, \Psi_3),$$

consistent with the relativistic description of space and time as a four-dimensional space incorporating three space components and one time component.

What meaning does one attach to the various components of a four-valued state $(\Psi_0, \Psi_1, \Psi_2, \Psi_3)$ for a charged particle? Amazingly, these four components represent *two* distinct charged particles: the electron and its antiparticle, the positron. In addition each particle has two new states peculiar to the quantum realm. Because of the intimate connection of these states with the angular momentum of the particle, these states are called *spin* states. Each spin state possesses only two eigenvalues: $+\hbar/2$ and $-\hbar/2$, the fundamental eigenvalues of angular momentum. The four-valued state represents an electron with two spin states and a positron with two spin states for a total of four states.

The most significant result of the quantum theory of the electron for the structure of the atom is the *Pauli Exclusion Principle.* It requires that no two electrons can be in the same state (where the spin is one of the quantities defining the state in addition to the three quantum numbers of the Casimir set).

The exclusion principle brings completion to the mechanical theory of the elements. Even though the ground state is the preferred state for a nucleus with an assembly of orbiting electrons, it can be occupied by two of them at most; and these two electrons must have opposite spins. Since two electrons (one of spin $+\hbar/2$ and one of spin $-\hbar/2$) can at most occupy each of the eigenstates Ψ_{nlm} for each unique combination of n, l, m, the pattern of eigenstates $1 + 4 + 9 + \cdots$ actually corresponds to electronic occupations of $2 + 8 + 18 + \cdots$ which is the basic pattern of the Periodic Table.

The emergence of the Periodic Table from the mechanics of the atom is one of the magnificent achievements of physics. It is the quantum counterpart to the equally magnificent emergence of the ellipse of Kepler from

Newtonian mechanics of the solar system. They are the motions—the one classical, the other quantum—of two bodies bound together by the inverse-square force.

The actual energy levels of a multi-electron atom will not quite follow the simple $2 + 8 + 18 + \cdots$ pattern of the two-body atom because of the mutual interaction among electrons in addition to their interaction with the nucleus. Just as the elliptical orbits of the planets about the sun are perturbed by one another's presence, so too, the pure quantum mechanical states of the electrons in the hydrogen atom are perturbed by one another's presence.

The atoms beyond hydrogen are *many-body* atoms. The approximate energy levels of multi-electron atoms, still using the ordering scheme of hydrogen, are shown in Fig. 5-7. It can be seen that there are *two blocks* of 8 states before the first appearance of the 18 states and then two blocks of 18 states consistent with the Periodic Table. Although the $n = 3$, $l = 2$ state (Ψ_{32m}) has a lower energy level than the $n = 4$, $l = 0$ state (Ψ_{400}) when electron interactions are neglected, the situation is reversed in an actual multi-electron atom.

This is because the inner electrons shield the positively charged nucleus making it appear to have a diminished charge to the outer electrons. Outer electron energy levels are therefore shifted upward from those which they would have if they interacted solely with the nucleus. In the $n = 2$ state, the $l = 1$ states are elevated above the ground state $l = 0$ because of the shielding effect of the two electrons in the ground state. The same effect shifts the $l = 1$ states for the $n = 3$ levels. At the $n = 3$ level the $l = 2$ states are further shifted because of the shielding by both the two $l = 0$ ground state electrons and the six $l = 1$ electrons. The cumulative effect of these shifts is to place the $n = 3$, $l = 2$ states slightly *above* the $n = 4$, $l = 0$ states so that they group themselves with the $n = 4$, $l = 0, 1$ states rather than the states of their own principal quantum number $n = 3$. This shifting effect creates two rows of 8 states before the onset of rows with 18 states. These rows of 18 states are also double because of the same effect as shown in Fig. 5-7.

The ground state energy of the primal element, $|\mathcal{E}_0| = k\mu^2/2\hbar^2$, is a fundamental atomic constant based upon the electron mass and the hydrogen nuclear mass. It has a value of 13.6 eV (the energy acquired by an electron if it is accelerated through a voltage difference of 13.6 V). Division of this energy by the mass of the hydrogen atom yields an energy scale of approximately 50,000 British thermal units (BTU) per pound or 100 million Joules (megajoules) per kilogram. This fundamental quantum mechanical energy level sets the energy scale of all chemical reactions. The energy levels of these eigenstates directly determine the chemical reactivity of the elements. It is the differences in energy levels that provide the basis for chemical changes in which electrons and nuclei rearrange themselves so as to achieve stable states of minimum energy.

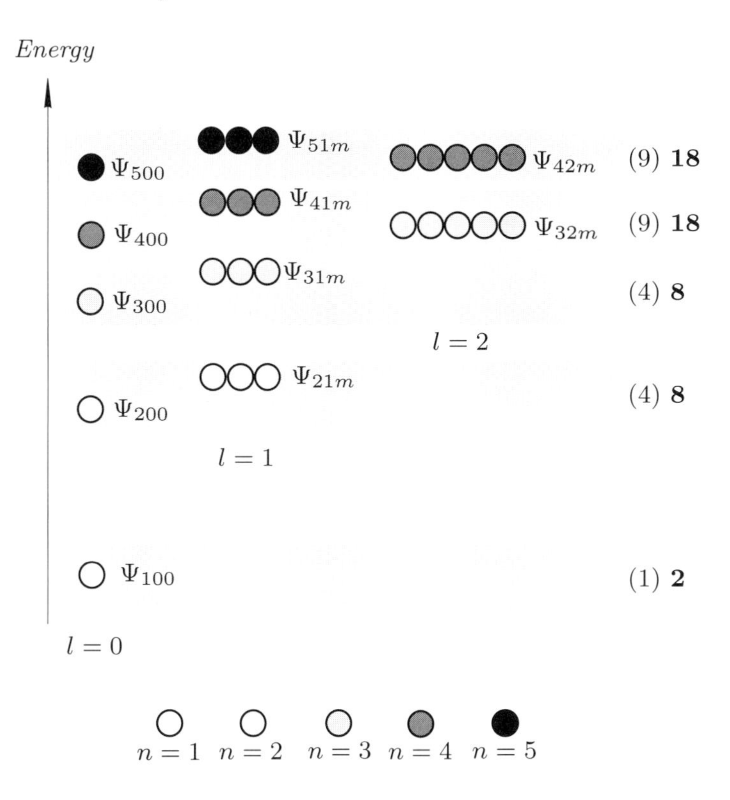

Figure 5-7. Energy States and the Periodic Table. The bands correspond to states with approximately the same energy level. States with different principal quantum numbers n are identified by different shades. The states are arranged according to the angular quantum number l. The Periodic Table groups states according to energy level; hence the numbers of states of the atom follow the number of states within each energy band which are $1 + 4 + 4 + 9 + 9 + \cdots$ and which correspond to electronic occupations of $2 + 8 + 8 + 18 + 18 + \cdots$.

The energy that may be released by a pound of dry wood if it is burned is about 10,000 BTU. In burning, the electrons orbiting the carbon and hydrogen nuclei in the wood interact with the electrons and nuclei of the oxygen in the air to form carbon dioxide and water molecules which end up with an excess of energy. The excess energy arises because the internal energy of the water and carbon dioxide molecules is less than the internal energy of the hydrogen and oxygen atoms. This excess energy is converted by collisions of the carbon dioxide and water molecules and by radiation to heat. The energy released when a beam of stone, wood, or steel is fractured is thousands of BTU per pound of the fractured portion. The energy contained in a pound of food and transformed into motion and heat by our bodies is several thousand BTU per pound. The chemical energy level of fat is a few thousand BTU per pound. The wild goose transforms fat into propulsive motion with its wings, flying about a thousand miles on several pounds of fat. All these chemical transformations are rooted in the quantum-mechanical ground state energy $|\mathcal{E}_0|$=13.6 eV.

It is the energy level differences of atomic states (rather than quantum number differences) that give each element its peculiar chemical properties. Hence, the clustered groups of states in Fig. 5-7 which correspond to the same energy level, rather than levels marked by the principle quantum number n of the two-body hydrogen atom, demark the chemical properties of the many-body elements. The full quantum-mechanical problem of multi-electron atoms is a many-body problem just like that of the solar system. It cannot be solved exactly like the single electron hydrogen atom.

Nonetheless, the hydrogen atom is the prototypical atom, the exactly soluble motion of quantum mechanics. Although the hydrogen atom is the only element for which an exact quantum-mechanical solution exists, mechanics nonetheless describes the motion of all the elements in their many-body complexity. Several powerful approximate methods exist for solving these many-body motions. Computational methods in which the operator equations are solved numerically have also yielded the revealing inner structure of both high atomic number elements and the molecules which they form. But these developments are beyond the domain of the two-body problem.

Reprise

A long journey spans the distance from the Ptolemaic world view to the quantum mechanics of the atom. Two milestones mark turning points on that journey: Newton's inverse-square law which produced the ellipse of Kepler from a fundamental law of motion and Bohr's quantum transformation of that same inverse-square law to create the atom. The harvest is extraordinarily rich. We now stand before one of the magnificent vistas of physics to which the Shaggy Steed of Physics has carried us: the heavens and the elements governed by a universal law of motion and the same symmetries. A complete set of isolating invariants formed from the conservation laws and completed by the eccentricity invariant for the inverse-square force of electricity and gravitation portrays this unity. Its underlying symmetries and invariants are those of four-dimensional rotations. From these invariants flow the fundamental orbit of heavenly bodies on the one hand and the fundamental pattern of the elements on the other.

Quantum mechanics, compelling as it is, can only be an approximation to a more complete theory of mechanics yet to be conceived. Since the probabilistic character of quantum mechanics arises because of the impossibility of observing half the invariants of motion exactly, half the invariants of classical motion rest upon an ultimately uncertain footing deep at the quantum scale. Yet the large-scale world in which we live and move possesses all these invariants! How do these uncertain invariants get assigned?

There is no more telling sign of the present incompleteness of quantum theory than that there is no way presently known to bridge the uncertainty gap between the quantum and the classical without paradox.*

We are always open to Ptolemaicism—seeing the simple in needlessly complicated ways. The centerpiece of quantum theory, the state function ("die Psi funktion" as Einstein always called it, denying it permanent significance beyond its symbolic form), faces us with this vulnerability. The central object of the theory *is not observable.* Rather, it is an intermediate construct on the way to the probability density $|\Psi|^2$ which is the observable.

Are these not Ptolemaic burdens, quantum versions of epicycles and deferents that we bear because we cannot yet see the way out of a needlessly obtuse perspective? Ptolemaicism, with its quite credible predictions of actual orbits, reminds us that accurate predictions of atomic structure afforded by quantum mechanics cannot alone guide us to the deep truth of the theory. Like Greek and medievalist, we too are limited by the vision of our era.

Why, one must finally ask, why this unique, omnipresent, inverse-square force in the large-scale world that closes orbits, makes electronic states degenerate, and creates elements with rich structure thereby giving the world pith and texture? Can there be a more dramatic portrait of the unfolding of the one into the many than that of charge and mass bound by the inverse-square force? For out of such nakedly simple attraction born out of a primal bloom there arise, in coalescing complexity, elements, molecules, planets, geological structures, atmospheres, and the entire evolution of living creatures—the vast portion of the creation that surrounds us. And if the world is splendid, this simple force mediating mass and charge is, indeed, a splendid stroke of creation.

No one has yet found a credible vision of the quantum that naturally merges into the classical or dispenses with the unobservable state function. But deeper insight into the pervasive nature of the inverse-square force has been found. It is Einstein's vision of general relativity. It is splendid.

---

* Unique perspectives on the paradoxes of quantum mechanics and an imaginative solution to the present incompleteness of the theory based upon *gravity* are offered by Roger Penrose in *300 Years of Gravitation*, S. W. Hawking and W. Israel (eds.), Cambridge U. Press, 17–48 (1987) and *The Emporer's New Mind*, Oxford University Press, Chap. 6 (1989).

CHAPTER 6

The Hidden Unity of Space and Time

We are approaching the limits of our journey with the two-body problem. The mechanics that cast asunder the ancient cosmos has borne us upon this journey. Looking back, one finds the static heavenly spheres gave way to a universal law sustaining all motion: the universe unfolds along the paths of least action of all its matter. The diversity and richness of the heavens and the elements are born out of the unity and simplicity of electrons and nuclei bound by gravity and electricity.

For two heavenly bodies the law of motion yields the classical ellipse of Kepler—and more: heavenly tori image heavenly motions in a simplicity and beauty more revealing than that of the spheres of antiquity. Extended into the quantum realm, the same law of motion describes the structure of the elements; and the prototypical structure of the elements is found to possess the same symmetries as the prototypical structure of the solar system.

The heavens and the elements are mathematically united by the same law of motion and the same underlying symmetries. The external beauty of the world is accompanied by an interior beauty of invariants and symmetries in which the perfection imagined in the heavenly spheres pervades the whole of space and time. The music of the spheres and the elements now resounds in the four-dimensional rotational symmetries embracing the solar system and the atom.

This vision is only a partial vision of a world layered in mystery. The portrait of the heavens and the elements drawn from the Galilean and Newtonian mechanics of gravity and electricity also shatters when pushed beyond its human-scale limits. Our portrayal of the heavens and the elements—like those of the past—is only a glimpse of a deeper and more embracing drama which continues to unfold.

In these final chapters the two-body problem reveals its limitations and in so doing it opens us to a wider world of physics. We shall first be confronted with the breakdown of Galilean symmetries and their displacement by the even more graceful symmetries of Einsteinian relativity. (Remarkably, the Einsteinian symmetries of space-time turn out to be the same as the symmetries of Galilean two-body motion when the motion is unbound.) Then we shall be introduced to the new features of motion which appear when more than two bodies are allowed to interact, the rich world of chaos and cosmos which appears in many-body motions.

Einsteinian Symmetry

Space and time are not quite what we have made of them. The position of a point in space, the instant of a point in time—these seem to be the bedrock of experience, the absolute ground upon which all descriptions of the dynamics of the world must be built. But they are not. They only appear to be for velocities which are small compared to the velocity of light. For the full range of velocities, the primal mechanical quantity is not position, nor is it time. The fundamental mechanical quantity of the world is a *velocity*, the velocity of light.

We say "of light," but it is more than this. It is the velocity with which both gravitational and electrical interactions are propagated from particle to particle. The electrical force is propagated from particle to particle at the speed of light. The gravitational force is propagated from particle to particle at the speed of light. The maximum velocity with which the subatomic forces can be propagated from particle to particle is the speed of light. The universe communicates at the speed of light.

At the dawn of the twentieth century it became clear that the speed of light is an invariant over all the events of the universe. The velocity of propagation of the gravitational or electrical force is measured as the velocity of light in all reference frames, no matter what their velocities. This is contrary to the Galilean view of space and time which has shaped our journey to the heavens and the elements. In the Galilean world space and time are *separately* homogeneous and isotropic according to the transformation law (3.2). Time flows absolutely and forces propagate infinitely fast for all observers.

Motion with velocities of the order of the speed of light is described as *relativistic* motion. At velocities which approach the speed of light, the finite speed of propagation of forces becomes noticeable. Whereas space and time are absolute in the Galilean world, the speed of light is the true absolute. Space and time are the relative notions. They are local properties of a particular reference frame.

Since the speed of light is such a large quantity compared to the human scale (approximately 186,000 miles/sec or 300,000 km/sec), the discovery that forces propagate at a finite speed did not occur until the late nineteenth and early twentieth centuries. If the speed of light were to be absolute over all reference frames, something else had to give. Time was this quantity; and the idea of time becoming a relative notion peculiar to each observer rather than an absolute quantity which ruled all motions proved to be a great stumbling block.

In the late nineteenth century H. A. Lorentz worked out the transformations of space and time which maintain the speed of light constant between any two reference frames. (These famous transformations now bear his name.) In so doing he found that the time coordinates of each reference frame are not the same. They shrink and expand with respect to one another depending upon the relative velocity between the two reference frames. This result was so confounding that Lorentz called one of the times the "absolute time" and the other the "local time." In reality neither time coordinate can claim a status different from the other. Both are on the same footing. There is no absolute; there are only local times. Albert Einstein was the first to truly affirm this conclusion as a profound revelation of nature and to foster its development into a watershed theory of twentieth century physics.

The implications for space and time which flow from forces which propagate at a finite speed may be drawn out in the following way. A gravitational or electrical force is initiated. Think of this force as a wave which propagates at the speed of light c. Let the spatial coordinates x_1, x_2, x_3 mark the wave front of the force and let t record the time in this reference frame. The wave front of the force propagates a distance ct in the time interval t. But this same distance between the initial and final positions of the wave front has the magnitude ${x_1}^2 + {x_2}^2 + {x_3}^2$. The quantity

$$s^2 = (ct)^2 - {x_1}^2 - {x_2}^2 - {x_3}^2 \tag{6.1}$$

must therefore be an invariant. This invariant of space and time which follows from the invariance of the velocity of light is known as the Lorentz interval.

The Lorentz interval s marks the interval between successive wave fronts of an electromagnetic or gravitational force. Since these fronts move at the invariant velocity of light, the Lorentz interval is observed to be identical in all reference frames. For any two reference frames $(\mathbf{x}, t)$ and $(\mathbf{x}', t')$, the Lorentz interval measured in each frame is the same:

$$s^2 = (ct)^2 - (x_i)^2 = (ct')^2 - (x_i')^2.$$

The most elementary symmetries of the physical world are in space and time themselves. In familiar three-space, reference frames in which the law of motion is invariant must be open to all origins, translational directions, and rotational orientations. These are the fundamental symmetries of homogeneity and isotropy.

In Galilean space–time homogeneity and isotropy are described by the ten symmetry parameters $\mathbf{a}$, $\boldsymbol{u}$, $\boldsymbol{\Omega}$, and τ which generate the Galilean relativity transformations

$$\begin{aligned} \mathbf{x}' &= \mathbf{x} + \delta\mathbf{a} + \delta\mathbf{u}t + \delta\boldsymbol{\Omega} \times \mathbf{x}, \\ t' &= t + \delta\tau. \end{aligned} \tag{6.2}$$

Space and time appear asymmetrically in Galilean relativity. Time is totally severed from the space coordinates in the second of the transformations (6.2). It is therefore absolute across all reference frames and plays no role in the isotropy of space. Rotational symmetry is only manifested in the space coordinates. This aesthetic flaw is directly connected with a physical flaw in which forces propagate infinitely fast.

Where is the logical defect in the Galilean symmetries that underlies their experimental and aesthetic flaws? It is in their ambiguous sense of *direction.* As described in Chap. 3, Galilean relativity postulates two sources of directional symmetry, one translational described by a uniform velocity (such as walking along a straight line) and another rotational described by an angular orientation (such as pointing at a star). In hindsight we now know there is only one basis for the notion of direction; and it is rotational. The direction which arises through the apparent translational nature of the velocity vector turns out to be rotational when the velocity of light is accepted as an invariant of nature. True translational symmetry is restricted to symmetry with respect to origins. This is appropriate; for translations have to do with the lack of a preferred origin rather than the lack of a preferred direction.

The flaws in Galilean relativity may be overcome by recognizing the hidden unity of space and time: they are bound together in a single four-dimensional space–time rather than in a three-dimensional space which stands separately from time. Instead of specifying a point in space with a three-component position vector which is a function of time, one describes events in space–time in terms of a single four-vector. Three components of this vector are the spatial components x_1, x_2, x_3. The fourth (or "zeroth") component, consistent with the dimensions of these spatial components, is the signal interval $x_0 = ct$. In this four-space the four components of space and time constitute the four-vector

$$x_\lambda = (x_0, x_1, x_2, x_3) = (ct, \mathbf{x}).$$

(A four-vector is distinguished with Greek suffixes, as in x_λ, $\lambda = 0, 1, 2, 3$, while the boldface notation and Latin suffixes, as in $\mathbf{x}$ and x_i, are reserved for three-vectors.)

The magnitude of a four-vector is symbolized as x^2. If it were a vector in Euclidean space, its magnitude would obey a Euclidean metric like that of Eq. (3.1):

$$x^2 = {x_0}^2 + {x_1}^2 + {x_2}^2 + {x_3}^2.$$

Such a Euclidean magnitude is contrary to the invariance of the Lorentz interval (6.1) required by the condition that forces propagate at the speed of light. The time component ${x_0}^2$ must have the opposite sign of the spatial components x_i^2. Space and time therefore constitute a non-Euclidean space in which the magnitude of a vector is composed from its components by*

$$x^2 = s^2 = {x_0}^2 - {x_1}^2 - {x_2}^2 - {x_3}^2. \tag{6.3}$$

Contrary to the position vector in the Galilean world, whose magnitude x^2 varies with time, the fundamental vector describing position in four-space—the "event" of space–time—has a magnitude which is an invariant.

The search for the simplest symmetries of space and time is reminiscent of the search of Copernicus and Kepler for a geometry of beauty and simplicity which reflected the solar system. The replacement of the Euclidean interval (3.1) by the Lorentz interval (6.3) in a non-Euclidean four-space is a visionary insight leading to a simplicity and beauty comparable to that of the Copernican replacement of the earth by the sun as the center of planetary motions.

The blank canvas of mechanics must be homogeneous and isotropic in four-dimensional space–time. Since the magnitude of the four-vector is invariant, the symmetry transformations may only shift the origin or change the direction of four-vectors. They may not alter their magnitudes. Transformations of vectors that change direction without changing magnitude are pure rotations. The symmetry transformations describing homogeneity and isotropy are therefore translations and rotations in four-dimensional space–time.

The symmetry group of relativistic motion is that of translations and four-dimensional rotations. Rotations of space in which one space coordinate is rotated into another are familiar enough. But the unification of space and time in a single four-space leads to a notion unthinkable in three-space: the rotation of a space coordinate into a time coordinate. Such a space–time rotation differs from a pure spatial rotation by having a rotation angle which is imaginary. These four-dimensional rotations come about in the following way.

In four-space there are six independent planes of rotation as described in Chap. 4. Three of these planes, (x_2, x_3), (x_3, x_1), and (x_1, x_2), contain the rotations among the spatial coordinates. The remaining planes, (x_0, x_1), (x_0, x_2), and (x_0, x_3), contain the rotations between the time and space coordinates. Four-dimensional rotations are described by six symmetry parameters which may be arranged as two three-vectors as described in Chap. 4. Rotations in the four-space of Lorentzian space–time are therefore described by six symmetry parameters which reside in two three-vectors and represent rotations in the six planes.

---

* There is a freedom in the overall sign of the Lorentz metric condition. One could equally well define the Lorentz interval as

$$x^2 = {x_1}^2 + {x_2}^2 + {x_3}^2 - {x_0}^2.$$

The subspace of rotations among the three spatial coordinates x_1, x_2, x_3 is Euclidean; all components x_i contribute terms x_i^2 to the metric with the same sign. The rotations in these planes are the same as those in Galilean space–time and are described by the three-vector rotational symmetry parameter $\boldsymbol{\Omega}$.

The rotations between the space and time coordinates differ from the purely spatial rotations because there is a sign difference between ${x_0}^2 = (ct)^2$ and any of the components ${x_1}^2$, ${x_2}^2$, ${x_3}^2$ in the metric condition. The rotations between space and time coordinates are hyperbolic rather than Euclidean. Whereas Euclidean rotations turn through real angles and are described by trigonometric functions, hyperbolic rotations turn through imaginary angles and involve hyperbolic functions. A hyperbolic rotation may be obtained from a Euclidean rotation by making the angle of rotation imaginary.

The rotations between the time and space coordinates are also generated by a three-vector symmetry parameter. What is it? The velocity symmetry parameter $\boldsymbol{u}$, shorn of its role as a generator of translations in Galilean space–time, turns out to be the generator of the rotations between the time and space coordinates. The velocity symmetry parameter may be expressed more informatively as a rotation angle $\boldsymbol{\phi}$ given in terms of $\boldsymbol{u}$ by

$$\tanh \phi_i = u_i/c.$$

For infinitesimal rotations, this expression shows $\delta\boldsymbol{\phi} = \delta\mathbf{u}/c$. The complete set of symmetry transformations in Lorentzian space–time are described by the pair of three-vectors, $\boldsymbol{\Omega}$ and $\boldsymbol{\phi}$. The pure spatial rotations are described by $\boldsymbol{\Omega}$ and the imaginary rotations between the time and space coordinates (which have come to be known as "boosts") are described by $\boldsymbol{\phi}$.

With directional symmetry completely described by rotations, the two cumbersome transformations (6.2) are swept away to be replaced by a single symmetry transformation of translations and rotations:

$$x'_\lambda = x_\lambda + \delta a_\lambda + \delta\Lambda_{\lambda\mu} x_\mu. \tag{6.4}$$

Translations are described by the four-vector δa_λ while the four-dimensional rotations—known as Lorentz transformations—are described by the four-matrix $\delta\Lambda_{\lambda\mu}$.

The rotation four-matrix $\delta\Lambda_{\lambda\mu}$ possesses only six (rather than sixteen) independent components. These six components consist of the six components of two three-vectors which are the symmetry parameters of four-dimensional rotations, $\delta\boldsymbol{\Omega}$ and $\delta\boldsymbol{\phi}$. The group of transformations of space and time is again described by ten symmetry parameters. Four of these describe translational symmetry, the components of the four-vector δa_λ. Six describe rotational symmetry, the two three-vectors $\delta\boldsymbol{\Omega}$ and $\delta\boldsymbol{\phi}$ whose components constitute the six independent components of $\delta\Lambda_{\lambda\mu}$:

$$\delta\Lambda_{\lambda\mu} = \begin{pmatrix} 0 & \delta\phi_1 & \delta\phi_2 & \delta\phi_3 \\ \delta\phi_1 & 0 & \delta\Omega_3 & -\delta\Omega_2 \\ \delta\phi_2 & -\delta\Omega_3 & 0 & \delta\Omega_1 \\ \delta\phi_3 & \delta\Omega_2 & -\delta\Omega_1 & 0 \end{pmatrix}.$$

One sees that the purely spatial rotations described by $\delta\boldsymbol{\Omega}$ form a three-matrix tucked into the lower right-hand corner while the space–time rotations described by $\delta\boldsymbol{\phi}$ border it.*

The two kinds of rotations contained within a Lorentz transformation may be illustrated in the following manner. If a reference frame x'_λ is obtained from a frame x_λ by a rotation about the x_3 axis through angle $\Omega_3 \equiv \theta$, the spatial coordinates x_1 and x_2 are transformed by the elementary rotation

$$\begin{aligned} x_1{}' &= x_1 \cos\theta + x_2 \sin\theta, \\ x_2{}' &= -x_1 \sin\theta + x_2 \cos\theta, \end{aligned} \tag{6.5}$$

and the coordinates $x_0 = ct$ and x_3 are unchanged. This is the pure spatial rotation part of a Lorentz transformation as shown in Fig. 6-1 *(a)*.

The hyperbolic rotation (or boost) of a space coordinate into a time coordinate is illustrated in Fig. 6-1 *(b)*. Let the four-vector x'_λ define an event in a given reference frame and let the four-vector x_λ define the event in a reference frame which translates along the x_3 axis with velocity $u_3 \equiv u$ relative to the first frame. The two coordinates x_1 and x_2 are unchanged in transformation between these two frames. Only ct and x_3 are changed. The hyperbolic rotation of ct into x_3 can be obtained by mimicking the Euclidean rotation of x_1 into x_2 but with an imaginary angle $\theta \to i\phi$. The trigonometric functions are transformed into hyperbolic functions and a space–time rotation is given by

$$\begin{aligned} ct' &= ct \cosh\phi + x_3 \sinh\phi, \\ x_3{}' &= ct \sinh\phi + x_3 \cosh\phi. \end{aligned} \tag{6.6}$$

The hyperbolic functions of the angular parameter ϕ may be expressed directly in terms of the velocity parameter as

$$\sinh\phi = \frac{u/c}{\sqrt{1-u^2/c^2}}, \qquad \cosh\phi = \frac{1}{\sqrt{1-u^2/c^2}}.$$

The Lorentz transformation (6.4) can therefore also be written

$$ct' = \frac{ct + (u/c)x_3}{\sqrt{1-u^2/c^2}}, \qquad x_3{}' = \frac{ut + x_3}{\sqrt{1-u^2/c^2}}. \tag{6.7}$$

* A finite rotation in four dimensions is obtained in a manner similar to that in three dimensions. One exponentiates the matrix $\Lambda_{\lambda\mu}$ corresponding to the infinitesimal rotation $x'_\lambda = x_\lambda + \delta\Lambda_{\lambda\mu} x_\mu$ to obtain the finite rotation $x'_\lambda = e^{\Lambda_{\lambda\mu}} x_\mu$.

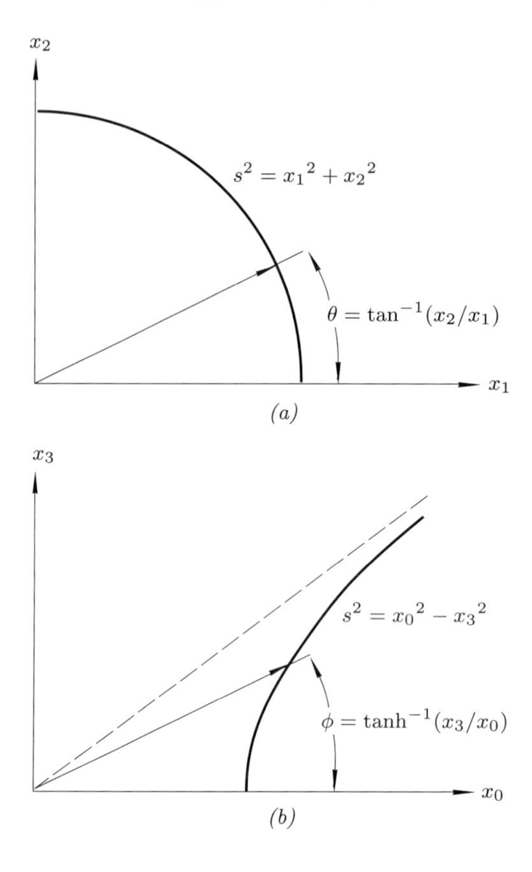

Figure 6-1. Euclidean and Hyperbolic Rotations. In the Euclidean rotation between two space coordinates *(a)*, the coordinates are related to the angle of rotation by $\tan\theta = x_2/x_1$. In the hyperbolic rotation between the space and time coordinates *(b)*, the coordinates are related to the angle of rotation by $\tanh\phi = x_3/x_0 = x_3/ct = u_3/c$.

These equations show what distressed H. A. Lorentz. According to the first of Eqs. (6.7), the time between events $\Delta t'$ experienced when one is at rest with respect to them is greater than the time Δt between the same events when one is in motion

$$\Delta t' = \Delta t/\sqrt{1 - u^2/c^2}.$$

Time passes more slowly in frames in motion relative to those which are fixed.

In the limit $u^2/c^2 \ll 1$ the Lorentz transformation (6.7) reduces to the Galilean transformation (6.2) and the velocity of light disappears:

$$x_3' = x_3 + ut, \qquad t' = t.$$

Relativistic Action

The classical action principle for relativistic motion follows that for Galilean relativity described in Chap. 3. It consists of the variation of the action due to the variations of the path and the variations due to the symmetry parameters. The action itself and its Lagrangian are different in relativistic motion revealing a profound geometric significance obscured in Galilean space–time.

The principle of least action in relativistic motion has a beautiful geometric form. The motion of a free particle takes place in such a way that its path from one event to another through four-dimensional space–time is the shortest possible; that is, its Lorentz interval $\int ds$ takes a shape in which its length is a minimum. Such minimum-length curves are known as geodesics. For example, the geodesics of a sphere are the great circles which lie upon its surface. The shortest distance between any two points on the surface of a sphere lies upon a great circle which intersects them.

The action principle reaches an epitome in relativistic space–time. It tells us that all motion is orchestrated by this simple law: in their movement between any two points in space–time particles take the shortest paths possible. In a deep sense, this is the only law of motion; for Einstein has shown that this principle is also true not only for a free particle but also for particles gravitationally and electrically interacting with one another. When interaction takes place, the metric of the space in which the path exists is no longer the simple Lorentz metric but is shaped by the presence of mass and energy.

The action is in essence the path length:

$$S = -mc \int ds.$$

The particle mass m and the speed of light c are coefficients which provide the proper dimensions. The negative sign is chosen because the interval minimizes S. The principle of least action with only path variations (symmetry parameter variations will be considered shortly) takes the form

$$\delta S_{\text{path}} = -mc\,\delta \int ds = -mc\,\delta \int \frac{ds}{dt}\,dt. \tag{6.8}$$

Since $S = \int L\,dt$, the Lagrangian for a free particle in relativistic motion is $L = dS/dt$ or

$$L = -mc\,ds/dt = -mc^2\sqrt{1 - \dot{x}^2/c^2}.$$

The momentum follows as

$$\mathbf{p} = \partial L/\partial \dot{\mathbf{x}} = m\dot{\mathbf{x}}/\sqrt{1 - \dot{x}^2/c^2}. \tag{6.9}$$

The total energy (or Hamiltonian) is given by $H = \mathbf{p} \cdot \dot{\mathbf{x}} - L$. The Lagrangian may be expressed in terms of the momenta through Eq. (6.9) as

$$L = -mc^2/\sqrt{1 + \mathbf{p}^2/m^2c^2},$$

where $\mathbf{p}^2$ denotes the magnitude of the three-momentum: $\mathbf{p}^2 = {p_1}^2 + {p_2}^2 + {p_3}^2$. The Hamiltonian then takes the form

$$H = c\sqrt{\mathbf{p}^2 + m^2c^2}. \tag{6.10}$$

The three components of the momentum and the energy may be gathered together as the four-momentum,

$$p_\lambda = (p_0, \mathbf{p}).$$

The fourth (or zeroth) component of the momentum turns out to be the bearer of the total energy of the particle $H = \mathcal{E}$. This can be seen by rearranging Eq. (6.10) in the form

$$(\mathcal{E}/c)^2 - {p_1}^2 - {p_2}^2 - {p_3}^2 = m^2c^2,$$

which corresponds to the metric condition for the four-momentum p_λ,

$${p_0}^2 - {p_1}^2 - {p_2}^2 - {p_3}^2 = p^2. \tag{6.11}$$

One finds the zeroth component of the four-momentum is fixed by the energy and its magnitude by the mass of the particle:

$$p_0 = \mathcal{E}/c, \qquad p = mc.$$

These same relationships with Eq. (6.9) show that the three-momentum is related to the three-velocity by

$$\mathbf{p} = (p_0/c)\dot{\mathbf{x}}. \tag{6.12}$$

The canonical coordinates of motion in Einsteinian relativity are the two four-vectors of position and momentum and there exists an intimate correspondence between the components and magnitudes of the two. Aside from factors of c, the zeroth component $x_0 = ct$ of the four-position is the time and its canonical mate is the energy $p_0 = \mathcal{E}/c$. The three-momenta $\mathbf{p}$ are the canonical mates of the space coordinates $\mathbf{x}$. The magnitude of the four-position is the Lorentz interval. It corresponds to the magnitude of the four-momentum which is the mass of the particle.

It is interesting to see the connection between the action as a path length in Einsteinian space–time (for which the Lagrangian is essentially the speed $-mc\,ds/dt$) and the Galilean action (for which the Lagrangian is $L = T - V$). In the Galilean limit $\dot{x}^2 \ll c^2$, the relativistic Lagrangian $L = -mc^2\sqrt{1 - \dot{x}^2/c^2}$ becomes

$$L = m\dot{x}^2/2 - mc^2$$

and is the same as the Lagrangian $L = T - V$ of a free particle in Galilean space–time where $T = m\dot{x}^2/2$ and the potential energy is $V = mc^2$. The condition that a particle minimize its Lorentz interval therefore reduces to the minimization of the Lagrangian $L = T - V$ in Galilean space–time.

When the action (6.8) is varied with respect to the symmetry parameters, invariants of the motion emerge. Translational symmetries for both space and time are now contained in a four-vector symmetry parameter, $\delta a_\lambda = (c\delta\tau, \delta\mathbf{a})$. The directional symmetries are now purely rotational and are contained in the symmetry parameters $\delta\boldsymbol{\Omega}$ and $\delta\boldsymbol{\phi} = \delta\mathbf{u}/c$.

The relativistic action when subjected to the symmetry variations is formally the same as that for the Galilean symmetries (**Note 1**),

$$\delta S_{\text{symmetries}} = -\Delta\boldsymbol{P}\cdot\delta\mathbf{a} - \Delta\boldsymbol{J}\cdot\delta\boldsymbol{\Omega} - \Delta\boldsymbol{N}\cdot\delta\mathbf{u} - \Delta H\delta\tau, \tag{6.13}$$

however there is a change in the meaning of the invariant $\boldsymbol{N}$ which corresponds to the velocity symmetry parameter $\boldsymbol{u}$. The conservation laws $\Delta() = 0$ lead to the ten invariants $\boldsymbol{P}$, $\boldsymbol{J}$, $\boldsymbol{N}$, and H. The four-vector $P_\lambda = \sum_\alpha p_{\lambda_\alpha}$ is the total four-momentum of the system. It consists of space and time components $P_\lambda = (P_0, \boldsymbol{P})$. The space components are the same as those for Galilean relativity whereas the time component, within factors of c, is the total energy

$$P_0 = \sum_\alpha p_{0_\alpha} = \mathcal{E}/c.$$

The angular momentum $\boldsymbol{J}$ is the same as that in Galilean relativity; but the mass-center invariant $\boldsymbol{N}$ is different. In Einsteinian relativity it is discovered to be more fundamentally the energy-center invariant. The center of energy $\boldsymbol{X}$ is defined by

$$\mathbf{X} = P_0{}^{-1}\sum_\alpha p_{0_\alpha}\mathbf{x}_\alpha.$$

The velocity of a particle is related to its momentum through its energy according to Eq. (6.12). This equation may be summed over all particles to become $P_0\dot{\boldsymbol{X}} = c\boldsymbol{P}$ and then integrated to give

$$\boldsymbol{N} = (\boldsymbol{P}X_0 - P_0\boldsymbol{X})/c, \tag{6.14}$$

where $X_0 = x_0 = ct$.

The relativistic invariant $\boldsymbol{N}$ is similar to the mass-center invariant $\boldsymbol{N}$ of Galilean relativity, except that the energy plays the role of the mass. The

Galilean limit occurs when $\mathbf{p}^2 \ll m^2c^2$ in which case $P_0 = \mathcal{E}/c \to mc$, and the energy-center invariant becomes the mass-center invariant

$$\boldsymbol{N} = (\boldsymbol{P}X_0 - P_0\boldsymbol{X})/c \to \boldsymbol{P}t - m\boldsymbol{X}.$$

Equivalence of Mass and Energy

The expression for the Hamiltonian or total energy (6.10) of a particle in relativistic motion is surprising in two ways. First, it is irrational. The Hamiltonian of a Newtonian particle is a rational function of the momentum: $H = p^2/2m$. The Hamiltonian of a relativistic particle depends upon the momentum through the radical $H = c\sqrt{\mathbf{p}^2 + m^2c^2}$. The rational Hamiltonian of a Newtonian particle occurs in the limit $\mathbf{p}^2 \ll m^2c^2$ for which expansion of the above radical gives

$$H \approx mc^2 + \mathbf{p}^2/2m.$$

Second, the relativistic Hamiltonian (6.10) introduces one of the most unanticipated connections ever to appear in physics. In the reference frame in which a free particle of energy $H = \mathcal{E}$ and mass m is at rest ($\boldsymbol{p} = 0$), one can immediately see that the particle possesses an internal energy contained by its mass given by a truly amazing relationship first discovered by Albert Einstein:

$$\mathcal{E} = mc^2.$$

The speed of light is invariant across all reference frames. This discovery of the abstract geometry of space and time in turn confronts us with an implication for the most concrete mechanical objects: *mass* and *energy* are equivalent.

The implications multiply. A uranium 235 nucleus is not equal to the mass of barium 141, krypton 92, and 2 neutrons, its fission products. That tiny difference in mass between the uranium nucleus and its fission products, about one-tenth the mass of a hydrogen atom, corresponds to 200 million electron-volts of energy. The ground state energy of the hydrogen atom, $|\mathcal{E}_0| = 13.6$ eV, is the energy scale of all chemical interactions. The nuclear energy scale is therefore ten million times the chemical energy scale. The destruction wrought by one pound of mass consumed in a nuclear reaction is equivalent to that induced by ten million pounds of chemical explosives. Nature opened to us the stunning ramifications of relativistic invariance for matter and motion. And with that revelation we built a bomb.

At this high peak on our journey through mechanics, we encounter the sad vista of physics, the mystery in matter comingled with mass destruction that is the hallmark of our age. This beautiful revelation of the inner behavior of nature was not affirmed with wonder and love of the natural world. It was used for an impious attack on the creation itself.

The Kepler–Einstein Coincidence

The relativistic space and time of Einstein possess four-dimensional rotational symmetry. Kepler motion in Galilean relativity also possesses four-dimensional rotational symmetry. The symmetries of these two quite different aspects of the physical world are the same, a source of some amazement.

The coincidence can be more clearly seen by examining the Poisson bracket algebra of the Lorentz invariants $\boldsymbol{J}$ and $\boldsymbol{N}$. The Poisson bracket in relativistic motion is built upon the four-vectors x_λ and p_λ:

$$[f, g] = \sum \left(\frac{\partial f}{\partial x_\lambda} \frac{\partial g}{\partial p_\lambda} - \frac{\partial g}{\partial x_\lambda} \frac{\partial f}{\partial p_\lambda} \right), \tag{6.15}$$

where it is understood that λ runs over all four coordinates and the sum is over all particles.

The Poisson bracket (6.15) shows that $\boldsymbol{J}$ and $\boldsymbol{N}$ have the relativistic algebra,

$$[J_i, J_j] = \epsilon_{ijk} J_k, \quad [N_i, J_j] = \epsilon_{ijk} N_k, \quad [N_i, N_j] = -\epsilon_{ijk} J_k / c^2, \tag{6.16}$$

which should be compared with the nonrelativistic algebra of the Kepler symmetries (3.48) in unbound motion for which the eccentrum becomes imaginary: ($\boldsymbol{E} \to i\boldsymbol{E}$):

$$[M_i, M_j] = \epsilon_{ijk} M_k, \quad [E_i, M_j] = \epsilon_{ijk} E_k, \quad [E_i, E_j] = -\epsilon_{ijk} M_k. \tag{4.10}$$

The Kepler–Einstein coincidence lies in the fact that, aside from the factor of $1/c^2$ in the third of Eqs. (6.16), the relativistic algebra is the same as that of the Kepler invariants $\boldsymbol{M}$ and $\boldsymbol{E}$ of Eqs. (4.10) for free motion with $\boldsymbol{M}$ playing the role of $\boldsymbol{J}$ and $\boldsymbol{E}$ playing the role of $\boldsymbol{N}$. The Poisson bracket $[N_i, N_j]$ vanishes identically for Galilean symmetry as shown in Table 3.51. However, for the Einsteinian symmetry it cyclically reproduces the angular momentum as shown in the third of Eqs. (6.16). Only in the limit $c^2 \to \infty$ does the relativistic algebra (6.16) give the Galilean result. This is because the rotational symmetry generated by velocity boosts which produces the invariant $\boldsymbol{N}$ approximates a translational symmetry in the Galilean limit.

It is important to observe that Eqs. (6.16) do not match the Kepler algebra for bound orbits for which the four-dimensional rotations are all real and the sign of the $[E_i, E_j]$ bracket switches. Imaginary rotations enter when the motion is free for which the Kepler constant, the eccentrum, and the angles of rotations involving these coordinates become imaginary. The

Kepler motion phase space in this case is non-Euclidean as is Einsteinian space–time.

The Kepler–Einstein coincidence does not imply a causal connection between relativity theory and Kepler motion. This is made clear both by the fact that there is no correspondence for bound orbits and, more importantly, by the fact that relativistic Kepler motion, in contrast to Galilean motion, *does not* possess four-dimensional rotational symmetry at all. We are going to shortly find that the invariance of the eccentricity is broken in relativistic motion and the symmetry algebra (4.10) does not exist. The coincidence of the symmetries rather reflects the many ways in which the same mathematical forms appear within the diversity of the physical world, an instance of what V. I. Arnol'd has called "the mysterious unity of all things."*

Relativistic Kepler Motion

As we move onward to describe the relativistic regime with two-body motion we shall be rewarded with stunning success followed by inescapable failure. In the relativistic crisis of two-body motion the two-body description of electricity will successfully reveal significant relativistic structure in the hydrogen atom. But the description of two-body gravitational motion will be flawed from the start. In the relativistic regime electricity and gravity are revealed to no longer be the same kinds of forces. Both the two-body problem for gravity and electricity will ultimately lead to contradiction.

Begin with a simple measure of the importance of the velocity of light. The maximum velocity in bound Kepler motion is the orbital speed of the ground state, $v_\phi = M/\mu r_0 = k/M$ where $r_0 = M^2/k\mu$ is the radius of the ground state orbit. The significance of the speed of light in classical Kepler motion can therefore be summed up in the parameter

$$\beta = v_\phi/c = k/Mc$$

which is the ratio of the maximum speed attainable in two body motion to the speed of light.

In quantum Kepler motion the ground state quantities may be constructed from the same classical quantities by replacing the angular momentum M by $\hbar$. The importance of the speed of light in the hydrogen atom is summed up in the parameter

$$\alpha = k/\hbar c \approx 1/137$$

which is the ratio of the maximum speed of the electron in the hydrogen atom to the speed of light. This natural physical constant is known as the

* V. I. Arnol'd, *Catastrophe Theory*, Third Edition, Chap. 16, Springer-Verlag, Berlin (1992).

fine-structure constant. The importance of the speed of light in the Kepler motion of the atom is a universal constant of nature, a fixed, small number.

Nature demands that relativistic effects in the atom be small. On the other hand, since β need not be a small quantity, nature has no such demand on the heavens. To the contrary, relativistic gravitational effects are the *sine qua non* of the large-scale cosmos. The difference between β and α, like that between h and $\hbar$, is a reflection of the decisive difference between the heavens and the elements.

A description of relativistic Kepler motion is formed from the union of the relativistic Hamiltonian for a free particle in the center of mass frame of motion and the inverse-square force potential $V = -k/r$ thereby creating the Hamiltonian and Hamilton-Jacobi equation for the relative motion:

$$\mathcal{H} = c\sqrt{\boldsymbol{p}^2 + \mu^2 c^2} - k/r = \mathcal{E}. \tag{6.17}$$

The corresponding quantum-mechanical energy operator for relativistic motion is obtained by substituting $\hat{p_i}^2 = -\hbar^2 \partial^2/\partial x_i^2$ for the squared momentum $\boldsymbol{p}^2$ in Eq. (6.17):

$$\hat{\mathcal{H}} = c\sqrt{-\hbar^2 \partial^2/\partial x_i^2 + \mu^2 c^2} - k/r. \tag{6.18}$$

These relativistic Kepler Hamiltonians embody a naked contradiction: mass and momenta are Lorentz invariant but the potential $V = -k/r$ is not.

The irrational operator $\sqrt{-\hbar^2 \partial^2/\partial x_i^2 + \mu^2 c^2}$ in Eq. (6.18) plunges relativistic quantum motion into ambiguity. The artistic openness of quantum mechanics is again before us as it was before P. A. M. Dirac who first puzzled over the operator (6.18).

What is one to make of the square-root of an energy operator with both positive and negative energy states? The ambiguity of sign with which the square-root presents us became an imaginative opening for Dirac. After a difficult struggle with several false steps, he ultimately interpreted the two possible signs of the square-root as signatures of two different particles: the electron and its antiparticle, the positron. The particle and its antiparticle are identical in mass but opposite in charge.

Dirac sought an equivalent set of operators that reproduce Eq. (6.18) but which were rational and linear. He found that such a set of operators could be conceived if they were four-by-four matrices obeying simple commutation rules. Moreover, the wave function Ψ to which the operators are applied should be conceived as a four-valued quantity, consistent with the four dimensions of space–time. Finally, the Dirac operators which imply Eq. (6.18) necessarily contain internal states which represent an internal angular momentum of the particle: the spin. However, these spin states possess only two eigenvalues, $\pm\hbar/2$.

In a brilliant stroke of disambiguation of the square-root of an operator Dirac was led to the discovery of antiparticles and the necessity of spin—both pillars of the microscopic world of elementary particles. The four components of the wave function can be thought of as representing two distinct particles—electron and positron—each with two spin components for a total of four components.

Discovery is an erratic and perplexing business. Dirac's ingenious revelation of the four-valued wave function upon which relativistic quantum operators act was so stunning that it could not be immediately comprehended. Two of the waves are associated with positive energies; but two correspond to negative energies. Both Dirac and his colleagues working on quantum mechanics in the late 1920's were initially devastated by the appearance of the negative energy states which they felt marred this otherwise beautiful theory. Dirac's first reaction was to declare them of no physical significance. Werner Heisenberg wrote Wolfgang Pauli in 1928 that "the saddest chapter of modern physics is and remains the Dirac theory."

Then, bit by bit, Dirac accepted his new child completely. He embraced the reality of the negative energy states which the mathematics proclaimed. But the struggle was not over. Dirac proposed that these negative energy states were *protons* (protons at that time were in want of theoretical explanation and the idea of whole new sets of antiparticles in nature was still too revolutionary). Dirac's proton proposal also blighted the theory; for the symmetry of the wave function requires that the negative energy states have the same mass as the positive energy states and the proton was empirically known to be a thousand times more massive than the electron. Finally, reassured by the experimental measurements of uniquely new particles with the electron's mass but opposite charge by Carl Anderson, the electron's antiparticle—the positron—was accepted. The physics community ushered in the inevitability of antiparticles as a deep aspect of nature.

Let us now draw out the orbits of relativistic, two-body motion. The relative energy $\mathcal{H}$ and the angular momentum $\boldsymbol{M}$ persist as invariants in the relativistic regime; hence the pair of vector Casimir sets

$$(\mathcal{H}, M, M_1), \qquad (\mathcal{H}, M, M_3),$$

continue to be appropriate sets of invariants in the relativistic regime and the motion will separate in their spherical-polar symmetry coordinates. But the third Casimir set contributed by the eccentricity invariant no longer exists because the eccentricity is no longer an invariant in the relativistic case and the motion no longer separates in parabolic coordinates. The portrait may only be drawn in spherical-polar coordinates.

Write $\boldsymbol{p}^2 = p_r^2 + M^2/r^2$ just as in nonrelativistic two-body motion. The Hamilton–Jacobi equation (6.17) can then be written as

$$(\mathcal{E} + k/r)^2 - c^2(p_r^2 + M^2/r^2 + \mu^2 c^2) = 0 \tag{6.19}$$

and may be compared with Eq. (4.33), its nonrelativistic counterpart. Relativistic motion forces us to rethink our notion of the total energy. It tells us that the energy of material bodies is enormous, even if their kinetic energy is slight; for the leading contribution to the energy is the rest energy μc^2. The formerly zero level of energy in Galilean relativity is now μc^2, the rest energy of the relative motion. Examination of Eq. (6.19) reveals that for $\mathcal{E} < \mu c^2$ the motion is bound, librating in the radial direction between the points $r_{\rm min}$ and $r_{\rm max}$ (the points where $p_r = 0$). There is a minimum or ground state energy $\mathcal{E}_0$ given by

$$\mathcal{E}_0 = \sqrt{1-\beta^2}\mu c^2 = \frac{2\sqrt{1-\beta^2}}{\beta^2}\left(k^2\mu/2M^2\right),$$

for which the orbit is a circle of radius

$$r_0 = \sqrt{1-\beta^2}\left(M^2/k\mu\right).$$

These results should be compared with the nonrelativistic ground state energy and orbit radius given by Eqs. (4.47) and (4.48) which are $\mathcal{E}_0 = -k^2\mu/2M^2$ and $r_0 = M^2/k\mu$. For $\beta \ll 1$, the expression for the ground state energy may be expanded as

$$\mathcal{E}_0 = \mu c^2 - \left(1 - \tfrac{1}{4}\beta^2 + \cdots\right)\left(k^2\mu/2M^2\right)$$

and, with rest energy accounted for, reduces to Eqs. (4.47) in the limit $\beta \to 0$. It can be seen that relativistic effects shift the ground state energy level upward and contract the radius of the ground state orbit.

The total action is $S(r,\phi,t) = -\mathcal{E}t + S_r(r) + S_\phi(\phi)$. The angular momentum $p_\phi = M$ is an invariant just as in nonrelativistic two-body motion. The form of the polar action

$$S_\phi(\phi) = M\phi$$

is also unchanged. The radial momentum is taken from Eq. (6.19) to construct the relativistic radial action $S_r(r) = \int p_r\, dr$:

$$S_r(r) = \int \sqrt{\left(\mathcal{E} + k/r\right)^2/\mu c^2 - M^2/r^2 - \mu^2 c^2}\, dr.$$

The trajectory equation $\partial S/\partial M = \partial S_r/\partial M + \phi = const$ leads to

$$\phi = \int \frac{M\, dr}{r^2\sqrt{\left(\mathcal{E} + k/r\right)^2/c^2 - M^2/r^2 - \mu^2 c^2}}.$$

These trajectories may be compared with their nonrelativistic counterparts in Eqs. (4.50). The integral is of similar structure to that of Eq. (4.50) involving a radical of inverse first and second powers of r. It is therefore expressible in trigonometric functions and the orbit is the *precessing* ellipse,

$$r(\phi) = \frac{r_*}{1 + e\cos[\sqrt{(1-\beta^2)}\phi]}, \tag{6.20}$$

shown in Fig. 6-2.

The libration points have a slightly different dependence on the invariants than in nonrelativistic motion. This can be seen by comparing the relativistic orbit (6.20) with the nonrelativistic orbit (4.52). In the nonrelativistic case the counterpart of the latus rectum r_* is $r_0 = M^2/k\mu$ in Eq. (4.52) and depends only upon the angular momentum; but the latus rectum of the relativistic orbit also depends upon the energy:

$$r_* = \frac{\mathcal{E}_0}{\mathcal{E}} r_0.$$

The pericentron and apocentron are given in terms of the latus rectum and eccentricity by

$$r_{\min} = r_*/(1+e), \qquad r_{\max} = r_*/(1-e),$$

and the eccentricity has the relativistic form

$$e^2 = \frac{(\mu c^2)^2}{\mathcal{E}^2}\left[1 - \frac{\mathcal{E}^2 - (\mu c^2)^2}{\mathcal{E}_0^2 - (\mu c^2)^2}\right],$$

where $\mathcal{E}_0$ is the relativistic ground state energy. In the limit $\beta \ll 1$, the ground state quantities, libration points, and eccentricity are identical to those for nonrelativistic motion when the rest energy level μc^2 is accounted for.*

The precessing ellipse (6.20) is not generally a closed orbit. It is an ellipse which rotates about the focus, forming an unending succession of rosettes for which the path is not retraced. When ϕ completes a cycle of 2π, the argument of $\cos[\sqrt{(1-\beta^2)}\phi]$ completes a cycle of $\sqrt{(1-\beta^2)}2\pi$. The precession of the angle for each cycle is

$$\Delta\phi = [1 - \sqrt{(1-\beta^2)}]2\pi.$$

---

* For $\beta \ll 1$, both $\mathcal{E}$ and $\mathcal{E}_0$ are dominated by the rest energy: $\mathcal{E} = \mu c^2 + \cdots$. The square of $\mathcal{E}$ has the expansion

$$\mathcal{E}^2 = (\mu c^2)^2 + 2\mu c^2(\mathcal{E} - \mu c^2) + \cdots,$$

with a similar expression for $\mathcal{E}_0^2$. Hence, as $\beta \to 0$ one finds $r_* \to r_0$ and

$$e^2 \to 1 - \frac{(\mathcal{E} - \mu c^2)}{(\mathcal{E}_0 - \mu c^2)},$$

which should be compared with the second of Eqs. (4.48).

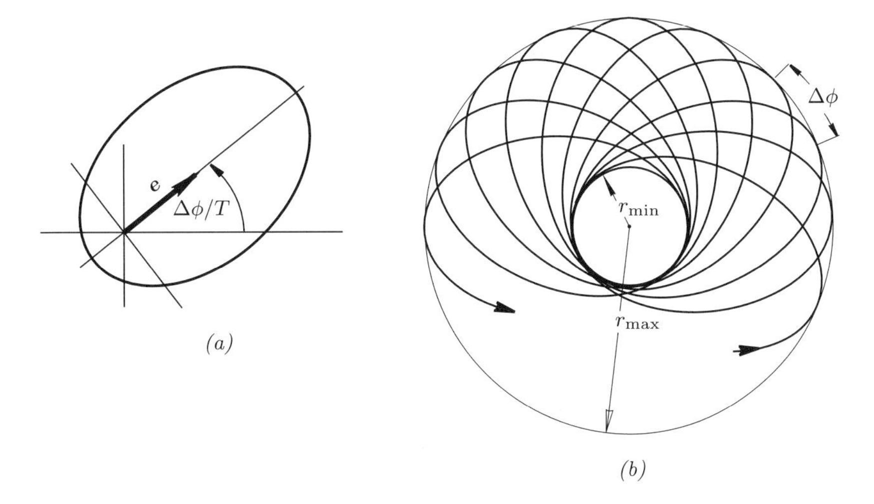

Figure 6-2. The Precessing Elliptical Orbit of Relativistic Motion. The eccentricity vector, drawn from the focus to the center *(a)*, is no longer invariant as in nonrelativistic motion. It spins at the precession rate $\Delta\phi/T$ about the focus of the ellipse where T is the period of the orbit. The orbit is generally open *(b)*; as time passes, it densely fills the annulus. The precession for each cycle is $\Delta\phi = [1 - \sqrt{(1-\beta^2)}]2\pi$ or for $\beta^2 \ll 1$, $\Delta\phi = \pi\beta^2$.

If the orbit is to close upon itself, the precession angle $\Delta\phi$ must be a rational fraction of 2π [$\Delta\phi = (l/m)2\pi$ where l, m are integers]. This may only occur for special values of β. For $\beta^2 \ll 1$, the precession angle is approximately given by

$$\Delta\phi = \pi\beta^2. \tag{6.21}$$

The eccentricity vector, the hidden invariant of nonrelativistic, two-body motion, is no longer an invariant. Consistent with an open orbit, there is no final isolating invariant for relativistic, two-body motion. The eccentricity vector, anchored to the ellipse by tip at the center and by tail at the focus, now spins at the precession rate $\Delta\phi/T$ where T is the period of the orbit.

The open orbit of a precessing ellipse is the classical manifestation of the breaking of the closed-orbit degeneracy by the Einsteinian symmetry of space–time. This degeneracy should also be relativistically broken on the quantum level resulting in uniquely different eigenvalues of the energy for each state of angular momentum. If the degeneracy were intact, all states of angular momentum would possess the same eigenvalue of the energy.

Relativistic, two-body, quantum motion described by the Hamiltonian operator (6.18) or, more precisely, Dirac's version of it, reveals that the degenerate energy levels of the hydrogen atom are indeed broken by relativistic effects. A given energy level is split into a bundle of nearly identical levels as shown in Fig. 6-3. Since these split levels result from relativistic effects, they must be dependent on the parameter summarizing those effects, the fine-structure constant α. The split energy levels are separated by gaps proportional to α^2. The split levels differ from one another by small amounts visible in the spectral pattern of radiation emitted by the atoms: about $\alpha^2 \approx 1/20,000$ of the principal energy level $\mathcal{E} = (-k^2\mu/2\hbar^2)/n^2$. These finely split levels correspond to different values of the angular momentum. The existence of different energy levels for each angular momentum quantum number l is the sign that the inverse-square force degeneracy of the hydrogen atom is broken by relativistic effects.

Arnold Sommerfeld, at a time in which a rigorous formulation of quantum mechanics was not yet fully born, extended the idea of Bohr that classical motions described by trajectories could be transformed into quantum motions by requiring that the trajectory support an integral number of particle waves. Sommerfeld generalized Bohr's idea by requiring that the actions generated in one period $J_i = \oint p_i dq_i$ be integral multiples of $\hbar$. In this way the actions of classical motions could be directly transformed into quantum-mechanical eigenvalues.

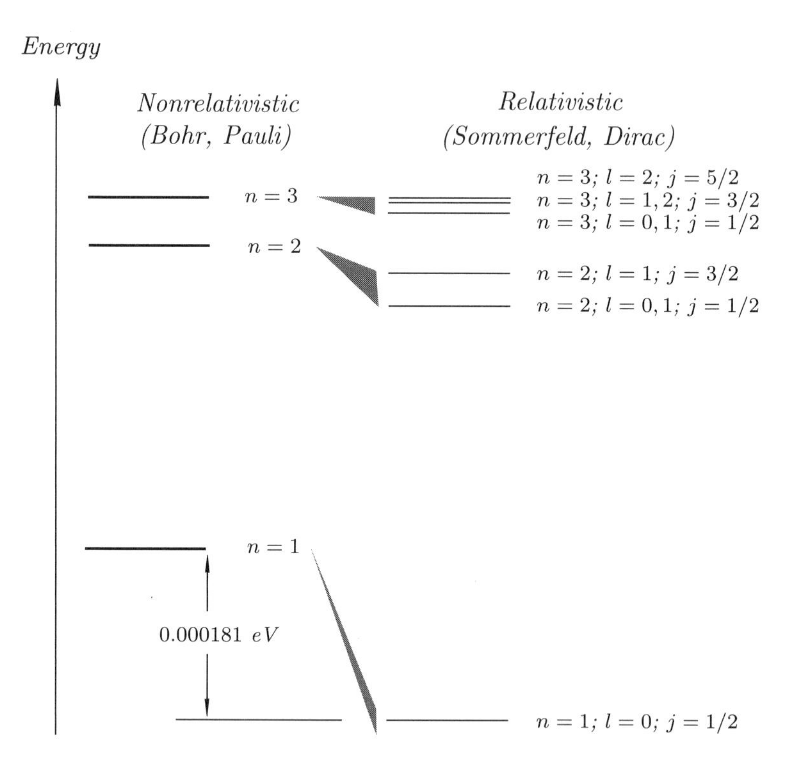

Figure 6-3. Relativistic Effects on Hydrogen Atom Energy Levels. To the left are the nonrelativistic energy levels postulated by Bohr and confirmed by Pauli. The split energy level shifts due to relativistic effects yielded by the Sommerfeld and Dirac theories are exaggerated by the factor α^2. The quantum number j which is the sum of the spin and orbital angular momentum quantum numbers is related to the quantum number l as $j = l \pm 1/2$ except for $l = 0$ when $j = 1/2$.

For the relativistic, two-body problem whose solution is a precessing ellipse, Sommerfeld's rule yields the energy levels

$$\mathcal{E} = -\frac{(k^2\mu/2\hbar^2)}{n^2}\left[1 + \frac{\alpha^2}{n}\left(\frac{1}{j} - \frac{3}{4n}\right)\right],$$

where j is the angular momentum quantum number.

Sommerfeld's calculation was surely inspired. The solution of the Dirac equation for the hydrogen atom, which inherently introduces spin, confirm Sommerfeld's result which knew nothing of spin. They show that the total angular momentum $\boldsymbol{J}$ of the atom is no longer just the orbital angular momentum $\mathbf{M}$. It is now given by

$$\boldsymbol{J} = \boldsymbol{M} + \boldsymbol{S}$$

and includes the intrinsic spin angular momentum of the electron $\boldsymbol{S}$. The eigenvalues of the spin $\boldsymbol{S}$ are $S = \pm\hbar/2$ so the total angular momentum eigenvalues are $j = l \pm 1/2$ for all $n > 1$ and $j = 1/2$ for $n = 1$. The only appearance of spin in the Sommerfeld calculation is in the interpretation of the quantum number j.

From an uncertain foundation in classical mechanics ignorant of spin, Sommerfeld built a structure to the exact result yielded by the Dirac equation, an equation built upon spin. This happy coincidence rests delicately upon a cancellation of energy shifts due to the coupling of the electron's internal spin with its orbital motion on the one hand and spin–spin coupling contained within the Dirac equation on the other. The Sommerfeld solution emerges as the classical survivor of quantum cancellations.

There are further splittings of the energy levels of the hydrogen atom unaccounted for in the Dirac equation which are proportional to higher powers of the fine-structure constant. These effects as well as the interactions of charged particles at high energy cannot be described by two-body interactions. Ultimately, two-body motion becomes a contradiction for high-energy interactions of electrons and positrons at relativistic speeds.

The Limits of Two-Body Motion

Relativistic, two-body motion illuminates much of the fine-structure of the hydrogen atom. Its success is reaped from the small value of the fine-structure constant. Let us now turn to the heavens and the gravitational two-body problem to see if the relativistic two-body problem illuminates and explains fine-structure in the heavens.

Mercury, the smallest planet, exhibits significant precession of its elliptical orbit. Most of this precession can be explained by nonrelativistic perturbations induced by Mercury's giant companions, Jupiter and Saturn.

But there is a stubborn discrepancy not accounted for by these planetary perturbations. The measured rate of precession of Mercury's ellipse is 574 sec of arc per century. Detailed nonrelativistic calculations of the perturbations of Mercury by the giant planets give a precession rate of 531 sec of arc per century. Somewhat over 40 sec of arc are missing and a search for those 40 sec of arc progressed through most of the late nineteenth and early twentieth centuries.

Does relativistic, two-body motion described by the Hamiltonian (6.17) have that additional 40 sec of arc? It does not. When one puts the numbers for Mercury's orbit into the precession, Eq. (6.21), only about 1/6 the required deficit is to be had. Relativistic, two-body motion yields the right dynamical structure but the wrong precession rate. Though its quantum counterpart well represents much of the fine electrical structure of the atom, the relativistic, two-body problem is an unreliable guide to the fine gravitational structure of the heavens. The contradiction contained in the two-body relativistic Hamiltonian (6.17) is a fundamental contradiction for gravity.

The relativistic two-body problem for gravity, in contrast to it's electrical counterpart, fails immediately as the general relativity theory of Einstein reveals. In the relativistic regime gravity and electricity are no longer structurally identical kinds of forces. Though both become inverse-square forces in the nonrelativistic limit, the electrical force ultimately has a vector character while the gravitational force has a tensor character, that is, its structure is like that of the product of two vectors.

Gravity requires a deepening of the idea of Lorentz invariance. Albert Einstein beautifully imagined such an extension of Lorentz invariance in his theory of general relativity in which gravity appears as the geometric response of space and time themselves to the presence of mass and energy. Indeed, the prediction of general relativity for the precession of the ellipse in the limit $\beta^2 \ll 1$ is

$$\Delta\phi = 6\pi\beta^2$$

and is to be compared with Eq. (6.21). Einstein's general relativity result yields an uncanny 43 sec of arc for the relativistic contribution to the precession of Mercury's ellipse each century.

The relativistic crisis of two-body motion is inescapable. The finite speed of propagation of forces and two-body motion inherently contradict one another. When one body suffers a change in position or velocity it transfers energy and momentum which cannot reach a second body until after a finite interval—the Lorentz interval. The energy and momentum have left the first body but they have yet to reach the second. Where do they reside? In additional bodies which are the bearers of the force. The energy and momentum which have left the first body but have yet to reach the second body are carried by force-bearing particles.

The particle which is the bearer of the force in the case of the electrical force is the photon. The force-bearing particle in the case of the gravitational force is the graviton. In the particle representation of a force a particle which interacts gravitationally with another emits a graviton which propagates at the speed of light to the other where it is absorbed. A similar picture exists for the interaction of two charged particles such as the proton in the atomic nucleus and an electron where the particle emitted and absorbed is the photon. For the motion of an electron about a charged nucleus the electrical force can be understood as a steady stream of virtual photons which are being continuously emitted and absorbed by the electron and the charged nucleus.

Reprise

Relativistic motion always involves more than two bodies. In the simplest motion of two charged or gravitating bodies a third entity—the force bearing particle—must necessarily be considered under relativistic conditions. The minimum number of bodies involved might seem to be three—two charged particles and one photon in the case of electricity or two mass particles and one graviton in the case of gravity. But this is not the case. Streams of unlimited numbers of force-bearing particles are required to represent the persistent nature of force. All interactions involve infinitely many photons or gravitons. Both the hydrogen atom and relativistically orbiting heavenly bodies are infinitely many-body motions. The split levels of Fig. 6-3 in the exact theory of quantum electrodynamics (which goes well beyond the Sommerfeld and Dirac theories) are actually split into even finer and finer sublevels. Each line is again another bundle of infinitely many fine lines.

At relativistic speeds quantum wave functions represent both particles and their antiparticles. At high particle kinetic energies $\mathbf{p}^2/2m \approx mc^2$, where m is the mass of a fundamental particle such as an electron, a plethora of particles are spontaneously created out of the vacuum. This is because the vacuum is not an abyss of nonexistence. Rather, it is an energetic sea of elementary particles and antiparticles spontaneously created and annihilated in a dynamical equilibrium in which the expectation value of all particles vanishes.

A high-energy particle projected into the vacuum, the primordial sea of particles and antiparticles, excites a rich variety of interactions which bring forth an array of elementary particles. The relativistic motion of all elementary particles is, from their source in the vacuum, a rich and complex many-body motion.

Chapter 7

The Manifold Universe

Relativistic speeds turn motion into an inherently many-body problem. But most motions at the human and planetary scale are not relativistic and therefore not inherently many-body. Solar system motions of macroscopic bodies including the atmosphere and oceans of a planet like the earth are well described nonrelativistically according to the mechanics of Galilean relativity. At the scale of the atom, the orbital speed of an electron is only a small fraction of the speed of light. The states of the hydrogen atom are well predicted by the nonrelativistic laws of quantum motion.

Yet even in this Galilean limit the isolated interaction of two bodies is an idealization, closely approximating many motions—including that of our planetary system and the hydrogen atom—but far removed from a description of others—such as the atmosphere and oceans of the earth and the complex molecules of life. Exact two-body motions are only an approximation because the world consists of more than two bodies. The fluids of the earth, formed of enormous numbers of atoms and molecules interacting with one another, sustain motion incomparably different from that of two bodies. Even on the level of Galilean mechanics two-body motion can carry us no farther.

The passage beyond the two-body problem to the many-body problem opens up qualitatively new features of motion. These features are inherent in the Hamiltonian mechanics flowing from the principle of least action; but it is impossible for them to appear in the motion of only two bodies bound by electricity and gravity. When more than two bodies interact classically they develop resonances; and resonances inherently harbor instabilities.

A resonance can be understood in the following way. A many-body motion may be regarded as a set of independent two-body motions which are allowed to interact. The integrable two-body motions, viewed in angle-action coordinates, consist of a set of independent oscillating modes, each oscillator possessing a characteristic frequency. In integrable motion the oscillators are independent and their amplitudes neither shrink nor grow; they are the invariants I of the motion.

When the independent modes are allowed to interact, the oscillators perturb one another and their amplitudes and frequencies change. Moreover, if one of the modes perturbs another at the same frequency and phase, a resonance develops in which a mode grows in amplitude much in the manner in which a swing is pumped in amplitude by pushing at the same frequency as the swing frequency. The amplitude of a resonant oscillator grows without bound: it is said to be unstable.

Many-body motion is rich in interacting oscillator modes which are free to closely approach one another in frequency thereby exciting a resonance, a situation impossible in the motion of only two bodies. An unstable resonance persists until it becomes strong enough to change either its frequency or the frequency of the perturbation thereby "detuning" the instability.

Resonances and their accompanying instability manifest themselves in phase space by creating a chaotic phase flow. The streamlines of a flow with unstable resonances form a complex web with tangled regions of disorder. This tangled disorder in the streamlines sends the trajectories emanating from neighboring points off on wildly different directions. Chaotic phase flows are therefore said to exhibit "sensitive dependence on initial conditions." In a chaotic region of phase space neighboring points, considered as initial conditions, exist which, no matter how close to one another they may be brought, belong to trajectories that diverge by uncontrollably large amounts. Resonances, instability, and sensitive dependence on initial conditions, are all manifestations of the same underlying phenomena in many-body motion perhaps best captured by the single word "chaos."

One can gain some understanding of a chaotic motion by placing it in context with motion at the opposite end of the spectrum—"cosmic" motion also known as regular motion. The most cosmic motion is one so highly structured by symmetry that it is both integrable and fully degenerate (it is separable and all its separable parts are in synchrony). Such is the motion of two celestial bodies. In analogy to the flow of a fluid, the orderly flow of two celestial bodies is laminar. Its streamlines are closed on a smooth torus. They exhibit regularity and high degree of order.

As one proceeds along the spectrum to motions of greater disorder, one next encounters quasiperiodic motions. These motions are integrable but nondegenerate. Since they are no longer in synchrony, they give the appearance of being chaotic; but their chaos is illusory because integrable motions possess symmetry sufficient to prevent the separate modes from interacting. Quasiperiodic motion cannot be resonantly excited. The apparent irregularity arises solely from the lack of synchrony between the various modes. The flows of such motions appear not to exhibit regularity (they densely fill the torus upon which they lie); but their streamlines have a highly ordered structure (they lie upon an invariant torus) and they do not exhibit sensitive dependence on initial conditions. They are therefore not chaotic.

Finally one encounters motion which is nonintegrable and hence cannot even be quasiperiodic. The symmetries of these motions are so broken they do not inhibit the separate coordinate modes from interacting. The motions are therefore subject to resonances. Their flows are not only irregular; the orderly patterns of the streamlines typical of integrable motion are lost. The streamlines, no longer confined to noninteracting tori, are tangled in a complex web, and the flow exhibits sensitive dependence on initial conditions. These are the essential characteristics of chaotic motion.

Contrary to the laminar flow of two celestial bodies, the flow of nonintegrable motion is generally turbulent like a storm-driven ocean. Two-body motion with inverse-square forces cannot be chaotic. But enter a third or more bodies and the motion is open to chaos.

Beyond Two Bodies: Chaos

The final encounter two-body motion offers us is a glimpse into the rich complexity of many-body motion. Any inquiry into the nature of many-body motion begins with a sobering theorem formulated at the end of the nineteenth century by Henri Poincaré and others which declares that all many-body motions from the three-body problem on *are not* integrable.

Poincaré found that the flows of many-body motions cannot be represented by transcendental functions (infinite series of algebraic polynomials), the most general functions of classical mathematical analysis. He showed that these series diverge for many-body motions. Although the trajectory flow cannot be represented as a transcendental function, it can be numerically computed. But even here deep questions about the precision and proper interpretation of the computed flow arise because of the sensitive dependence on initial conditions.

Although the many-body problem is not integrable, Poincaré discovered that the most significant properties of many-body motion could be revealed as *perturbations* of integrable two-body motions. In fact, most of the crucial and interesting properties of many-body motions are those which are close to integrable motions. These perturbed motions are called *near-integrable* motions.

An example of a near-integrable motion is that of our own solar system. Each planet and the sun, considered in isolation, constitute an integrable two-body motion. Since the sun is so much more massive than any single planet, the effects of the planets on one another's two-body interaction with the sun may be regarded as a small perturbation. The resulting motion is a near-integrable motion.

A second example is the hydrogen atom in the presence of a magnetic field. The electron not only interacts with the nucleus; it also interacts with the magnetic field. This is a many-body interaction in which the magnetic field represents the effects of numerous electrons flowing in the magnet circuits. Rather than consider the hydrogen electron interacting with each of these electrons, one considers the magnet electrons organized into an ensemble which generates the field. The interaction is then reduced to the interaction between the magnetic field, the hydrogen atom electron, and the nucleus. The motion may again be represented as a near-integrable motion about the integrable two-body states of the hydrogen atom.

Viewed in phase space, one has a flow upon the invariant tori of the integrable two-body motions. These are described by the action variables $J = (J_1, J_2, \ldots, J_s)$ and their canonical mates, the angle variables $\alpha = (\alpha_1, \alpha_2, \ldots, \alpha_s)$. When allowed to interact as many-bodies, the actions J are no longer invariant and the tori are perturbed by one another's presence.

The action $S(\alpha, J)$ is the ultimate description of near-integrable motion. If one knows the action, one can find all the other dynamical variables. One begins with a known integrable motion with Hamiltonian $H_0(J)$ and known action $S_0 = \alpha_j J_j$. It is to be the basis of an approximation of the motion whose Hamiltonian differs from it by a small perturbation $H'(\alpha, J)$:

$$H(\alpha, J) = H_0(J) + H'(\alpha, J).$$

The action of the many-body motion is

$$S(\alpha, \bar{J}) = \alpha_j \bar{J}_j + S'(\alpha, \bar{J}),$$

where $\bar{J}$ are new action coordinates (which differ from J since the $\bar{J}$ are no longer invariant after the perturbation is made) and $S'(\alpha, \bar{J})$ is the response of the action to the Hamiltonian perturbation $H'(\alpha, \bar{J})$. Since the Hamiltonian is specified by the actual configuration of the many-bodies, the perturbation Hamiltonian $H'(\alpha, \bar{J})$ is a known function. The question is therefore this: what is the action perturbation $S'(\alpha, \bar{J})$ induced by the Hamiltonian perturbation $H'(\alpha, \bar{J})$?

Poincaré* found that one could obtain the crucial answers to this question in a powerfully generic manner. The actual flow clearly depends upon the particular form of the Hamiltonian, that is, the number of bodies and the particular form of the force law through which they interact. But it is a result of no small significance that the generic properties of the motion are independent of the particular form of the Hamiltonian. Poincaré's method which has come to be known as canonical perturbation theory is summarized in **Note 8**.

Poincaré makes a near-integrable motion integrable by constructing a canonical transformation which takes the motion to new coordinates $(\bar{\alpha}, \bar{J})$ in which the new Hamiltonian is a function only of the action coordinates, $\bar{H} = \bar{H}(\bar{J})$ only. This is accomplished by averaging the Hamiltonian over the angle variables (**Note 8**). Averaging over the angle variables is the

---

* Poincaré, H., *Les Methods nouvelles de la méchanicque céleste*, vols. 1–3, Gauthiers-Villars, Paris (1892–99); also available as a reprint from Dover Books, New York (1957). Poincaré's method has come to be formalized as canonical perturbation theory and is presented in standard works, for example, Chap. 2 of A. J. Lichtenberg and M. A. Lieberman, *Regular and Stochastic Motion*, Springer-Verlag, New York (1992).

key to the solution of the motion which is otherwise nonintegrable, a procedure whose justification, if not validity, rests upon the smallness of the perturbation.

The generating function of this canonical transformation is the action $S(\alpha, \bar{J})$. It consists of the identity transformation $\alpha_i \bar{J}_i$ and a perturbation induced by the many-body interaction $S'(\alpha, \bar{J})$. The Hamiltonian under the averaging procedure is mandated to have the form

$$\bar{H}(\bar{J}) = H_0 + \langle H'(\bar{\alpha}, \bar{J}) \rangle,$$

where $\langle\,\rangle$ indicates an average over all the angle coordinates. This is the crucial step which makes near-integrable motion "integrable." The Hamiltonian may be expanded in a Taylor's series in the deviations of the new actions from the old, $J' \equiv \bar{J} - J$,

$$\bar{H}(\bar{J}) \approx H_0 + \frac{\partial H_0}{\partial J} J' + H'(\alpha, J).$$

Since $\omega \equiv \partial H_0/\partial J$ and $J' = \partial S'/\partial\alpha$ according to the first of Eqs. (3.30), comparison of these two expressions for $\bar{H}(\bar{J})$ reveals that the perturbed action S' is directly related to the Hamiltonian deviation from the average

$$\Delta H'(\bar{\alpha}, \bar{J}) \equiv H'(\bar{\alpha}, \bar{J}) - \langle H'(\bar{\alpha}, \bar{J}) \rangle,$$

as

$$\omega \frac{\partial}{\partial \bar{\alpha}} S'(\bar{\alpha}, \bar{J}) = -\Delta H'(\bar{\alpha}, \bar{J}). \tag{7.1}$$

To this order of approximation it is permissible to replace α by $\bar{\alpha}$ in $S'(\alpha, \bar{J})$ and $H'(\alpha, \bar{J})$. [A more detailed exposition of the steps leading to Eq. (7.1) may be found in **Note 8**.]

For s pairs of conjugate position and momentum coordinates, the angular dependence of the action perturbations may be represented as an s-fold Fourier series in the angle coordinates with Fourier coefficients which depend upon the action coordinates

$$S'(\bar{\alpha}, \bar{J}) = \sum_m S'_m(\bar{J}) e^{i(\bar{\alpha}_1 m_1 + \bar{\alpha}_2 m_2 + \cdots + \bar{\alpha}_s m_s)}.$$

The quantities $m = (m_1, m_2, \ldots, m_s)$ take all the positive and negative values of the integers.

The deviation of the Hamiltonian perturbation from the average has a similar Fourier representation,

$$\Delta H'(\bar{\alpha}, \bar{J}) = \sum_m \Delta H'_m(\bar{J}) e^{i(\bar{\alpha}_1 m_1 + \bar{\alpha}_2 m_2 + \cdots + \bar{\alpha}_s m_s)}.$$

By substituting the Fourier expansions for $S'(\bar{\alpha}, \bar{J})$ and $\Delta H'(\bar{\alpha}, \bar{J})$ into Eq. (7.1), one finds the action for near-integrable many-body motion is

$$S(\bar{\alpha}, \bar{J}) = \bar{\alpha}_j \bar{J}_j + \sum_m i \frac{\Delta H'_m(\bar{J}) e^{i(\bar{\alpha}_1 m_1 + \bar{\alpha}_2 m_2 + \cdots + \bar{\alpha}_s m_s)}}{m_1\omega_1 + m_2\omega_2 + \cdots + m_s\omega_s}. \tag{7.2}$$

The action (7.2) describes the behavior of many-body motions quite acceptably for a wide range of initial conditions (which fix the s invariants J), for perturbations which are not too large, and over time scales which are not too long. It does nicely for the small perturbations of our planetary system due to the mutual interactions of the planets.

To a first approximation, one may describe the solar system as a set of integrable two-body motions, one for each planet and the sun as described in Chap. 4. In the next approximation, the mutual interaction of the planets is included as the small perturbation. Individual angular momenta and eccentricities of the planets cease to be invariant in this approximation; however, the sum of the eccentricity vectors is again an invariant because of the averaging of Poincaré's method over the angle variables of all the planets. The sum of the angular momenta of the planets remains invariant because the total angular momentum is always invariant. The solar system as a whole therefore possesses invariant angular momentum and eccentricity vectors.

The individual Keplerian orbits of each planet are modified in the following way. The major axis undergoes small oscillations about its two-body value. The eccentricity vector not only undergoes small oscillations about its two-body value; it is also set slowly spinning so that each Keplerian orbit acquires a precession. This effect is reflected in the time a planet spends in the vicinity of the sun and can induce effects on the planet's climate with the same frequency signature. The period of precession of the earth's eccentricity vector is of the order of ten thousand years. The earth's ice ages are driven in large measure by the precession of its eccentricity vector.

These many-body effects are not qualitatively different from the kinds of motions characteristic of two bodies. But there are other kinds of many-body motion contained within Eq. (7.2) which are radically different from the oscillating and precessing trajectories of the kind we have seen thus far. There exist motions for which Eq. (7.2) predicts a catastrophic phenomenon. The Fourier amplitudes of the action have the denominator $m_1\omega_1 + m_2\omega_2 + \cdots + m_s\omega_s$ where the $\omega_i = \partial H_0/\partial I_i$ are the natural frequencies of the unperturbed motion and the m_i may take the values of all the positive and negative integers. This denominator may vanish:

$$m_1\omega_1 + m_2\omega_2 + \cdots + m_s\omega_s = 0. \tag{7.3}$$

When it does, a resonance has appeared.

Since the m_i are integers, the denominator (7.3) cannot vanish unless the motion is degenerate, that is, the ω_i are rationally commensurate as they indeed are for both pure Kepler and Hooke motion. But even if the motion is nondegenerate, sets of $m = (m_1, m_2, \ldots, m_s)$ can still be found which closely satisfy the resonance condition (7.3) because abundant rational numbers m_i/m_j lie arbitrarily close to any irrational ω_j/ω_i. It is the resonances which prevent series expansions like Eq. (7.2) from converging and hence prevent the representation of many-body motion in transcendental functions.

Although the elementary perturbation procedure utilized above cannot be used to calculate the motion at resonance, resonance is not an artifact of the perturbation procedure. It is a deep-seated aspect of many-body motion revealed by more powerful secular perturbation theory and by direct numerical calculation of the trajectories using the Hamilton equations.

Resonances signal a qualitative transition in motion. They open it to chaos. Resonances are also pervasive in many-body motion because of the dense intertwining of rational and irrational numbers. Since Fourier modes $m = (m_1, m_2, \ldots, m_s)$ always exist which closely satisfy the resonance condition (7.3), chaotic behavior is densely intertwined with the regular motion of the invariant tori in near-integrable motion; yet regular motion predominates in the sense that the irrationals predominate the set of real numbers. (If one selects any real number at random, the probability that that number is rational is zero.)

Many-body motion is therefore not generally pure chaos. It is, in the vicinity of a near-integrable motion, a mixture of both regular and chaotic motion. But for large perturbations corresponding to many-body regimes far from integrable motions, the motion becomes fully chaotic.

Chaotic Kepler Motion

The qualitatively new behavior introduced by many-body motion may be illustrated with the two intimately related motions through which the two-body problem has led us, those of Kepler and Hooke. The Hamiltonian

$$\mathcal{H} = p^2/2\mu - k/r + \kappa\rho^2/2, \tag{7.4}$$

where ρ is the projection of the position vector onto the plane $x_3 = 0$,

$$\rho^2 = {x_1}^2 + {x_2}^2 = r^2 - {x_3}^2,$$

combines both motions. Aside from a constant term not included here, this is the Hamiltonian for the hydrogen atom in the presence of a magnetic field aligned with the polar axis x_3. We are going to view this motion as a classical rather than quantum motion. The orbits of interest are therefore the outer orbits with quantum numbers $n \gg 1$ for which the motion is well approximated as a classical motion.

In the limit in which the electron is free of the magnetic field and interacts only with the nucleus one has Kepler motion with the Hamiltonian

$$\mathcal{H} = p^2/2\mu - k/r.$$

The motion possesses a full set of isolating invariants for which the eccentrum $\boldsymbol{E}$ completes the conservation law invariants $\mathcal{H}$ and $\boldsymbol{M}$. The trajectory for bound motion is an ellipse perpendicular to the angular momentum vector whose center lies at a focus of the ellipse as shown in Fig. 7-1 *(a)*. The fundamental frequency of the motion is the Kepler frequency $\omega_K = \sqrt{8|\mathcal{H}|^3/k^2\mu}$ given by Eq. (4.55).

In the limit in which the electron is free of the nucleus and interacts only with the magnetic field aligned with the polar axis x_3 the relative motion is that of the Hooke Hamiltonian

$$\mathcal{H} = p^2/2\mu + \kappa\rho^2/2.$$

The motion is that of a particle bound by the linear force but confined by the magnetic field to the plane perpendicular to the x_3 or polar axis. (It is a unique property of the magnetic force that it cannot act on a charged particle in its own direction; hence the potential energy term in the Hamiltonian is independent of x_3: it is expressed in terms of ρ^2 rather than r^2.) The coefficient κ depends upon the strength of the applied magnetic field. This motion also possesses a full set of isolating invariants with the Hooke invariant matrix A completing the conservation law invariants. The motion is integrable and the trajectory is an ellipse lying in the plane perpendicular to the polar axis whose center is the center of the ellipse as shown in Fig. 7-1 *(b)*. The fundamental frequency of the uncoupled Hooke motion is $\omega_H = \sqrt{\kappa/\mu}$.

If the electron is now considered to be the hydrogen atom electron, the Hamiltonian describes both interaction with the nucleus and interaction with the magnetic field and is given by Eq. (7.4). If the motion were strictly two-body, the full position vector r^2 would appear in the place of ρ^2 and the combined Kepler–Hooke Hamiltonian would be separable. The confinement of the Hooke term to the plane is the effect of the magnetic field—all those many-body electrons interacting with the hydrogen atom electron.

Another way of describing this situation is that the Kepler inverse-square force potential $1/r$ possesses complete rotational symmetry about all three axes but the Hooke potential $\rho^2 = r^2 - {x_3}^2$ possesses rotational symmetry only about the x_3 axis. As a result, the Hamiltonian (7.4) is not integrable; it is, at root, a many-body Hamiltonian.

Even though there does not exist a coordinate system in which the motion is separable, the separability of Kepler motion in parabolic coordinates and the presence of the projection ρ^2 in the Hooke term, a naturally occurring quantity in these coordinates, makes the parabolic coordinates ξ,

η, ϕ the most revealing coordinates in which to view it. They are defined in Chap. 4 by

$$x_1 = \rho\cos\phi, \qquad x_2 = \rho\sin\phi,$$
$$x_3 = (\xi - \eta)/2, \qquad r = (\xi + \eta)/2,$$

with the projection ρ given by

$$\rho^2 = r^2 - {x_3}^2 = \xi\eta.$$

From Eq. (4.38), the Hamiltonian (7.4) has the parabolic coordinate form

$$\mathcal{H} = \frac{4}{\xi+\eta}\left(\xi p_\xi^2/2\mu + \eta p_\eta^2/2\mu\right) + \frac{1}{\xi\eta}p_\phi^2/2\mu - \frac{2k}{\xi+\eta} + \frac{\kappa}{2}\xi\eta. \tag{7.5}$$

If one chooses the Kepler ellipse semi-major axis a as length scale and the inverse Kepler frequency ${\omega_K}^{-1} = T/2\pi$ as time scale, the energy $\mathcal{H} = \mathcal{E}$ takes the nondimensional value $\mathcal{E} = -\frac{1}{2}$. The Hamiltonian (7.5) in nondimensional variables now scaled to these reference quantities takes the form

$$\mathcal{H} = \frac{4}{\xi+\eta}\left(\xi p_\xi^2/2 + \eta p_\eta^2/2\right) + \frac{1}{\xi\eta}p_\phi^2/2 - \frac{2}{\xi+\eta} + \frac{\epsilon^2}{2}\xi\eta = -\frac{1}{2}. \tag{7.6}$$

The natural measure of the importance of Hooke motion compared to Kepler motion is the ratio of the Hooke and Kepler frequencies,

$$\epsilon = \sqrt{k^2\kappa/8|\mathcal{H}|^3} = \omega_H/\omega_K,$$

and is the natural measure of perturbation strength. The ratio of the potential energy in Hooke motion to that in the Kepler motion is ϵ^2. For $\epsilon \to 0$, the motion is pure Kepler. For $\epsilon \to \infty$ the motion is pure Hooke. The motion is integrable in both these limits. One expects the maximum departure from integrable motion for $\epsilon \approx 1$.

The phase space of the Hamiltonian (7.5)–(7.6) is six-dimensional; but the motion possesses two invariants, the energy $\mathcal{H}$ and the polar angular momentum $p_\phi = M_3$. (The invariance of the remaining angular momentum components and the eccentricity $\mathbf{e}$ are destroyed by the presence of the magnetic field.) This means that a portion of the motion is separable—the polar mode corresponding to the invariant M_3 and it allows a reduction in the order of the Hamiltonian system of two (**Note 6**). This integrable mode has the polar action J_ϕ and angle variable $\alpha_\phi \equiv \phi$. The action is the invariant $J_\phi = M_3$ and the phase velocity is $\omega_\phi = \partial\mathcal{H}/\partial J_\phi = M_3/\xi\eta$ so that the separable mode is described by

$$\phi = (M_3/\xi\eta)t + const, \qquad J_\phi = M_3. \tag{7.7}$$

The remaining nonseparable modes constitute the reduced phase space which is four-dimensional. The remaining invariant, the energy $\mathcal{H}$, defines a three-dimensional surface in this space upon which the trajectories must wind. The known invariants therefore reduce the manifold of the flow from the full phase space to a three-dimensional space. This space turns out to be a three-sphere (**Note 6**).

The integrable mode may be removed from the Hamiltonian by choosing coordinates aligned with the natural directions of Kepler motion, the three orthogonal directions $\mathbf{M}$, $\mathbf{e}$, and $\boldsymbol{M} \times \mathbf{e}$ illustrated in Fig. 4-11. In spherical-polar coordinates it is appropriate to align $\boldsymbol{M}$ with the polar axis. In parabolic coordinates it is appropriate to align $\mathbf{e}$ with the polar axis in which case $M_3 = 0$ and the Hamiltonian becomes

$$\mathcal{H} = \frac{4}{\xi + \eta}\left(\xi p_\xi^2/2 + \eta p_\eta^2/2\right) - \frac{2}{\xi + \eta} + \frac{\epsilon^2}{2}\xi\eta = -\frac{1}{2}. \tag{7.8}$$

With $M_3 = 0$, the electron cannot orbit the x_3 axis. The polar mode phase velocity ω_ϕ in Eq. (7.7) correspondingly vanishes and the motion lies in the plane $\phi = const.$ The case of pure Kepler motion ($\epsilon = 0$) is that of an ellipse lying in this plane *parallel* to the x_3 axis as shown in Fig. 7-1 *(a)*. (This is in contrast to the orientation of the Kepler ellipse in the plane perpendicular to x_3 used throughout Chaps. 4 and 5 in spherical-polar coordinates because one is now dealing with parabolic coordinates in which $\mathbf{e}$ rather than $\boldsymbol{M}$ is aligned with x_3 and hence $M_3 = 0$.)

The Kepler ellipse is a circle when $e = 0$ and collapses into a straight-line vibration along the x_3 axis whenever the angular momentum M completely vanishes. (M is no longer an invariant so this can occur in the course of the motion.) As was shown in Chap. 4, the vibratory motion to which the Kepler ellipse degenerates when the angular momentum vanishes corresponds to $e = 1$. This is a collisional trajectory and it is distinct from the rotational Kepler motion because it passes through the origin. The Kepler motion therefore consists of rotations ($e < 1$) and vibrations ($e = 1$).

Pure Hooke motion occurs in the limit $\epsilon \to \infty$ and consists of an ellipse lying in the plane perpendicular to the plane of Kepler motion shown in Fig. 7-1 *(b)*. For $M_3 = 0$, the rotational modes are prohibited and the Hooke motion consists only of the straight-line vibration perpendicular to the polar axis (and to the Kepler vibration) as indicated in Fig. 7-1 *(c)*. The uncoupled motions for $M_3 = 0$ are Kepler rotations and mutually perpendicular Kepler and Hooke vibrations.

The combined motion takes place in the plane $\phi = const$ as indicated by Eq. (7.7). The motion is a competition between rotational and vibratory modes. The hydrogen nucleus attempts to draw the electron into a Kepler rotation in a plane parallel to the polar axis or a vibration along the polar axis; the magnetic field attempts to draw it into a Hooke vibration perpendicular to this axis. The magnetic force acting on the electron on the Kepler ellipse is a maximum when the orbit is circular ($e = 0$) and vanishes for the Kepler vibration ($e = 1$).

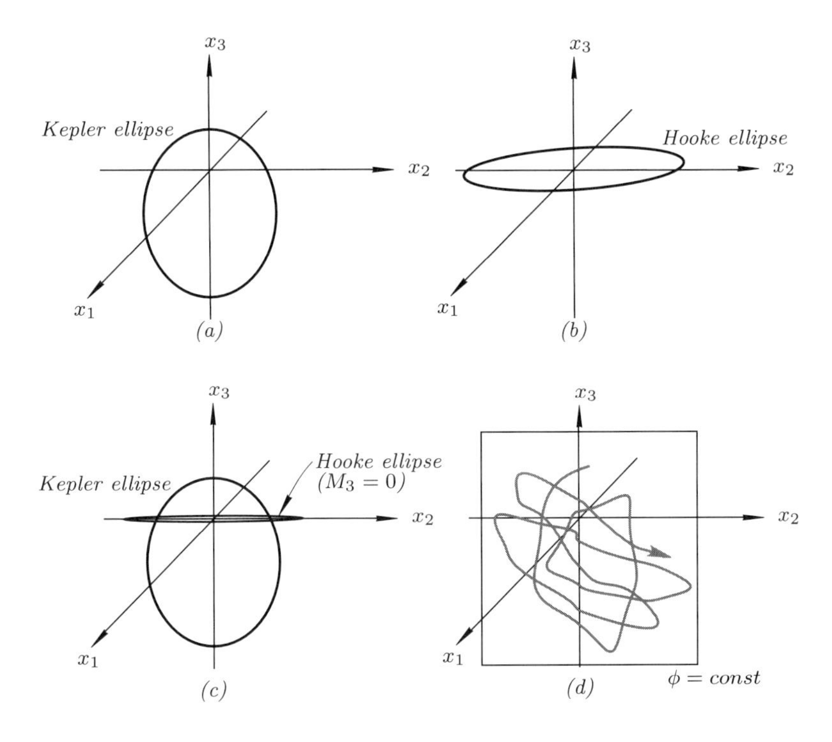

Figure 7-1. Combined Kepler and Hooke Motion. (a) Pure Kepler motion consists of rotations ($e < 1$) lying in a plane parallel to the polar axis and vibrations ($e = 1$) along the polar axis. *(b)* Pure Hooke motion consists of a Hooke ellipse perpendicular to the polar axis. *(c)* For $M_3 = 0$ the Hooke ellipse degenerates to a straight-line vibration perpendicular to the Kepler ellipse and the combined motion *(d)* takes place in the plane $\phi = const.$

The Hamiltonian acquires an even simpler form if, instead of ξ and η, squares are used as coordinates (**Note 6**):

$$\sigma^2 \equiv \xi = (r + x_3), \qquad \tau^2 \equiv \eta = (r - x_3). \tag{7.9}$$

The corresponding momenta are

$$p_\sigma = \partial L/\partial \dot{\sigma} = 2\sqrt{\xi} p_\xi, \qquad p_\tau = \partial L/\partial \dot{\tau} = 2\sqrt{\eta} p_\eta,$$

and Eq. (7.8) reduces to

$$\mathcal{H} = \frac{1}{\sigma^2 + \tau^2}\left(p_\sigma^2/2 + p_\tau^2/2\right) - \frac{2}{\sigma^2 + \tau^2} + \frac{\epsilon^2}{2}\sigma^2\tau^2 = -\frac{1}{2}. \tag{7.10}$$

The Hamiltonian (7.10) is not separable because of the term in ϵ^2 and the trajectories must be obtained by numerical integration of the Hamilton equations. Small perturbations from pure Kepler motion may be described by Poincaré's theory (7.2).

Poincaré found that a motion could be more simply grasped not by directing one's attention to the trajectory as a whole but rather to a specially selected set of its points. These points are created by the trajectory on a plane of observation which Poincaré termed a *surface of section* shown in Fig. 7-2. This is a surface in phase space which lies transverse to the flow. Each time the trajectory pierces the surface of section it generates a point. Poincaré found that the behavior of points in the surface of section characterized the trajectory as a whole.

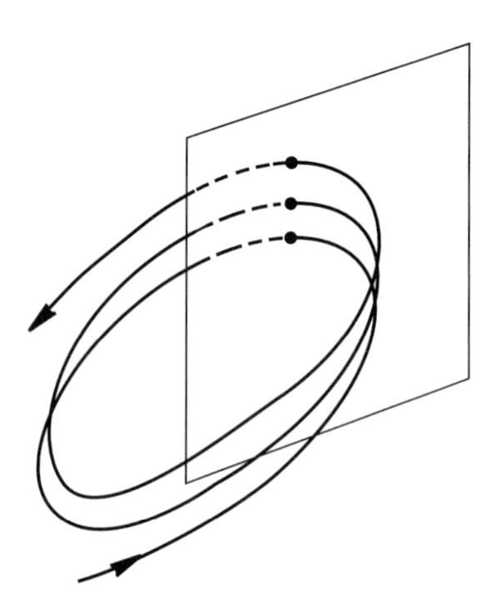

Figure 7-2. Poincaré Surface of Section. The surface of section is a surface in phase space which lies transversely to the trajectory. It is intersected by successive crossings of the complete trajectory.

For coupled Hooke and Kepler motion, an appropriate surface of section is a plane perpendicular to the flow defined by a conjugate coordinate pair such as the (τ, p_τ) plane at $\sigma = 0$. This surface of section is shown in Fig. 7-3. The circular boundary corresponds to the limiting curve $p_\sigma = 0$ in the surface $\sigma = 0$ from Eq. (7.10):

$$\tau^2 + p_\tau^2 = 4.$$

Flows which degenerate to a point in the Poincaré section are important features of a motion and are known as the fixed points of the flow. Five fixed points lie in the surface of section of the hydrogen atom in a magnetic field. The center point $(0, 0)$ corresponds to the Kepler vibration (one may think of the electron vibrating along a line perpendicular to the page). The points $(\pm\sqrt{2}, 0)$ correspond to the rotational Kepler orbits (the electron emerges from one point and re-enters the section at the opposite point). The points $(0, \pm\sqrt{2})$ correspond to the Hooke vibration (each is the turning point of the $e = 1$ collapsed Hooke ellipse).

The integrable motion about which the perturbation of the combined motion is made is a pure Kepler ground state—the circle corresponding to

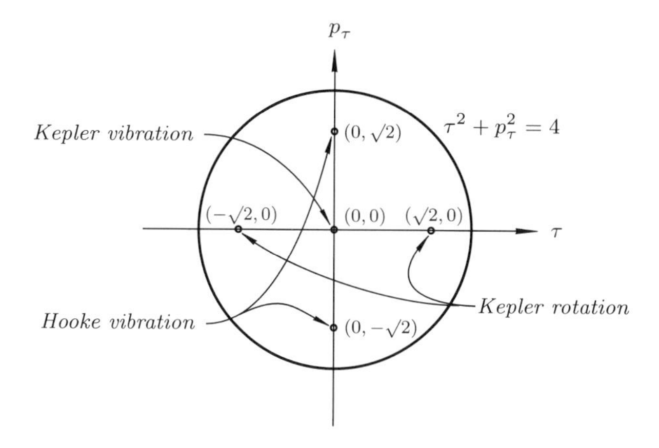

Figure 7-3. Fixed Points of the Hydrogen Atom in a Magnetic Field. The points are exhibited in the (τ, p_τ) plane at $\sigma = 0$. The center point $(0, 0)$ corresponds to the straight-line Kepler vibration for which $e = 1$ (the electron vibrates along a line perpendicular to the page). The points $(\pm\sqrt{2}, 0)$ correspond to the rotational Kepler orbits (the electron emerges from one point and re-enters the section at the opposite point). The points $(0, \pm\sqrt{2})$ correspond to the vibratory Hooke motion (each is the turning point of the $e = 1$ Hooke ellipse).

the rotational fixed points on the horizontal axis of Fig. 7-3. The combined motion may be determined by numerically computing the trajectories from the Hamilton equations for the Hamiltonian (7.10). These trajectories, captured in the Poincaré section, appear in Fig. 7-4.

Once the perturbation, however small, is made, a significant shift in the motion occurs. This can be seen in Fig. 7-4 *(a)*. The surface of section of the unperturbed motion consists of only the two rotational fixed points on the horizontal axis; but under even the smallest perturbation, the trajectories escape from these fixed points and range over the entire surface of section. With the invariance of the eccentricity broken, the orbiting electron is free to explore all eccentricity values ranging from rotational modes close to its original circle $e \approx 0$ to the vibratory mode $e = 1$. A small perturbation introduced by the magnetic field $\epsilon \ll 1$ does not lead to a small perturbation in e. This global shift of the motion under a small perturbation is a result of the complete degeneracy of the hydrogen atom.

A perturbed trajectory no longer continually pierces the surface of section in isolated fixed points; instead the orbit ticks off a multiple point pattern like that illustrated in Fig. 7-4. This pattern entwines three kinds of point sets: a continuous closed curve (including the vibratory fixed points for which the curve has degenerated to a point), a chain of closed loops called *islands* which are continuously revisited, and a chaotic spattering of points which are never revisited.

A densely filled out curve in the surface of section corresponds to an *open* trajectory. This is because a closed orbit in phase space cannot create a closed curve in the Poincaré section. It pierces the surface of section in a countable number of disconnected points. But an open orbit in phase space can create a closed curve because it pierces the Poincaré section in an infinite number of points densely filling out the curve. Such curves are known as *invariant* curves. They can be seen in Fig. 7-4 *(a)* for which $\epsilon = 0.19$, a perturbation strength still close to the integrable Kepler motion. Each invariant curve in Fig. 7-4 *(a)* corresponds to a particular trajectory generated by a particular set of initial conditions, all of which correspond to the same value of ϵ.

The fixed points of the motion are resonant points because they correspond to closed trajectories. The two rotational resonances $(\pm\sqrt{2}, 0)$ of the circular Kepler orbit in Fig. 7-4 *(a)* are *intersected* by invariant curves. The vibratory Kepler resonance which appears as the single point $(0, 0)$ and the vibratory Hooke resonance which appears as the pair of points $(0, \pm\sqrt{2})$ are *encircled* by invariant curves. The vibratory fixed points are stable to small perturbations in the sense that orbits close to them when perturbed maintain the same structure.

The two invariant curves which intersect the rotational fixed points $(\pm\sqrt{2}, 0)$ separate the vibratory modes from the rotational modes and are known as the *separatrices* of the motion. The rotational fixed points are unstable. Orbits near them are destroyed and different kinds of trajectories come into being once the perturbation is made.

The unstable fixed points and the separatrices emanating from them are the seats of chaotic behavior; it is along the separatrices that chaos first develops as can be seen in Fig. 7-4 *(b)*. These are indicated by both the island chains and the spattering of isolated points which have appeared in the place of the separatrices. Spattered points, unlike the points which make up the invariant curves, cannot be ordered. They do not define a curve. It is this lack of order which makes such trajectories chaotic.

The different island chains correspond to different combinations of m_1 and m_2 which closely satisfy the resonance condition (7.3). A chain of small islands surrounding its central resonant points corresponds to the same trajectory. The successive crossings of this trajectory jump from island to island, beginning at one island, moving successively over this island chain, and returning to the first where the sequence repeats.

Chaotic trajectories are the outcome of resonances and their unstable separatrices. Many-body resonances are pervasive because a rational number exists arbitrarily close to any irrational number. Resonant motions are therefore densely entwined with nonresonant motions in near-integrable motion. Thin layers of chaotic points comingled with island chains are actually distributed throughout the surface of section for all levels of perturbation strength. These resonance layers and their island chains are very thin for small perturbations and disappear into the invariant curves on the scale of the pictorial accuracy of the figure. (Imagine looking at an invariant curve through a microscope and discovering that it is not constructed from a solid

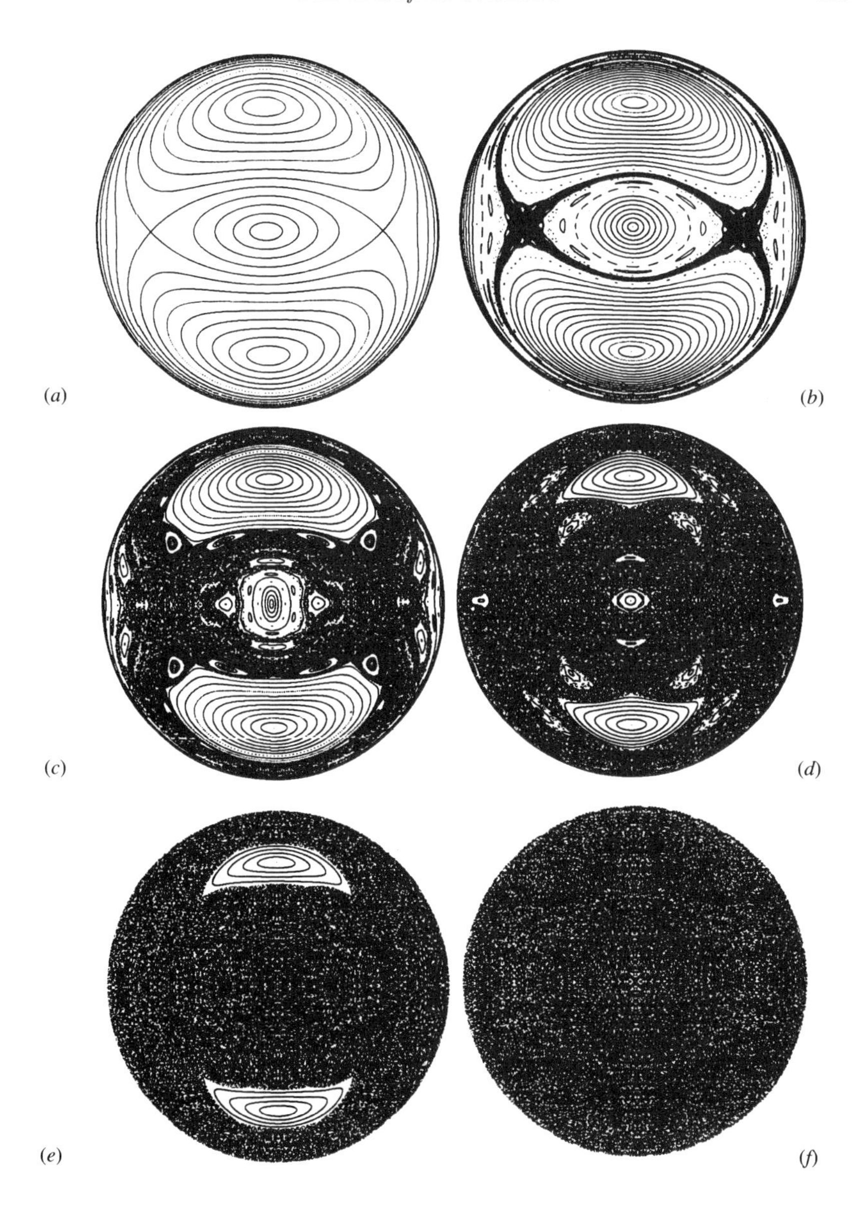

Figure 7-4. Surfaces of Section in the (τ, p_τ) Plane for Chaotic Kepler Motion. In *(a)* the perturbation strength is $\epsilon = 0.19$. Invariant curves encircle the primary resonances shown in Fig. 7-3. In *(b)* ($\epsilon = 0.22$), chaotic trajectories indicated by the spattered points and originating near the the separatrices have appeared. They become more numerous in *(c)* ($\epsilon = 0.24$) and *(d)* ($\epsilon = 0.26$) for which island chains are also in evidence. Invariant curves are progressively destroyed as perturbation strength increases to $\epsilon = 0.30$ in *(e)* and finally completely engulfed for $\epsilon = 0.38$ resulting in ergodic motion *(f)*. [Coutesy H. Freidrich and D. Wintgen, *Physics Reports* **183**, No. 2, 37–79 (1989)]

line but rather from infinitely fibred braids of finer and finer island chains.) For larger perturbation strengths, the island chains and the chaotic points are clearly visible in Figs. 7-4 *(c)–(e)*.

The stable vibratory fixed points are well isolated from the chaotic regions by their invariant curves for small perturbation strengths. The chaotic layers are isolated from one another and chaotic trajectories in one layer cannot intersect those in another.

As the perturbation strength is increased, chaos invades more and more of the phase space by destroying the invariant curves separating the islands as shown in Fig. 7-4 *(d)* and *(e)*. The chaos progressively encroaches upon the stable fixed points and finally engulfs them as in Fig. 7-4 *(f)*. When the last invariant curve is destroyed the chaos becomes global, filling the entire phase space. Motions whose flows fill entire regions of phase space are said to be *ergodic*. The trajectories which began as the fully degenerate trajectories of the unperturbed Kepler orbit filling a manifold of one dimension now fill the nonintegrable phase space of three dimensions.

Chaos and the Quantum

Classical many-body motion is inherently open to chaos as illustrated by the outer quasi-classical orbits of the hydrogen atom in a magnetic field. The complexity of chaotic motion and the smearing of the particle flow over the phase space gives chaotic motions a random or stochastic character. Quantum motion also has a stochastic character; but it originates in a distinctly different source in the quantum uncertainty principle.

It might be expected that quantum motions with their inherent uncertainty and many-body motions with their inherent chaos when combined would produce intense stochastic behavior. In fact, the opposite is true. Quantum effects suppress classical chaos. This is because chaos feeds on phase space filling the largest scales down to the finest scales. Quantum effects limit the otherwise classically limitless fineness of phase space which chaotic orbits can enter.

One can think of the uncertainty principle as a quantum symmetry. It manifests itself in the destructive and constructive interference of quantum waves and structures the flow in volumes of phase space on scales fixed by the Planck constant. Since phase space is built up of canonical pairs (q_i, p_i) whose product represents an action, those regions of phase space on scales Δq_i, Δp_i whose volume is $V \approx \prod_i^s \Delta q_i \Delta p_i \approx \hbar^s$ will be governed by quantum rather than classical mechanics.

The interference of the waves in the quantum regions blocks some orbits and reinforces others thereby reimposing order upon chaotic motion which has broken the constraints of its classical symmetries. This can only occur for the chaotic motions which have reached the quantum level. The larger scales of phase space are free of the uncertainty principle and the chaos operating on these scales is not suppressed by the quantum. This effect occurs in the hydrogen atom in a magnetic field. The outer orbits which are quasi-classical are chaotic. As one progresses inward to the lower orbits, the phase flow progressively becomes a field of quantum waves which suppress the chaotic flow. For the hydrogen atom this is the scale of the lowest lying orbits.

Cosmos and Chaos

The hydrogen atom in a magnetic field illustrates the manner in which many-body motions evolve chaotically. Chaos manifests itself in the atom and the elements. Chaos also manifests itself in the heavens. The solar system is our cosmos, its slow evolution still strongly imprinted by the initial conditions of its formation. Yet it too is open to chaotic behavior. The motion of the planet Pluto appears to be chaotic on a time scale of tens of millions of years. The large asteroid belt in the vicinity of the inner planets is rich in chaotic behavior generating particle orbits with large eccentricities which send meteorite material into the Earth and Mars.

Since resonances are the seat of chaotic motion, it is chaos which is the generator of the regular rings of Saturn and the gaps in the asteroid belts. In an interesting twist, the "regular" rings are the former abodes of chaotic matter. Chaos sculpts regularity in a dynamical *bas relief* in the following way.

Orbiting material in each planetary ring or asteroid belt has a unique orbital frequency associated with the ring or belt. The motion of this material is perturbed by the giant planets and resonances develop for material whose orbital frequency is a rational multiple of the orbital frequency of the planet. As we saw in the case of the hydrogen atom in a magnetic field, small perturbations in energy can nonetheless lead to large perturbations in the eccentricity. Rings regularly ordered by rational numbers exist because chaos generated by the resonant motion of the bodies in that ring with the orbital frequency of the planet sends them off on wildly eccentric orbits which collide with other planets thereby removing them and delineating the ring or belt. The adjacent ring or belt containing matter is nonresonant.*

The motion of two celestial bodies is the exemplary cosmic motion. Its orderly flow unfolds from and is completely specified by the initial state. In the eighteenth and nineteenth centuries this fact was extended to an even more sweeping observation based upon the classical equations of motion. The Marquis de Laplace proclaimed that given the initial state of all the particles of the universe, the equations of motion determine the future course of the world. Everything to be is contained in the initial state of mechanics—and indeed, by the same Laplacian argument—in the present state; but it is only visible to an omniscient being.

---

* Wisdom, J., "Chaotic Dynamics of the Solar System," *ICARUS*, **72**, 241–275 (1987).

The creation was understood as a great mechanical system, a complex set of interacting trajectories projected from their initial conditions and bound by the principle of least action. The Creator set up the law of motion and the initial conditions and then abandoned the machine to run along on its own.

The uncertainty of quantum motions overthrew this deterministic picture of the physical world. Quantum mechanics tells us that any initial condition cannot be specified exactly; hence the motion evolving from those initial conditions is not fully contained within the the initial state in the sense of Laplace.

Quantum motions are indeterministic in principle because the initial state cannot be endowed with all the information required to fully specify the motion. Classical many-body motions are deterministic in principle; but this determinism leads to chaos. The notions of cosmos and chaos are therefore as important in describing the evolution of the classical equations of motion as that of their determinism, a distinction unknown to Laplace.

A cosmic evolution of deterministic equations, such as that governing two heavenly bodies, fits appropriately with Laplace's vision. A deterministic mechanics which evolves chaos does not. Classical chaotic evolution, like quantum uncertainty, undermines the imperial role of initial conditions. It does so because of the sensitive dependence of chaotic evolution to initial conditions.

The role of chaos in nature may be illuminated, not by focusing on the connection between a chaotic motion and its initial state, but rather on the large-scale connections between different chaotic motions. From this perspective, chaos is not formless. It is the building block of form and structure on new scales. This means that many-body motion perceived from the local scale of distances separating the bodies in any near-integrable motion is chaotic; but on the global scale of an ensemble of chaotic motions as a whole, the combined motion of all bodies possesses a coherent form and structure. Cosmos gives way to chaos out of which new cosmos arises.

Metamorphosis

In the limit in which myriads of atoms and molecules interact together, individual chaotic motions of the particles coalesce to form a continuum—a solid, liquid, or gaseous medium. Individual molecular motions lose their significance and the motion of the continuum as a whole becomes the dynamically interesting object. Individual microscopic particle motions are chaotic; but, remarkably, the motion of the continuum as a whole need not be chaotic.

The continuum of solids, liquids, and gases shows that the integrable mechanics exemplified by two-body motion is reborn from chaos—a rebirth as revealing as the apparition of Prince Conn's companion from the remains

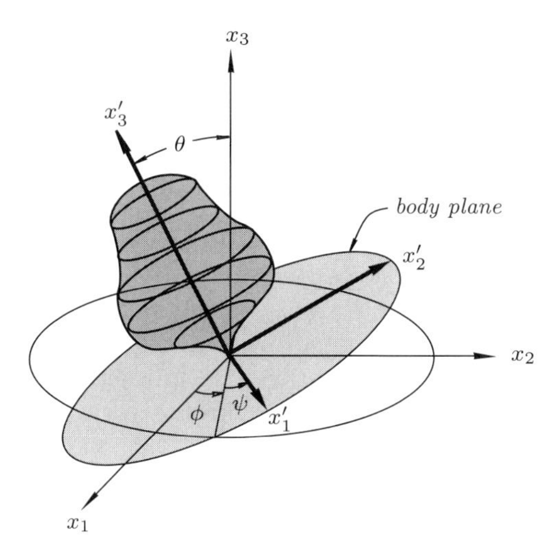

Figure 7-5. The Rigid Body. The motion of every point within a rigid body is completely determined by the motion of the reference frame $\mathbf{x}'$ embedded in it. The relative motion of this reference frame possesses only three degrees of freedom such as the three angles $\phi(t)$, $\psi(t)$, $\theta(t)$ (known as Euler angles). The motion of these angles therefore fixes the relative motion of every point within the body.

of the shaggy steed. This metamorphosis is revealed in the destruction of cosmic two-body motion in a continuum and its replacement by the chaotic many-body motion of atoms or molecules within a rigid body. The rigid body is a continuum of internal molecular motion, often chaotic, which as a whole is cosmic: the motion of a rigid body is integrable.

Because of its rigidity, the positions of all points within a rigid body are fixed with respect to a reference frame (x'_1, x'_2, x'_3) embedded within it as shown in Fig. 7-5. Their motion is therefore completely determined by the motion of this reference frame. The specification of the motion of the three angles which fix the orientation of this reference frame with respect to the center-of-mass frame (x_1, x_2, x_3) is sufficient to describe the relative motion of all the points in the body. The relative motion of a rigid body therefore has only three degrees of freedom and two Casimir sets of three invariants each,

$$(\mathcal{H}, M, M_3), \qquad (\mathcal{H}, M, M_1),$$

just like the vector representation of two bodies.

One sees immediately that generic rigid body motion is integrable since its Casimir sets contain three invariants which match the three degrees of freedom of the relative motion. Moreover, since there is more than one independent Casimir set, rigid body motion is degenerate. But since it does not possess three independent sets, it is not fully degenerate unless additional symmetries are imposed. The phase flow of a generic rigid body is open and densely covers its torus.

The conservation laws are the only deep laws of motion for the continuum. Since there are many more degrees of freedom than there are conservation laws, the internal motions of the particles seemingly defy description. But there is a saving feature in the many-body limit of the continuum—the ergodic behavior of the chaotic regime: individual particles are free to explore the entire phase space of the motion subject only to the constraint of the conservation laws. As a result, statistical methods of description and analysis of the internal motions becomes feasible.

One no longer speaks of individual particle motions in the continuum. One speaks instead of the mechanics of the continuum. The mechanical quantities inherent in the conservation laws (linear momentum, angular momentum, energy) also describe the continuum because the global conservation laws remain universally valid.

The global conservation laws are in themselves incomplete to describe the continuum. New mechanical concepts unique to the continuum and which have no significance for an individual molecule come into being. These include temperature, pressure, stress, heat, and entropy. These macroscopic continuum quantities, rather than the individual microscopic trajectories of particles, more properly describe the internal many-body motion of the continuum.

The rigid body is the leading illustration of the new level of cosmos which arises at the macroscopic scale of the continuum underlain by microscopic chaos. The rigid body is always a cosmic motion. But on the new level of structure and dynamics of the continuum, deformable solids as well as liquids and gases also acquire a new cosmic simplicity and new levels of chaos.

Another example of the interplay of cosmos and chaos is offered by the ultimate many-body medium: the fluid. In a fluid composed of a continuum of atoms and molecules, the new cosmos which arises can be ordered. One of the most basic structures of the fluid continuum is the vortex: a smoke ring shown in Fig. 7-6. Although the internal motion of individual molecules of the air and smoke are chaotic, the ensemble motion is ordered into tightly bound spirals.

But the cosmos of the continuum has its own life-cycle. It too dissolves into chaos. The motion of the fluid smoke ring is highly ordered as it first emerges; but it is a vulnerable order which will grow unstable and break up into a new kind of chaos proper to the macroscopic level: turbulent motion in which the fluid becomes highly disorganized. One encounters a new disintegration of the cosmos of the continuum into the chaos of turbulence. The appearance of chaotic orbits in many-body motion on the molecular level is recapitulated on the macroscopic level where ordered ensemble motion gives way to chaotic motion on the macroscopic level.

Nature surrounds us with ascending and descending rounds of chaos and cosmos. Cosmic two-body motion on the molecular level gives way to the chaos of many-body motion which in turn gives way to the cosmos of the

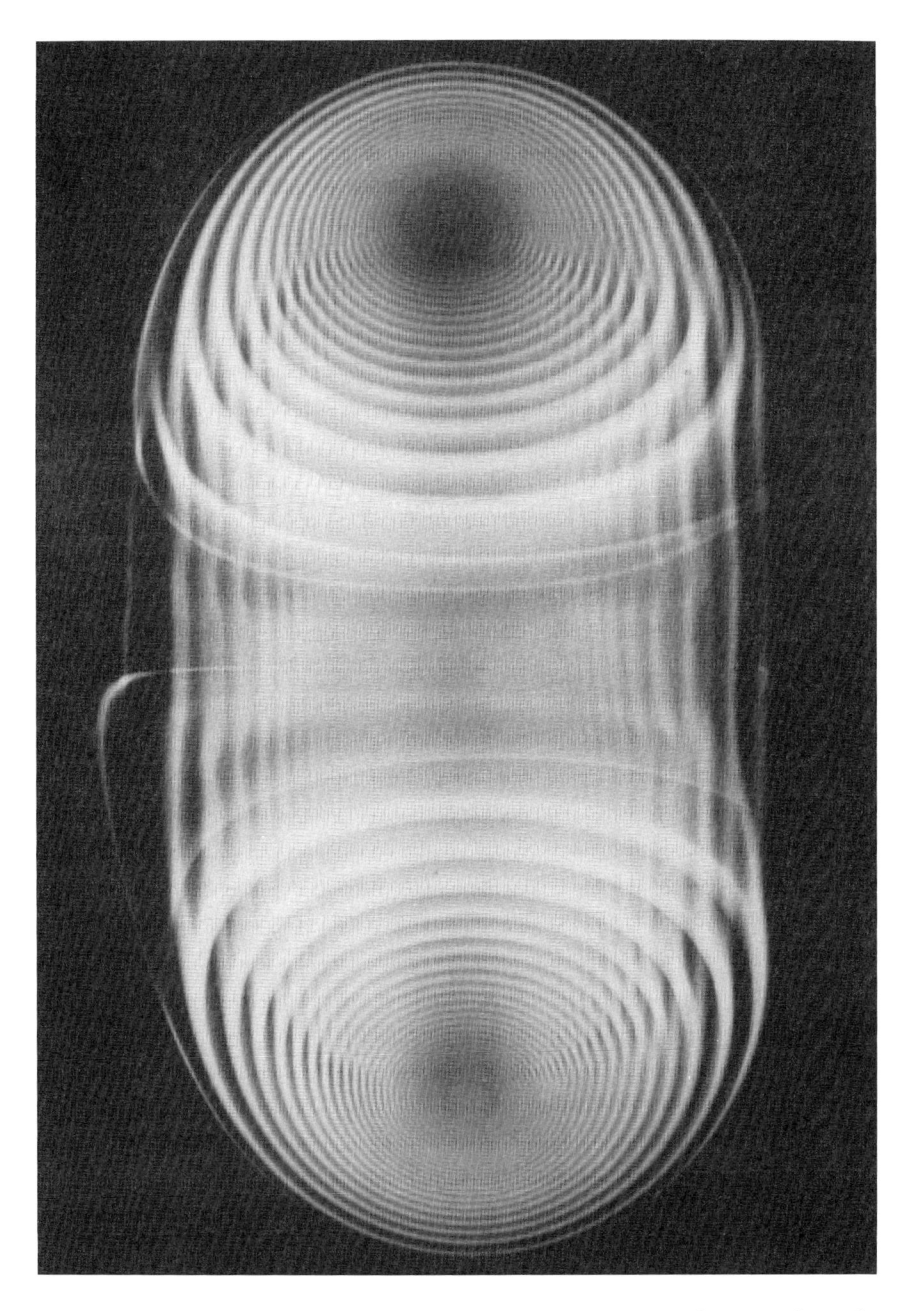

Figure 7-6. Ordered Motion Built upon Molecular Chaos. The vortical motion of a smoke ring. The internal motion of the individual atoms and molecules of the air is chaotic; but the macroscopic motion which the air as a whole undergoes is highly ordered. [Courtesy R. H. Magarvey and C. S. Maclatchy, *Canadian J. Phys.* **42**, 678–683, (1964)]

continuum. But the continuum is open to instabilities which destroy its order and once again re-open the way to the chaos of turbulence.

The round continues to higher levels of scale. The earth and sun with their turbulent oceans and atmospheres are mere mass points on the scale of the entire solar system. On solar system scales these two point masses approximately execute two-body, cosmic motion. But on galactic scales embracing myriads of stars, the universe behaves like a continuous, gravitating fluid in which earth and sun are chaotic, many-body participants, much like atoms or molecules in a gas or liquid.

Perhaps the most remarkable aspect of the natural world is the manner in which each level of cosmos is interleaved with other levels so as to develop its own integrity and largely follow its own course. The disparate scales of the elementary forces establish the basic levels of structure of the universe. For example, a single water molecule consists of nuclei of protons and neutrons binding a constellation of orbiting electrons. The dynamics of the electrons are not strongly coupled to the internal nuclear states of the protons and neutrons which make up the hydrogen and oxygen nuclei to which they are bound. The electronic states have organization different from the organization of the nucleus. The elecrical force establishes a unique scale of structure from that of the nuclear forces. But if only the nuclear, gravitational, and electrical forces differentiated matter in the universe, the world would be a monotonous one indeed.

Might it not be that the transitions from regular to chaotic motion are the generators of the complex textures of the physical world? The chaotic possibilities inherent in Hamiltonian many-body mechanics provide for a manifold universe. Chaos makes determinism perishable. If it did not do so, large aggregations of microscopic constituents would simply be big versions of the same microscopic structure.

Water again illustrates the point. A many-body ensemble of water molecules possesses a variety of forms. Solid water manifests itself as crystaline ice. The crystal lattice is immobile: it is a highly cosmic arrangement in which the nuclei of the water molecules are locked in fixed positions.* The ensemble is a large-scale replication of its constituent water molecules. The patterns of snowflakes directly exhibit the bond angles and symmetries of the underlying angularly shaped water molecules of which they are composed.

In the liquid phase the lattice is destroyed. The water molecules break free of their confinement in the lattice, melting into a chaotic, many-body motion: the solid becomes a liquid droplet. In contrast to the ordered arrangement in the immobile solid the molecular many-body motion within the liquid droplet is chaotic. The overall structure of a water droplet is not a

* It is interesting to note that it is the quantum nature of matter that makes the ordered lattice of the solid phase possible. The classical many-body motion of point particles does not possess any mechanism for phase changes. Classical particles cannot form a crystal lattice.

large replica of the water molecules chaotically jostling about within it. The droplet has shape, structure, and mechanical organization distinct and dramatically different from that of the individual water molecules which make it up. This sharp difference between the liquid droplet and its molecules (in contrast to the strong similarities between the ice crystal and its molecules) is a result of the chaotic motion of the molecules in the liquid phase.

The earth is not a big molecule. It is a decidedly different, unique structure built from molecules. Chaos limits the mechanical organization which resides in the delicate determinism of one level while a new mechanical organization and an equally perishable determinism coexists at another level.

Continue to yet higher levels of the universe. The wind currents of a thunderstorm containing massive numbers of water droplets do not turn sensitively on the motion of a single water droplet within the storm's downdraft. The structure and motion of a thunderstorm globally shaped by its massive anvil cloud are uniquely different from the structure and motion of an individual water droplet flying about in the turbulent cascade within it. Yet the storm itself, arising out of the moist unstable atmosphere owes its existence to the water vapor which has formed these droplets.

The global motion of the earth's atmosphere is not the result of a single thundershower. The atmosphere is rich in fluid structures—cyclones, frontal systems, jet streams, boundary layers, cumulus towers—of which a thunderstorm is only one among many. Chaos provides the means of differentiation through which each uniquely different fluid structure acquires its own identity. Operating on the microscopic motions of atoms and molecules bound by electrical forces up through macroscopic turbulence driven by gravity and the heat of the sun, the chaos inherent in Hamiltonian mechanics limits the determinism of each level thereby delineating one structure from another.

Continue beyond the planetary level. The spiral motions of galaxies are coupled in only the most tenuous way to the motions of a single star let alone to the atmosphere of one of the planets of that star. Yet all these levels are interleaved with one another and each is the stuff of its underlying constituents from the largest galactic scale down to the protons, neutrons, and electrons of the atomic and nuclear level.

The world sustains exceedingly complex motions such as those of the atmosphere, oceans, and galaxies. The general equations of motion of the continuum are not integrable (indeed, *cannot* be integrable if they are to faithfully describe the complexity of the world). But these equations do contain some surprisingly simple solutions—exact integrable solutions no less—which frequently appear in nature. (Nature contains structures of pure simplicity as well as tangled complexity!)

One class of these integrable motions are those for solitary waves—or *solitons* as they have come to be called. These are exact wave-like solutions of the fluid equations which consist of a solitary wave which propagates without change of shape. The first soliton was worked out by Lord Rayleigh in 1875 to describe a dramatic solitary wave observed in an English canal by the engineer and naval architect Scott Russell. Russell chased the wave on horseback for the better part of two miles until it escaped him. It was thirty feet long, a few feet in maximum height, and moved unchanged in shape at a speed of nearly ten miles per hour. Another example of a solitary wave embedded in a turbulent planetary atmosphere is the great red spot of Jupiter.

The transformation of the cosmic motion of two-bodies and its apparition in the many-body motion of the continuum is an engaging round in the cycles of motion. Chaotic evolution severs mechanics from a rigid capitulation to initial conditions. The products of the initial conditions in chaotic molecular motion are the raw materials for cosmos at a larger scale. Chaotic evolution supports the rich structure of the world in which new levels of order involving complex molecular structures including the molecules of life and living organisms come into being.

Mechanical law is, without exception, validated as the necessary condition of all motion; but in its inherently quantum and chaotic evolution, it introduces both uncertainty and complexity into mechanics. It governs the ground of motion while leaving open the kinds of structures which evolve and play upon that ground. We ourselves illustrate this fact. On the small scale of the organic molecules which make up the fluids and cells of our bodies, motion is one of many-body chaos; but on the large scale of our tissues, organs, and nerves built from myriads of these molecules, we are highly structured creatures. The world sustains large-scale order and structure built upon an underlying classical chaos and quantum uncertainty. How it does so is an inviting mystery.

Finale

The Shaggy Steed of Physics reveals the unity and simplicity of the heavens and the elements while at the same time it confronts us with the inexhaustible complexity and mystery in which such simplicity resides. The universe seems to be embraced by ultimate cosmological limits: an origin in the Big Bang, expansion, and possibly recompression back to the origin from whence it began. Within this envelope there flows a bountiful creation with rich dynamics on a vast series of smaller scales. Mechanical law binds the elemental fabric of the universe giving it warp and weft. But the fabric seems otherwise free to flow and gather shape into the more complex structures of creation from galaxies, stars, planets, and the geological structures up through the flow of being to complex living creatures.

The many-body motion of the universe manifests a freedom from the tyranny of initial conditions while it is still very much a cosmos sustained by the law of motion. The creation seems to spring anew at each stage of its complexity, appropriately incorporating its historical structure from the past but with open possibilities for the future. This evolution, the unfolding of the old and the knitting together of the new, is shrouded in a mystery in which the law of motion is the ground. But that which breathes life into the law of motion and each unfolding event transcends initial conditions. Mechanics is not the determinant of the universe but the vivified space and time of an ongoing creation.

People

The story of the physical world is told by people. It has a human face. Here is a collection of brief biographies of some of the principal figures in the story of the heavens and the elements. My selection is not meant to be complete; many of overshadowing importance are not represented—Copernicus and Bohr just to name two. I have chosen those with a particular connection to the law of motion and the Kepler problem and its topology and symmetries. They are Europeans for the most part—but they represent the entire continent eastward to Siberia and westward to the British Isles and Ireland, northward to Scandanavia and south to Italy. They came from scholarly, clerical, merchant, or banking backgrounds with a notable exception: Isaac Newton was born into a farm family.

Some, like Newton and Kepler, are household words. Others, like Heinz Hopf, are little known. Still others like Emmy Noether have only recently been celebrated. Jacob Hermann has lain in obscurity for two centuries; however the late twentieth century is discovering him.*

The contributions for which some of the individuals are remembered were brief moments. Emmy Noether's work on the symmetries of the variational principal was a youthful discovery. Most of her life was devoted to the problems of abstract algebra. Wolfgang Pauli is more notably remembered for the neutrino and the exclusion principle than his elegant use of the eccentricity invariant in the construction of the hydrogen spectrum. Jacobi's canonical transformation theory is a key aspect of both classical and quantum mechanics; but he himself was most devoted to his theory of elliptic integrals. On the other hand, Mendeleev was passionately occupied with the Periodic Table for his entire life, Kepler was consumed with the problem of the planetary orbits, and P. A. M. Dirac's creative energies rarely strayed from the central problems of quantum mechanics.

---

* A number of works on Hermann are fostered by the Bernoulli Edition, Basel. See F. Nagel, "A Catalog of the Works of Jacob Hermann (1678–1733)," *Historia Mathematica*, **18**, 36–54 (1991) and "The Life and Work of Jacob Hermann" in the same issue.

Johannes Kepler
1571 Weil der Stadt, Germany–1630 Regensburg, Germany
(Courtesy AIP Emilio Segrè Visual Archives)

From an early age, Kepler was passionately dedicated to discovery of the geometry of the heavens. It was Kepler's happy fate that his imaginative powers found their way to a close (if often strained) collaboration with Tycho Brahe, a skilled observer. Following Tycho's death, Kepler was the recipient of the finest astronomical data of the time. The first fruit of Kepler's fertile imagination was quite misguided. Kepler found that if one nests the classical solids—cube, tetrahedron, dodecahedron, icosahedron, and octahedron—the ratios of the spheres enclosed between them are fair approximations to the ratios of the orbits of the planets. While a brilliant *tour de force*, the discrepancies were not insignificant and the solids lacked dynamical significance; Kepler lost interest in the scheme. Nonetheless, it was this kind of inspired exploration based upon Copernican heliocentricity that ultimately led him to the ellipse and its area and period law 20 years later.

Dmitry Ivanovich Mendeleev
1834 Tobolsk, Siberia—1907 St. Petersburg
(Courtesy AIP Emilio Segrè Visual Archives, W.F. Meggers Collection)

Mendeleev was a devoted student of chemistry for his long and productive life with wide ranging interests from theoretical chemistry to cheesemaking, mining, and petroleum. Yet, the periodic arrangement of the elements overshadowed all other interests. As Kepler sought the underlying unity of the heavens, Mendeleev sought the underlying unity of the elements. The crucial step came in an inspired moment in 1869 when, working on his treatise *Osmovy khimii*, it suddenly struck him that, contrary to the established theory of types, he should place metals—copper and silver—next to the alkalis. This led him to use atomic weight as a basis for ordering the elements. Mendeleev then made a deck of cards, each face an element, and began arranging them in various groups. This allowed him to recognize periodicity. Subsequent refinements revealed errors in established atomic weights and gaps that led to the prediction of elements not yet discovered. Although the Periodic Table was first resisted, the subsequent discovery of the predicted elements, particularly gallium and germanium, ushered in its total acceptance by 1880.

Isaac Newton
1643 Woolsthorpe Lincolnshire, England—1727 London

A growing wave of mathematics and mechanics spurred by the Copernican revolution and the revival of learning in the renaissance reached an apotheosis in Isaac Newton, a child not of aristocratic or scholarly origins (as was the case of most mathematicians and scientists), but of yeomen. Many of Newton's most important discoveries came about during a seminal two year period in his early twenties when, having been exposed to the works of Euclid, Galileo, Kepler, Hooke, Descartes, Walis, and Schooten, and with the Cambridge University closed because of the plague, he was free to take up his own thoughts in seclusion. Newton's abundant gifts produced the differential calculus, the foundations of classical mechanics, and epochal discoveries in optics, mathematical analysis, and geometry. The Kepler ellipse, the area law, and the period law all received their explanation in the Newtonian synthesis of the equations of motion and the inverse-square law of gravity.

Paul Adrien Maurice Dirac
1902 Bristol, England—1984 Miami, Florida
(Photo by Börtzells Tryckeri, Courtesy AIP Emilio Segrè Visual Archives)

P. A. M. Dirac was a gifted mathematical inventor who saw how quantum mechanics rises from classical mechanics, yet transcends it. Dirac did not know of the Bohr atom when he arrived at Cambridge in 1923; yet he quickly began contributing to the mathematical structure demanded by quantum phenomena, discovering the connection between the Poisson bracket and the commutator of Heisenberg's matrix representation of observables. Then, with careful attention to its classical antecedent, Dirac found the equation governing the evolution of the matrix elements which had eluded Heisenberg in the operator $i\hbar d\hat{A}/dt = \hat{A}\hat{H} - \hat{H}\hat{A}$. He then went on to discover spinors in describing the relativistic electron and antimatter implied by the quantum in relativistic space–time. Dirac conceived the many-time formulation of relativistic quantum mechanics and laid the foundations of the Feynman path integral thereby opening the way to quantum electrodynamics. Newton synthesized the foundations of classical mechanics. In fitting kinship, Dirac, who did the equivalent for quantum mechanics, filled the chair at Cambridge held by Newton.

Leonhard Euler
1707 Basel—1783 St. Petersburg

A watershed of classical mathematics was reached under Euler in number theory, analysis, geometry, and topology (Euler gave the solution to the famous problem of the seven bridges of Königsberg). In so doing, he conceived virtually all the modern notation of mathematics. Euler was an important contributor to problems of physics and engineering including celestial mechanics, elasticity theory, and the mechanics of fluids. Maupertuis, the president of the Berlin academy of which Euler was the most fruitful member, had begun to formulate problems in mechanics as problems of minimizing path integrals of functions such as the velocity of a particle. Euler was the first to systematize Maupertuis' ideas in a calculus of variations and to create general methods for their solution. He showed how minimization of the integral of a function between two fixed points could be reduced to the solution of a differential equation, the now famous Euler–Lagrange equation.

Joseph-Louis Lagrange
1736 Turin—1813 Paris

Lagrange was born into a family of civil servants and destined for the law; but his natural gifts turned him to mathematics. In a youthful discovery he found that he could express Euler's equation for the function minimizing a functional integral (the equation that would come to be known as the Euler–Lagrange equation) in a neat and compact form using integration by parts (illustrated in Chap. 3). Recognizing the beauty of this method, Euler recommended him to Maupertuis who was eager to encourage research in variational methods. Lagrange then generalized Euler's variational calculus so that it could handle all the problems of mechanics. He refined the choice of the integrand of the functional appropriate to mechanics, introducing the "Lagrangian," $L = T - V$. (Maupertuis and Euler had used what amounted to only the kinetic energy which then required additional side conditions.) Lagrange went on to make important contributions to a wide range of mathematical subjects including the general theory of small perturbations of the solar system.

William Rowan Hamilton
1805 Dublin—1865 Dunskirk Observatory, Ireland

The celebrated equations that now bear his name were, for Hamilton, an adjunct of a more fundamental interest in optics. Like Jacobi, Hamilton knew the formal connection between point mechanics and geometrical optics for which Fermat's principle corresponds to the action principle for particles and light-ray surfaces of constant phase (which Hamilton called his "characteristic function") are analogous to level surfaces of action. He solved a variety of optical problems with the characteristic function including a prediction that a light-ray incident on a biaxial crystal should emerge as a hollow cone. Its demonstration two months later caused a sensation in the scientific community. Hamilton's use of the eccentricity invariant of the Kepler problem occurred in relation to his application of the characteristic function to problems of celestial mechanics. Hamilton's other beautiful discovery was that of quaternions, a four-dimensional generalization of the two-dimensional complex variable. The algebra of these quantities is the same as that of four-dimensional rotations, an amazing coincidence with the symmetry algebra of Kepler motion of which Hamilton was not aware.

Carl Gustav Jacob Jacobi
1804 Potsdam—1851 Berlin

After schooling himself in the works of Lagrange and Euler while a student at Berlin, Jacobi secured a position at Königsberg where a voluminous outpouring of work chiefly centered on elliptic integrals followed. Jacobi brought to further perfection the methods begun by Euler and refined by Lagrange by developing the most general transformations of the Euler–Lagrange equations (now known as canonical transformations) and showed that the trajectories could be obtained by solving a first order partial differential equation for the generating function of a canonical transformation as an alternative to solving the Hamilton equations (see *Action Redux*, Chap. 3). The methods of both Jacobi and Hamilton showed the strong formal similarity between the mechanics of point particles and the optics of waves. They therefore showed the way in which quantum mechanics and classical mechanics merge and, remarkably, presaged the development of de Broglie matter waves, the Schroedinger equation, and ultimately the Feynman path integral.

Jacob Hermann
1678 Basel—1733 Basel
(Courtesy Bernoulli Edition, Basel; original in private hands)

Jacob Hermann was born into the mathematical ferment sparked by the famous Bernoulli family centered in Basel. He became an early champion of the calculus of Leibniz, winning the favor of the master. Teaching positions in central Europe were scarce (and the Bernoullis who filled them numerous) so Hermann was forced to seek a post in Padua, Galileo's old haunt. It was in Padua that he completed his most important work in celestial mechanics including the discovery of the hidden invariant of Kepler motion culminating in his magnum opus, the *Phoronomia*. He then moved to St. Petersburg where he worked freely with Euler and Daniel and Nicholas II Bernoulli on problems of algebra and motion. Hermann was well remembered and his works seriously studied in the eighteenth century along with those of the Bernoullis and Leibniz (his work was highly regarded by Immanuel Kant). He then seems to have been forgotten for 200 years but is now being rediscovered by the twentieth century.

Wolfgang Pauli
1900 Vienna—1958 Zurich
(Courtesy AIP Emilio Segrè Visual Archives)

Gifted by native intelligence and family background (his father was a distinguished professor of Chemistry at Vienna and Ernst Mach was his godfather), Pauli was one of physics' child prodigys. At the age of 19 he wrote the article on Einstein's theory of relativity for Felix Klein's *Encyclopädie der Mathematischen Wissenschaften*, a brilliant exposition which remained definitive for over half a century. Pauli remained an important and often severe critic of Einstein's field theories; but his own work was centered on the problems of quantum mechanics. He was the first to present the complete solution of the quantum motion of the hydrogen atom for which he used the eccentricity invariant he had learned from Lenz. The puzzling behavior of atoms in a magnetic field was solved by Pauli's discovery of the exclusion principle and the important role of spin in the atom: the Lorentz invariance of space–time requires that electrons possess spin and no two electrons can be in the same state. His last notable discoveries were the neutrino, required for conservation of energy in radioactive decay via the weak nuclear force, and the celebrated spin–statistics theorem.

Amalie Emmy Noether
1882 Erlangen, Germany—1935 Bryn Mawr, Pennsylvania

"Emmy Noether," wrote Herman Weyl in her obituary, "was born into a Jewish family distinguished for the love of learning and championed by the great Göttingen mathematician, David Hilbert." Nonetheless, she was barred from academic standing because she was a woman. Emmy Noether was still able to gather around her a notable group of students including the Dutch mathematician B. L. van der Waerden. She traveled extensively to other institutions and had an international array of colleagues including the distinguished Russian mathematician P. S. Alexandroff. Her discovery of the connection between the symmetries and invariants of the variational calculus was stimulated by Felix Klein during her early years at Göttingen. She was summarily dismissed and the Göttingen Algebraic School disbanded when the Nazis came to power. She emigrated to the United States where she lived out the last year and a half of her life at Bryn Mawr College.

Heinz Hopf
1894 Breslau, Germany—1971 Zollikon, Switzerland
(Courtesy G. Alexanderson, from the Polya Picture Album)

Heinz Hopf was inspired by the power of algebraic methods to reveal important properties of geometry and topology in attending lectures on the work of L. E. J. Brouwer. Following his studies at Breslau he passed through the vibrant mathematical circle at Göttingen where he became acquainted with Emmy Noether and began a lifelong correspondence with P. S. Alexandroff. One of Hopf's great discoveries, completed in 1931, was that of the topological relationships between the one-, two-, and three-dimensional spheres. (These relationships have at their core the manner in which each higher dimensional sphere is foliated with those of lower dimension.) The Hopf spheres are the basis of the symmetries of Kepler motion, though Hopf approached them as purely mathematical objects and was not aware of their physical connection to the motion of the solar system and the hydrogen atom.

Hedrick Antoon Lorentz
1853 Arnheim, Netherlands—1928 Haarlem, Netherlands
(Algeneen/Rijksarchief, The Hague, Courtesy AIP Emilio Segrè Visual Archives)

Maxwell's electromagnetic field equations received their connection with the sources of these fields (the Lorentz force law) in the work of Lorentz whose substantive work was on the theory of the electron. Lorentz was a gifted lecturer and writer who became the leading statesman of the late nineteenth century world of physics; he also inspired great affection and was beloved by generations of students and colleagues, most notably Einstein. Lorentz discovered the transformations of space–time required for the invariance of the Maxwell equations leading, incomprehensively, to the relativity of time and Lorentz contraction of bodies. Einstein used Lorentz transformations as the definitive insight into the geometric structure of space–time; but Lorentz, to the end of his life, regarded them as a mathematical artifice that would ultimately be reinterpreted in classical concepts based on absolute time.

Henri Poincaré
1854 Nancy—1912 Paris

Poincaré is a pivotal figure for both physics and mathematics in the twentieth century because he sparked a development immensely enriching both: the role of topology. It should not be forgotten Poincaré had independently of Lorentz and Einstein found the transformations of Maxwell's equations which preserve their invariance and, more importantly, recognized their geometric and group-theoretic significance. Poincaré's geometric and topological insights were of a general and abstract nature embracing all manifolds, not just those of space–time. Thus, the phase space was also deeply interesting to Poincaré and he initiated the systematic exploration of the connections between the topology and stability of mechanical systems, introducing mappings of manifolds as representations of the motions of mechanical systems. Topological features of mapped manifolds then reveal crucial qualitative properties of motion such as their stability. Poincaré showed, among other things, that one can derive this information from a study of the fixed points of mappings, a powerful concept which has grown to abundant use in the late twentieth century.

Notes

Note 1
Variational Principle

The law of motion has a basis in a variational principle. All motion may be gathered up in a single mechanical quantity, the action S. The trajectories which matter takes to get from one point in space and time to another are those for which the action takes the least possible value between the two points.*

The statement that the action shall take a minimum value between any two points in space and time is simple. The final laws of motion which result from this statement are also simple. But the pathway of deduction from one to the other is not. The following is a guided tour along this deductive path.

Let $L = L(\mathbf{x}_\alpha, \dot{\mathbf{x}}_\alpha)$ be the Lagrangian function of the rectangular positions $\mathbf{x}_\alpha$ and velocities $\dot{\mathbf{x}}_\alpha$ of a system of particles $\alpha = 1, 2, \ldots$. The action is defined as the integral of the Lagrangian over the time path of the system between any two time instants t_1 and t_2:

$$S = \int_{t_1}^{t_2} L(\mathbf{x}_\alpha, \dot{\mathbf{x}}_\alpha, t)\, dt. \tag{1.1}$$

To find the functions $\mathbf{x}_\alpha(t)$ for which S is stationary, the functions $\mathbf{x}_\alpha(t)$ and $\dot{\mathbf{x}}_\alpha(t)$ are given variations which displace them to $\mathbf{x}'_\alpha(t)$ and $\dot{\mathbf{x}}'_\alpha(t)$ while still allowing them to pass through fixed end points $(\mathbf{x}_\alpha)_1$ and $(\mathbf{x}_\alpha)_2$. The time t is also given a variation which displaces it to t'.

The displaced coordinates $(\mathbf{x}'_\alpha, \dot{\mathbf{x}}'_\alpha, t')$ are not only generated by the path variations $\delta\mathbf{x}_\alpha$, $\delta\dot{\mathbf{x}}_\alpha$; they are also generated by variations in the symmetry parameters describing the homogeneity and isotropy of space and time: $\delta\mathbf{a}$, $\delta\mathbf{u}$, $\delta\boldsymbol{\Omega}$, $\delta\tau$. The path variations are described with respect to a given reference frame. The symmetry parameter variations correspond to the variations of the reference frame. The displaced coordinates are given in terms of these variations by

$$\begin{aligned} \mathbf{x}'_\alpha &= \mathbf{x}_\alpha + \delta\mathbf{x}_\alpha + \delta\mathbf{a} + \delta\mathbf{u}t + \delta\boldsymbol{\Omega} \times \mathbf{x}_\alpha, \\ \dot{\mathbf{x}}'_\alpha &= \dot{\mathbf{x}}_\alpha + \delta\dot{\mathbf{x}}_\alpha + \delta\mathbf{u}, \\ t' &= t + \delta\tau. \end{aligned} \tag{1.2}$$

* More precisely, the action must be *stationary* which, over finite segments of time, allows not only minima, but maxima and inflection points. However, for infinitesimal segments, the action is always a minimum.

The variation of S now takes the form

$$\begin{aligned}\delta S = &\int_{t_1}^{t_2} \delta L(\mathbf{x}_\alpha, \dot{\mathbf{x}}_\alpha)\, dt \\ &+ \int_{t_2}^{t_2+\delta t_2} L(\mathbf{x}_\alpha, \dot{\mathbf{x}}_\alpha)\, dt - \int_{t_1}^{t_1+\delta t_1} L(\mathbf{x}_\alpha, \dot{\mathbf{x}}_\alpha)\, dt.\end{aligned} \tag{1.3}$$

The two terms on the second line of Eq. (**1.3**) appear because of the variation of the time.

The variation $\delta L(\mathbf{x}_\alpha, \dot{\mathbf{x}}_\alpha)$ may be expressed to first order in the displacements of the coordinates as

$$\delta L(\mathbf{x}_\alpha, \dot{\mathbf{x}}_\alpha) = \sum_\alpha \left(\frac{\partial L}{\partial \mathbf{x}_\alpha} \cdot (\mathbf{x}'_\alpha - \mathbf{x}_\alpha) + \frac{\partial L}{\partial \dot{\mathbf{x}}_\alpha} \cdot (\dot{\mathbf{x}}'_\alpha - \dot{\mathbf{x}}_\alpha) \right)$$

and the variations $\mathbf{x}'_\alpha$, $\dot{\mathbf{x}}'_\alpha$, t' expressed in terms of $\delta\mathbf{x}_\alpha$, $\delta\dot{\mathbf{x}}_\alpha$, δt and the symmetry parameters through Eq. (**1.1**). The variation of S given by Eq. (**1.3**) then becomes to terms first order in the variations:

$$\begin{aligned}\delta S = &\int_{t_1}^{t_2} \sum_\alpha \left(\frac{\partial L}{\partial \mathbf{x}_\alpha} \cdot \delta\mathbf{x}_\alpha + \frac{\partial L}{\partial \dot{\mathbf{x}}_\alpha} \cdot \delta\dot{\mathbf{x}}_\alpha \right) dt \\ &+ \int_{t_1}^{t_2} \sum_\alpha \frac{\partial L}{\partial \mathbf{x}_\alpha} \cdot (\delta\mathbf{a} + \delta\mathbf{u}t + \delta\boldsymbol{\Omega} \times \mathbf{x}_\alpha) + \int_{t_1}^{t_2} \sum_\alpha \frac{\partial L}{\partial \dot{\mathbf{x}}_\alpha} \cdot \delta\mathbf{u}\, dt \\ &+ \int_{t_2}^{t_2+\delta t_2} L(\mathbf{x}_\alpha, \dot{\mathbf{x}}_\alpha)\, dt - \int_{t_1}^{t_1+\delta t_1} L(\mathbf{x}_\alpha, \dot{\mathbf{x}}_\alpha)\, dt.\end{aligned} \tag{1.4}$$

The first term on the second line of Eq. (**1.4**) vanishes since, as described in Chap. 3, the Lagrangian of an isolated system of particles satisfies the conditions

$$\sum_\alpha \frac{\partial L}{\partial \mathbf{x}_\alpha} = 0, \qquad \sum_\alpha \mathbf{x}_\alpha \times \frac{\partial L}{\partial \mathbf{x}_\alpha} = 0.$$

To the approximation of first order in the variations, the last two terms in Eq. (**1.4**) reduce simply to the product of the Lagrangian and time variations at the end points of the path: $(L\delta t)|_{t_2} - (L\delta t)|_{t_1}$.

The term involving $(\partial L/\partial \dot{\mathbf{x}}_\alpha) \cdot \delta\dot{\mathbf{x}}_\alpha$ on the first line of Eqs. (**1.4**) may be transformed into one involving $\delta\mathbf{x}_\alpha$ through inegration by parts as in the transition from Eq. (3.9) to Eq. (3.10). Substitution of these results into the general variation (**1.4**) and integration of the term $d/dt(\partial L/\partial \dot{\mathbf{x}}_\alpha \cdot \delta\mathbf{x}_\alpha)$ leads to

$$\begin{aligned}\delta S = &-\int_{t_1}^{t_2} \sum_\alpha \left(\frac{d}{dt}\frac{\partial L}{\partial \dot{\mathbf{x}}_\alpha} - \frac{\partial L}{\partial \mathbf{x}_\alpha} \right) \cdot \delta\mathbf{x}_\alpha\, dt \\ &+ \sum_\alpha \frac{\partial L}{\partial \dot{\mathbf{x}}_\alpha} \cdot \delta\mathbf{x}_\alpha \Big|_{t_1}^{t_2} + \int_{t_1}^{t_2} \sum_\alpha \frac{\partial L}{\partial \dot{\mathbf{x}}_\alpha}\, dt \cdot \delta\mathbf{u} \\ &+ (L\delta t)|_{t_2} - (L\delta t)|_{t_1}.\end{aligned} \tag{1.5}$$

It is important to observe that the boundary term in the integration by parts, $\sum_\alpha (\partial L/\partial \dot{\mathbf{x}}_\alpha) \cdot \delta \mathbf{x}_\alpha |_{t_1}^{t_2}$, does not now vanish when the time t as well as the paths $\mathbf{x}_\alpha$ are varied. The requirement that the varied paths all pass through the same fixed end points is that $\mathbf{x}'_\alpha|_{t_1+\delta t_1} = \mathbf{x}_\alpha|_{t_1}$, *not* $\mathbf{x}'_\alpha|_{t_1} = \mathbf{x}_\alpha|_{t_1}$. Thus the path variation at an end point, such as $\delta \mathbf{x}_\alpha|_{t_1}$, does not vanish as it does in Eqs. (3.9) and (3.10) where the times are held fixed and symmetry parameter variations are not considered.

The variation at t_1 such as $\delta \mathbf{x}_\alpha|_{t_1}$ contains a contribution due to the variation of the time at the end point given by

$$\delta \mathbf{x}_\alpha|_{t_1+\delta t_1} = \delta \mathbf{x}_\alpha|_{t_1} + (\dot{\mathbf{x}}_\alpha \delta t)|_{t_1}.$$

A similar connection exists between $\delta \mathbf{x}_\alpha|_{t_2+\delta t_2}$ and $\delta \mathbf{x}_\alpha|_{t_2}$. These results and the condition $\delta t_1 = \delta t_2 = \delta \tau$ from the third of Eqs. (**1.2**) may be incorporated into Eq. (**1.5**) resulting in the action variation

$$\begin{aligned} \delta S = &-\int_{t_1}^{t_2} \sum_\alpha \left(\frac{d}{dt}\frac{\partial L}{\partial \dot{\mathbf{x}}_\alpha} - \frac{\partial L}{\partial \mathbf{x}_\alpha} \right) \cdot \delta \mathbf{x}_\alpha \, dt \\ &+ \Delta\Big(\sum_\alpha \frac{\partial L}{\partial \dot{\mathbf{x}}_\alpha} \cdot \delta \mathbf{x}_\alpha \Big) + \int_{t_1}^{t_2} \sum_\alpha \frac{\partial L}{\partial \dot{\mathbf{x}}_\alpha} dt \cdot \delta \mathbf{u} \\ &+ \Delta\Big(L - \sum_\alpha \frac{\partial L}{\partial \dot{\mathbf{x}}_\alpha} \cdot \dot{\mathbf{x}}_\alpha \Big) \delta \tau, \end{aligned} \qquad (\mathbf{1.6})$$

where Δ indicates the difference of values between end points of the path of any quantity before which it stands: $\Delta() \equiv ()_2 - ()_1$.

The path variations $\delta \mathbf{x}_\alpha$ at the end points which appear in the term

$$\Delta \left(\sum_\alpha \partial L/\partial \dot{\mathbf{x}}_\alpha \cdot \delta \mathbf{x}_\alpha \right)$$

on the second line of Eq. (**1.6**) do not vanish in the presence of symmetry parameter variations. Since the end points of the trajectories are fixed as $\mathbf{x}'_\alpha|_{t_1+\delta t_1} = \mathbf{x}_\alpha|_{t_1}$ and $\mathbf{x}'_\alpha|_{t_2+\delta t_2} = \mathbf{x}_\alpha|_{t_2}$, the end point variations of the paths must be canceled by the variations of the symmetry parameters or, according to Eqs. (**1.2**),

$$\delta \mathbf{x}_\alpha|_{t_1} = -\delta \mathbf{a} - \delta \mathbf{u} t_1 - \delta \boldsymbol{\Omega} \times \mathbf{x}_\alpha|_{t_1},$$

with a similar condition for the end point at t_2. Substitution of these conditions into Eq. (**1.6**) yields the action variation

$$
\begin{aligned}
\delta S = &-\int_{t_1}^{t_2}\sum_\alpha\left(\frac{d}{dt}\frac{\partial L}{\partial\dot{\mathbf{x}}_\alpha}-\frac{\partial L}{\partial\mathbf{x}_\alpha}\right)\cdot\delta\mathbf{x}_\alpha\,dt \\
&-\Delta\left(\sum_\alpha\frac{\partial L}{\partial\dot{\mathbf{x}}_\alpha}\right)\cdot\delta\mathbf{a} \\
&-\left[\Delta\left(\sum_\alpha\frac{\partial L}{\partial\dot{\mathbf{x}}_\alpha}t\right)-\int_{t_1}^{t_2}\sum_\alpha\frac{\partial L}{\partial\dot{\mathbf{x}}_\alpha}\,dt\right]\cdot\delta\boldsymbol{u} \\
&-\Delta\left(\sum_\alpha\mathbf{x}_\alpha\times\frac{\partial L}{\partial\dot{\mathbf{x}}_\alpha}\right)\cdot\delta\boldsymbol{\Omega} \\
&-\Delta\left(\sum_\alpha\frac{\partial L}{\partial\dot{\mathbf{x}}_\alpha}\cdot\dot{\mathbf{x}}_\alpha-L\right)\delta\tau.
\end{aligned}
\tag{1.7}
$$

The variation of the action δS thus results from the trajectory variations $\delta\mathbf{x}_\alpha$, the origin shift $\delta\mathbf{a}$, the space translation $\delta\boldsymbol{u}$, the space rotation $\delta\boldsymbol{\Omega}$, and the time shift $\delta\tau$. These variations are arbitrary and independent of one another. Hence, their coefficients must vanish if the variation δS in Eq. (**1.7**) is to vanish.

A vanishing variation δS corresponding to the variations $\delta\mathbf{x}_\alpha$ results in the equations of motion

$$
\frac{d}{dt}\frac{\partial L}{\partial\dot{\mathbf{x}}_\alpha}-\frac{\partial L}{\partial\mathbf{x}_\alpha}=0.
$$

The last four lines of Eq. (**1.7**) are in the form of *conservation laws*: a dynamical quantity is unchanged between any two arbitrary end points of the path of motion. The first conserved quantity corresponds to the invariance of the law of motion to arbitrary shifts of the origin of the space coordinates:

$$
\Delta\left(\sum_\alpha\frac{\partial L}{\partial\dot{\mathbf{x}}_\alpha}\right)=0. \tag{1.8}
$$

The second corresponds to invariance to velocity space translations

$$
\Delta\left(\sum_\alpha\frac{\partial L}{\partial\dot{\mathbf{x}}_\alpha}t\right)-\int_{t_1}^{t_2}\sum_\alpha\frac{\partial L}{\partial\dot{\mathbf{x}}_\alpha}\,dt=0. \tag{1.9}
$$

The third corresponds to invariance to space rotations:

$$
\Delta\left(\sum_\alpha\mathbf{x}_\alpha\times\frac{\partial L}{\partial\dot{\mathbf{x}}_\alpha}\right)=0. \tag{1.10}
$$

The fourth corresponds to invariance to shifts in the origin of the time coordinate:

$$
\Delta\left(L-\sum_\alpha\frac{\partial L}{\partial\dot{\mathbf{x}}_\alpha}\cdot\dot{\mathbf{x}}_\alpha\right)=0. \tag{1.11}
$$

The Lagrangian and its derivatives which appear in these expressions are recognizable as the total momentum $\boldsymbol{P}$, angular momentum $\boldsymbol{J}$, and energy H of the system:

$$\begin{aligned}\boldsymbol{P} &= \sum_\alpha \frac{\partial L}{\partial \dot{\mathbf{x}}_\alpha},\\ \boldsymbol{J} &= \sum_\alpha \mathbf{x}_\alpha \times \frac{\partial L}{\partial \dot{\mathbf{x}}_\alpha},\\ H &= \sum_\alpha \frac{\partial L}{\partial \dot{\mathbf{x}}_\alpha} \cdot \mathbf{x}_\alpha - L.\end{aligned} \qquad (\mathbf{1.12})$$

Since the momentum is $\boldsymbol{p}_\alpha = \partial L/\partial \dot{\mathbf{x}}_\alpha = m_\alpha d\mathbf{x}_\alpha/dt$, the second term on the third line of Eq. (**1.7**) has the integral

$$\int_{t_1}^{t_2} \sum_\alpha \frac{\partial L}{\partial \mathbf{x}_\alpha}\,dt = \sum_\alpha (m_\alpha \mathbf{x}_\alpha)_2 - \sum_\alpha (m_\alpha \mathbf{x}_\alpha)_1 = \Delta(m\boldsymbol{X}), \qquad (\mathbf{1.13})$$

where $m = \sum_\alpha m_\alpha$ is the total mass of the system and $\boldsymbol{X}$ is the coordinate of the center of mass defined by $m\boldsymbol{X} = \sum_\alpha m_\alpha \mathbf{x}_\alpha$. The third line of Eq. (**1.7**) therefore involves the conservation law for the *mass-center* $\mathbf{N}$ defined as

$$\mathbf{N} = \sum_\alpha \frac{\partial L}{\partial \dot{\mathbf{x}}_\alpha} t - \sum_\alpha m_\alpha \mathbf{x}_\alpha = \boldsymbol{P}t - m\boldsymbol{X}. \qquad (\mathbf{1.14})$$

The general variation (**1.7**) in these variables corresponding to the symmetry parameter variations is then

$$\delta S_{\text{symmetries}} = -\Delta \boldsymbol{P}\cdot\delta\mathbf{a} - \Delta\mathbf{N}\cdot\delta\mathbf{u} - \Delta\boldsymbol{J}\cdot\delta\boldsymbol{\Omega} - \Delta H\delta\tau. \qquad (\mathbf{1.15})$$

The overarching condition that action shall be a minimum leads to two kinds of results: equations of motion and conservation laws The conservation laws are a result of a vanishing action variation due to the symmetry parameter variations given by Eq. (**1.15**). These flow from the symmetries of the Lagrangian and reveal the invariants of motion. For Galilean relativity (**1.2**) these invariants are $\boldsymbol{P}$, $\mathbf{N}$, $\boldsymbol{J}$, and H. They are ten in number and correspond to the ten symmetry parameters of the Galilean group.

Note 2
Closed Orbits of Two-Body Motion

The bound motion of two bodies takes place in an effective potential:

$$W(r) = M^2/2\mu r^2 + V(r). \qquad (\mathbf{2.1})$$

The potential $V(r)$ may be quite general. However, of this general family of potentials, only two potentials generate orbits that are closed: the attractive inverse first power potential $V = -k/r$ (corresponding to the inverse-square force $\boldsymbol{f} = -k\mathbf{r}/r^3$) and the attractive quadratic potential $V = \kappa r^2/2$ (corresponding to the linear force $\boldsymbol{f} = -\kappa \mathbf{r}$).

If the motions are bound, the potential $W(r)$ which corresponds to $V(r)$ must possess a local minimum $W(r_0) = \mathcal{E}_0$ at the point $r = r_0$ and a positive second derivative $W''(r) > 0$ as shown in **Fig. 2-1**. Turning points of the radial motion ($\dot{r} = 0$) correspond to the intersections of the potential curve $W(r)$ with the energy $\mathcal{E}$. These intersections are the roots of the condition

$$\mathcal{E} = W(r) = M^2/2\mu r^2 + V(r). \tag{2.2}$$

The radial motion of the orbit consists of a libration "to and fro" between these two turning points designated $r = r_{\text{min}}$ and $r = r_{\text{max}}$. For $r_{\text{min}} = r_{\text{max}} = r_0$, the radial motion vanishes and the orbit is a circle of radius r_0.

An orbit will be closed if the angle $\Delta\phi$ through which the position vector turns from $r = r_{\text{min}}$ to $r = r_{\text{max}}$ and back again is a rational fraction of 2π: $\Delta\phi = 2\pi m/n$ where m and n are integers. In that event, after n revolutions of the angle ϕ, the radial coordinate will have completed m libration periods corresponding to m complete rotations of the position vector. The position vector will then be back precisely at the point from which the motion started bringing closure to the orbit. The behavior of the angle $\Delta\phi$ reveals the closed and open nature of the orbit.

The orbit angle is given by Eq. (3.50) and the angle corresponding to one libration period is

$$\Delta\phi = 2\int_{r_{\text{min}}}^{r_{\text{max}}} \frac{M/r^2}{\sqrt{2\mu(\mathcal{E} - W)}}\, dr. \tag{2.3}$$

The angle integral can also be expressed in terms of the perpendicular momentum $p_\perp = M/r$ of Eq. (4.4) in terms of which it becomes

$$\Delta\phi = 2\int_{p_{\perp\text{min}}}^{p_{\perp\text{max}}} \frac{dp_\perp}{\sqrt{2\mu(\mathcal{E} - W)}}, \tag{2.4}$$

with

$$W(M/p_\perp) = p_\perp^2/2\mu + V(M/p_\perp).$$

The libration limits of the perpendicular momentum follow directly from Eq. (**2.2**) and are the roots $p_\perp = p_{\perp\text{min}}$ and $p_\perp = p_{\perp\text{max}}$ in

$$\mathcal{E} = p_\perp^2/2\mu + V(M/p_\perp). \tag{2.5}$$

First examine the orbits close to the minimum energy point $r = r_0$. Such orbits are near-circular orbits. At the minimum energy point the first derivative $W'(r_0)$ vanishes. According to Eq. (**2.1**), a vanishing first

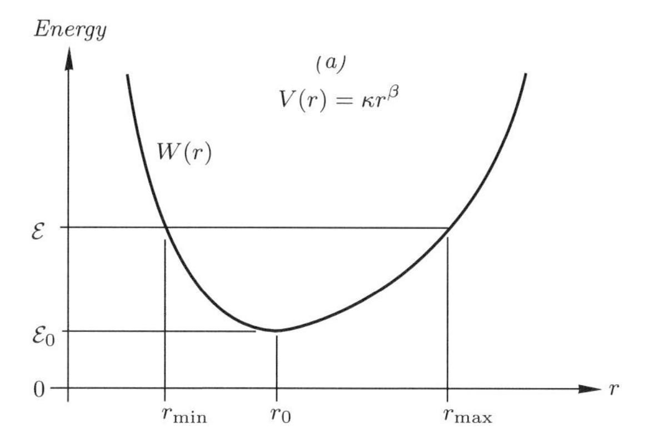

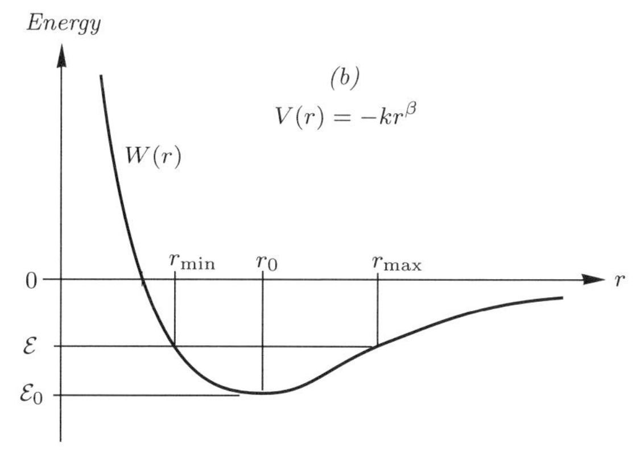

Figure 2-1. Effective Potentials for Two-Body Motion. Bound motions take place centered about an effective potential $W(r)$ which is convex with a positive second derivative $W'' > 0$. The points $r_{\min}$ and $r_{\max}$ correspond to the intersections of the potential energy curve with the total energy $\mathcal{E}$. The radial motion consists of a libration "to and fro" between $r_{\min}$ and $r_{\max}$. The point r_0 corresponds to the minimum energy point $W(r_0) = \mathcal{E}_0$ where the first derivative $W'(r_0)$ vanishes. In *(a)* the potential $W(r)$ for the attractive direct power-law potential $V(r) = \kappa r^{\beta}$ for $\kappa > 0, \quad \beta > 0$ is illustrated. In *(b)* $W(r)$ for the attractive inverse power-law $V(r) = -kr^{\beta}$ potential corresponding to $\beta < 0$ is illustrated. Bound motions only exist for $\mathcal{E} < 0$ in this case.

derivative of $W(r)$ imposes a condition on the first derivative $V'(r)$. The conditions at the minimum point are

$$W'(r_0) = 0, \qquad r_0 V'(r_0) = M^2/\mu {r_0}^2.$$

The potential may be expanded about the minimum point as

$$W(r) = \mathcal{E}_0 + \tfrac{1}{2} W''(r_0)(r - r_0)^2 + \cdots .$$

The libration point condition possesses a similar expansion:

$$\mathcal{E} = W(r_{\max}) = \mathcal{E}_0 + \tfrac{1}{2} W''(r_0)(r_{\max} - r_0)^2 + \cdots .$$

[Note that linear terms in $W'(r_0)(r-r_0)$ do not appear in these expansions since $W'(r_0)$ vanishes at the minimum energy point.]

If one keeps only terms up to second order in $\mathcal{E} - W$, the energy is symmetric about the minimum point with $r_{\text{max}} - r_0 = -(r_{\text{min}} - r_0) = \sqrt{2(\mathcal{E} - \mathcal{E}_0)/W''}$. The angle (**2.4**) for near-circular orbits is then

$$\Delta\phi = 4\sqrt{\frac{M^2/\mu r_0{}^2}{r_0{}^2 W''}} \int_{r_0}^{r_{\text{max}}} \frac{r_0\, dr}{\sqrt{(r_{\text{max}} - r_0)^2 - (r - r_0)^2}}$$

which is

$$\Delta\phi = 2\pi\sqrt{\frac{M^2/\mu r_0{}^2}{r_0{}^2 W''}}.$$

The second derivative W'' and the term $M/\mu r_0{}^2$ may be eliminated in favor of derivatives of the potential $V(r_0)$ through Eq. (**2.1**):

$$M^2/\mu r_0{}^2 = r_0 V', \qquad W'' = 3V'/r_0 + V'',$$

resulting in the angle

$$\Delta\phi = 2\pi\sqrt{\frac{V'}{r_0 V'' + 3V'}}.$$

For arbitrary potentials, the angle depends upon the size of the circular orbit r_0 about which the near-circular orbits exist. If the angle is to be independent of the size of the orbit, the potential $V(r)$ must satisfy the differential equation

$$\frac{rV'' + 3V'}{V'} = \alpha,$$

where $\alpha > 0$ is a constant. The solutions of this differential equation are

$$V(r) = kr^\beta, \quad \beta \neq 0; \qquad V(r) = k\ln r, \quad \beta = 0. \tag{2.6}$$

The exponent $\beta > -2$ is related to the constant α as $\beta = (1-2\alpha)/\alpha$. Only power-law and logarithmic force laws yield angles which are independent of the size of the orbit. The corresponding angle for all near-circular orbits which are independent of size is

$$\Delta\phi = \frac{2\pi}{\sqrt{\beta + 2}}. \tag{2.7}$$

Since the logarithmic potential $\beta = 0$ leads to an angle $\Delta\phi = \sqrt{2}\pi$ which is an irrational multiple of π, closed orbits can exist only for power-law potentials $V = kr^\beta$. But this still leaves a wide range of values of the exponent $\beta > -2$. Since these conditions pertain to orbits which are near-circular, they are necessary rather than sufficient. To find sufficient conditions, orbits in other limits which deviate from circularity must be examined.

Consider all direct power-laws $V = \kappa r^\beta$, $\beta > 0$. These orbits deviate from circularity by the maximum amount in the limit $\mathcal{E} \to \infty$. In this limit, the roots of the turning point condition (**2.5**) become

$$p_{\perp\max}^2/2\mu \to \mathcal{E}, \qquad \kappa M^\beta/p_{\perp\min}^\beta \to \mathcal{E},$$

and the integrand $1/\sqrt{2\mu(\mathcal{E}-W)}$ will vanish except in the regions $p_\perp \approx p_{\perp\min}$ and $p_\perp \approx p_{\perp\max}$. Since these regions are well separated from one another, the angle integral may be broken into two parts and the lower and upper limits extended to zero and infinity respectively:

$$\Delta\phi = 2\int_0^{p_{\perp\max}} \frac{dp_\perp}{\sqrt{(p_{\perp\max}^2 - p_\perp^2)}} - 2\int_{p_{\perp\min}}^{\infty} \frac{dp_\perp}{\sqrt{2\mu\kappa M^\beta(p_{\perp\min}^{-\beta} - p_\perp^{-\beta}\ \ )}}.$$

The integrals may be rearranged to

$$\Delta\phi = 2\int_0^1 \frac{d\xi}{\sqrt{1-\xi^2}} - 2\frac{p_{\perp\min}^{(\beta+2)/2}}{\sqrt{2\mu\kappa M^\beta}}\int_1^\infty \frac{d\xi}{\sqrt{1-\xi^{-\beta}}} \qquad (\mathbf{2.8})$$

with the substitution $\xi = p_\perp/p_{\perp\max}$ in the first and $\xi = p_\perp/p_{\perp\min}$ in the second. In the limit $\mathcal{E} \to \infty$, there results $p_{\perp\min} \to 0$. The second integral of Eq. (**2.8**) therefore makes a vanishing contribution and the first becomes

$$\Delta\phi = 2\int_0^1 \frac{d\xi}{\sqrt{1-\xi^2}} = \pi. \qquad (\mathbf{2.9})$$

The two conditions (**2.9**) and (**2.7**) may be combined to yield the single condition $\pi = 2\pi/\sqrt{\beta+2}$ or $\beta = 2$. Only the direct power force law $V = \kappa r^2$ yields closed orbits.

Consider now the inverse-power potentials for $V = -kr^\beta$ for which $-2 < \beta < 0$. In contrast to the direct power-law potentials which are positive for bound motions, the inverse power-law potentials must be negative. The circular orbit for this potential occurs at the minimum energy point $\mathcal{E}_0 \leqslant 0$. The orbit which deviates most from circularity is that for $\mathcal{E} = 0$. For

$$\mathcal{E} = W(M/p_\perp) = p_\perp^2/2\mu + kM^\beta p_\perp^{-\beta},$$

the roots of the turning point condition at $\mathcal{E} = 0$ are

$$p_{\perp\min} = 0, \qquad p_{\perp\max} = (2\mu k M^\beta)^{1/(2+\beta)}. \qquad (\mathbf{2.10})$$

The angle integral is then

$$\Delta\phi = 2\int_0^{p_{\perp\max}} \frac{dp_\perp}{\sqrt{2\mu\left(kM^\beta p_\perp^{-\beta} - p_\perp^2/2\mu\right)}}.$$

The term kM^β may be eliminated in favor of $p_{\perp\max}$ through the second of Eqs. (**2.10**) and the angle integral simplified to

$$\Delta\phi = 2\int_0^1 \frac{d\xi}{\sqrt{\xi^{-\beta} - \xi^2}} = 2\pi(2+\beta). \qquad (\mathbf{2.11})$$

Combining Eq. (**2.11**) with Eq. (**2.7**), one finds $2\pi(2+\beta) = 2\pi/\sqrt{2+\beta}$ and the condition $\beta = -1$. The only inverse power potential which yields a closed orbit is the potential $V = -k/r$. Closed orbits for two-body motion therefore exist only for the potentials $V = \kappa r^2$ and $V = -k/r$ where k and κ are positive coefficients.

Note 3
The Eccentricity Invariant

Two-body motion for which the force between the bodies is the inverse-square force $f = -k/r^2$ possesses an isolating invariant beyond the conservation law invariants. This invariant is the eccentricity

$$\mathbf{e} = \mathbf{p} \times \boldsymbol{M}/k\mu - \mathbf{r}/r. \qquad (\mathbf{3.1})$$

The eccentricity vector lies in the plane of motion perpendicular to the angular momentum vector. Its magnitude $e^2 = 1 + 2M^2\mathcal{H}/k^2\mu = 1 - M^2/h^2$ where $h \equiv \sqrt{k^2\mu/(-2\mathcal{H})}$ is the Kepler constant appears in the family of conic sections in (r, ϕ) coordinates:

$$r(\phi) = r_0/(1 + e\cos\phi).$$

The properties of the eccentricity vector may be revealed through the Poisson bracket. If $\mathbf{e}$ is an invariant, its Poisson bracket with $\mathcal{H}$ must vanish. To show this, form the bracket of Eq. (**3.1**) with $\mathcal{H}$ and use the property of Poisson brackets (which follows from the rules of differentiation) $[AB, C] = A[B, C] + B[A, C]$ to obtain

$$[\mathbf{e}, \mathcal{H}] = \mathbf{p} \times [\boldsymbol{M}, \mathcal{H}]/k\mu + [\mathbf{p}, \mathcal{H}] \times \boldsymbol{M}/k\mu - [\mathbf{r}/r, \mathcal{H}].$$

Since $\boldsymbol{M}$ is an invariant, it satisfies $[\boldsymbol{M}, \mathcal{H}] = 0$ and the bracket $[\mathbf{e}, \mathcal{H}]$ reduces to

$$[\mathbf{e}, \mathcal{H}] = [\mathbf{p}, \mathcal{H}] \times \boldsymbol{M}/k\mu - [\mathbf{r}/r, \mathcal{H}]. \qquad (\mathbf{3.2})$$

The Poisson bracket of any dynamical variable with $\mathcal{H}$ generates its time evolution. Hence $[\mathbf{p}, \mathcal{H}]$ generates Newton's law of motion in Hamilton's form $[\mathbf{p}, \mathcal{H}] = \dot{\mathbf{p}} = \boldsymbol{f}$. Represent the force as a general central force $\boldsymbol{f} = f(r)\mathbf{r}/r$ and the angular momentum as $\boldsymbol{M} = \mathbf{r} \times \mathbf{p}$. Equation (**3.2**) then becomes

$$[\mathbf{e}, \mathcal{H}] = \mathbf{r} \times (\mathbf{r} \times \dot{\mathbf{r}})f(r)/kr - [\mathbf{r}/r, \mathcal{H}]. \qquad (\mathbf{3.3})$$

Both the double cross-product $\boldsymbol{r} \times (\boldsymbol{r} \times \dot{\boldsymbol{r}})$ and the Poisson bracket $[\mathbf{r}/r, \mathcal{H}]$ appearing in Eq. (**3.3**) are actually different representations of the total time derivative of the unit position vector $\boldsymbol{r}/r$. The Poisson bracket of the unit position vector with the Hamiltonian is manifestly so:

$$\frac{d}{dt}(\mathbf{r}/r) = [\mathbf{r}/r, \mathcal{H}].$$

To see the connection between the double cross-product and the time derivative of the unit vector, expand it as

$$\Big(\mathbf{r} \times (\mathbf{r} \times \dot{\boldsymbol{r}})\Big)_i = (x_i x_j - r^2 \delta_{ij})\dot{x}_j. \tag{3.4}$$

The time derivative of the unit vector may be formally calculated from $x_i/r = x_i/\sqrt{x_j x_j}$ and, aside from sign and the scalar factor r^3, yields the same result,

$$\frac{d}{dt}(x_i/r) = -(x_i x_j - r^2 \delta_{ij})\dot{x}_j/r^3. \tag{3.5}$$

One therefore finds the elegant identity

$$\frac{d}{dt}(\mathbf{r}/r) = -\boldsymbol{r} \times (\mathbf{r} \times \dot{\boldsymbol{r}})/r^3. \tag{3.6}$$

Combining the results of Eqs. (**3.4**)–(**3.6**) with Eq. (**3.3**), one obtains

$$[\mathbf{e}, \mathcal{H}] = -\Big(1 + r^2 f(r)/k\Big)\frac{d}{dt}(\mathbf{r}/r). \tag{3.7}$$

For an attractive inverse-square central force $f(r) = -k/r^2$, the right-hand side of Eq. (**3.6**) vanishes revealing the invariance of the eccentricity vector.

Let us now consider the Poisson brackets of the eccentricity. The Poisson bracket of any two components of the eccentricity is

$$\begin{aligned}[e_i, e_j] =& [(\boldsymbol{p} \times \boldsymbol{M})_i/k\mu, (\boldsymbol{p} \times \boldsymbol{M})_j/k\mu] - [(\boldsymbol{p} \times \boldsymbol{M})_i/k\mu, x_j/r] \\ &+ [(\boldsymbol{p} \times \boldsymbol{M})_j/k\mu, x_i/r] + [x_i/r, x_j/r].\end{aligned} \tag{3.8}$$

The last term in Eq. (**3.8**) vanishes. The remaining terms may be established by direct calculation; but some simple theorems for Poisson brackets simplify the work. The Poisson bracket of a vector product $\boldsymbol{A} \times \boldsymbol{M}$ and a vector $\boldsymbol{B}$ can be shown to be

$$[(\boldsymbol{A} \times \boldsymbol{M})_i, B_j] = \epsilon_{ikl}[A_k, B_j]M_l - A_j B_i + (\mathbf{A} \cdot \mathbf{B})\delta_{ij}. \tag{3.9}$$

One may use this result again to show that the Poisson bracket of a vector product with the angular momentum is

$$[(\boldsymbol{A} \times \boldsymbol{M})_i, (\boldsymbol{A} \times \boldsymbol{M})_j] = \epsilon_{ikl}\epsilon_{jmn}[A_k, A_m]M_l M_n - A^2 \epsilon_{ijk} M_k. \tag{3.10}$$

Using (**3.9**) with $\boldsymbol{A} = \mathbf{p}$, one finds

$$[(\mathbf{p} \times \boldsymbol{M})_i/k\mu, x_j/r] - [(\mathbf{p} \times \boldsymbol{M})_j/k\mu, x_i/r] = 2\epsilon_{ijk} M_k/k\mu r. \qquad \mathbf{(3.11)}$$

The result (**3.10**) may be used with $\boldsymbol{A} = \mathbf{p}$, and the fact that $[p_k, p_m] = 0$ to show that

$$[(\mathbf{p} \times \boldsymbol{M})_i, (\mathbf{p} \times \boldsymbol{M})_j] = -p^2 \epsilon_{ijk} M_k.$$

The combination of these results yields

$$[e_i, e_j] = -(p^2/k^2\mu - 2/kr)\epsilon_{ijk} M_k = -(2\mathcal{H}/k^2\mu)\epsilon_{ijk} M_k,$$

where $\mathcal{H} = p^2/2\mu - k/r$ is the total energy. The Poisson bracket $[e^2, e_i]$ is simply $2e_j[e_j, e_i]$. Hence, the eccentricity Poisson brackets are

$$[e_i, e_j] = -(2\mathcal{H}/k^2\mu)\epsilon_{ijk} M_k, \qquad [e^2, e_i] = (4\mathcal{H}/k^2\mu)\epsilon_{ijk} e_j M_k.$$

The Poisson brackets of the eccentricity and angular momentum follow directly from the cyclic reproduction properties of the angular momentum:

$$[e_i, M_j] = \epsilon_{ijk} e_k, \qquad [M^2, e_i] = 2\epsilon_{ijk} e_j M_k, \qquad [e^2, M_i] = 0. \qquad \mathbf{(3.12)}$$

The eccentricity $\boldsymbol{e}$ may be expressed in terms of its dimensional form, the eccentrum $\boldsymbol{E}$, which has the same dimensions as the angular momentum:

$$\boldsymbol{E} = h\boldsymbol{e}.$$

The components of $\boldsymbol{E}$ are real for bound motions ($\mathcal{H} < 0$) and imaginary for free motions ($\mathcal{H} > 0$). The Poisson brackets for $\boldsymbol{E}$ and $\boldsymbol{M}$ are

$$[M_i, M_j] = \epsilon_{ijk} M_k, \qquad [E_i, M_j] = \epsilon_{ijk} E_k, \qquad [E_i, E_j] = \epsilon_{ijk} M_k. \quad \mathbf{(3.13)}$$

As described in Chap. 4, the two three-vectors $\boldsymbol{M}$ and $\boldsymbol{E}$ represent four-dimensional rotations. Four-dimensional rotations may also be represented by an antisymmetric four-matrix. Because of its antisymmetry, such a matrix has only six (rather than sixteen) independent components; and these six components are the six components of the two three-vectors $\boldsymbol{M}$ and $\boldsymbol{E}$. An antisymmetric matrix $F_{\mu\lambda}$ can be built from the two three-vectors as $F_{0j} = -F_{j0} = E_j$ for $j = 1, 2, 3$, and $F_{ij} = \epsilon_{ijk} M_k$ for $i, j, k = 1, 2, 3$. The matrix $F_{\mu\lambda}$ is a unification of the eccentrum and angular momentum in a single mechanical object:

$$F_{\mu\lambda} = \begin{pmatrix} 0 & E_1 & E_2 & E_3 \\ E_1 & 0 & M_3 & -M_2 \\ E_2 & -M_3 & 0 & M_1 \\ E_3 & M_2 & -M_1 & 0 \end{pmatrix}. \qquad \mathbf{(3.14)}$$

The Poisson brackets of the matrix $F_{\mu\lambda}$ may be found from Eqs. (**3.13**) and are the matrix form of the cyclic reproduction property of rotational symmetry:

$$[F_{\mu\lambda}, F_{\lambda\nu}] = F_{\mu\nu}. \tag{3.15}$$

The Poisson bracket (**3.13**) or (**3.15**) whose cyclic structure indicates a rotational symmetry show that the eccentricity and the angular momentum invariants $\boldsymbol{E}$ and $\boldsymbol{M}$ or, more compactly, the matrix $F_{\mu\lambda}$ represent the group of rotations in four dimensions just as the angular momentum is a representation of the group of rotations in three dimensions.

The matrix representing rotations in four-space is actually made up of two kinds of vectors. An ordinary vector changes sign if the coordinates are inverted, that is, $\mathbf{x} \to -\mathbf{x}$. Such vectors have come to be called polar vectors. However, a vector which arises as the cross-product of two polar vectors has the interesting property of not changing sign if the coordinates are inverted (the product of two vectors which change sign does not itself change sign). Such vectors are called axial vectors.

In the language of parity, a polar vector has odd parity whereas an axial vector has even parity. A matrix representing rotations in four-space is made up of a polar vector $\boldsymbol{E}$ and an axial vector $\boldsymbol{M}$. The eccentrum $\boldsymbol{E}$ is a polar vector because the term $\boldsymbol{p} \times \boldsymbol{M}$ *does* change sign upon coordinate inversion since $\boldsymbol{M}$ is an axial vector. The vector product $\boldsymbol{p} \times \boldsymbol{M}$ is the product of a polar and an axial vector. Hence, the invariants of the two-spheres supporting the four-dimensional rotational symmetry of the motion are divided into even ($\boldsymbol{M}$) and odd ($\boldsymbol{E}$) parity.

The matrix $F_{\mu\lambda}$ exists independently of any particular coordinate system. Although the components representing a matrix such as $F_{\mu\lambda}$ will change in the passage from one set of coordinates to another, a matrix possesses certain quantities which are invariant to coordinate transformations. For $F_{\mu\lambda}$, the matrix invariants are $M^2 + E^2$ and $\boldsymbol{M} \cdot \boldsymbol{E}$. These are precisely the Casimir invariants which appear in the subsidiary conditions (4.8) when $\boldsymbol{e}$ is expressed in terms of $\boldsymbol{E}$:

$$M^2 + E^2 = h^2, \qquad \boldsymbol{M} \cdot \boldsymbol{E} = 0. \tag{3.16}$$

The reader familiar with the electromagnetic field will recognize the matrix $F_{\mu\lambda}$ in the case of unbound motions as the image of the electromagnetic field-strength tensor

$$F_{\mu\lambda} = \begin{pmatrix} 0 & iE_1 & iE_2 & iE_3 \\ iE_1 & 0 & B_3 & -B_2 \\ iE_2 & -B_3 & 0 & B_1 \\ iE_3 & B_2 & -B_1 & 0 \end{pmatrix} \tag{3.17}$$

with the role of the electric field $i\boldsymbol{E}$ played by the eccentrum $\boldsymbol{E}$ and that of the magnetic field $\boldsymbol{B}$ played by the angular momentum $\boldsymbol{M}$.

The electromagnetic field equations are invariant to the transformations of Lorentzian space–time (6.4). There is an elegant representation of Lorentzian space–time proposed by Herman Minkowski in which the field-strength tensor and the electromagnetic field equations may be expressed. In Minkowski's representation the time component of the four-vector of an event is made imaginary as $x_0 = ict$. The imaginary time component of the four-vector $x = (ict, x_1, x_2, x_3)$ then turns the Euclidean magnitude of a vector $|x|^2 = x_\lambda x_\lambda$ into the Lorentzian magnitude (save for the overall sign):

$$|x|^2 = -(ct)^2 + {x_1}^2 + {x_2}^2 + {x_3}^2.$$

Maxwell's equations of electromagnetic theory can be written in Minkowski space as

$$\frac{\partial F_{\mu\lambda}}{\partial x_\mu} = J_\lambda,$$

where J_λ is the four-current of charge.

It is well known that the invariants of the electromagnetic field-strength tensor are $E^2 - B^2$ and $\boldsymbol{E}\cdot\boldsymbol{B}$. These invariants match the Casimir invariants for unbound motions of the two-body problem given by Eq. (**3.16**) when $\boldsymbol{E}$ and $\boldsymbol{M}$ appearing in Eq. (**3.14**) are related to the electromagnetic field variables as

$$\boldsymbol{E} \to i\boldsymbol{E}, \qquad \boldsymbol{M} \to \boldsymbol{B}.$$

Note 4
Orbit Integrals of Kepler Motion

The trajectory equations of Kepler motion drawn from the action result in the integral for the orbit in space relating r and ϕ given by Eq. (4.50) and that for the orbit in time relating r and t given by Eq. (4.51).

The orbit in space (4.50) is integrable in elementary trigonometric functions:

$$r = r_0/(1 + e\cos\phi). \tag{4.1}$$

This is the general equation for a conic section (**Fig. 4-1**) with the origin of coordinates r, ϕ at one of the foci. The rectangular coordinates of the conic,

$$x = r\cos\phi, \qquad x_2 = r\sin\phi,$$

are given by

$$x_1 = \frac{r_0\cos\phi}{1 + e\cos\phi}, \qquad x_2 = \frac{r_0\sin\phi}{1 + e\cos\phi}. \tag{4.2}$$

It is readily observed from Eqs. (**4.2**) that in these coordinates *all* conics cross the x_2 axis at the same points $\phi = \pm\pi/2$ for which $x_1 = 0$, $x_2 = \pm r_0$ independently of e. They also all cross the x_1 axis at $x_1 = r_{\min} = r_0/(1+e)$; but this point depends on e and is different for different conics.

The bound orbits which correspond to the circle and ellipse cross the x_1 axis at two points. The first which is closest to the focus at the origin is at $\phi = 0$ and corresponds to $x_1 = r_{\min}$ or

$$(x_1)_{\min} = r_{\min} = r_0/(1+e). \tag{4.3}$$

The second which is farthest from the focus at the origin is at $\phi = \pi$ for which

$$(x_1)_{\max} = r_{\max} = r_0/(1-e). \tag{4.4}$$

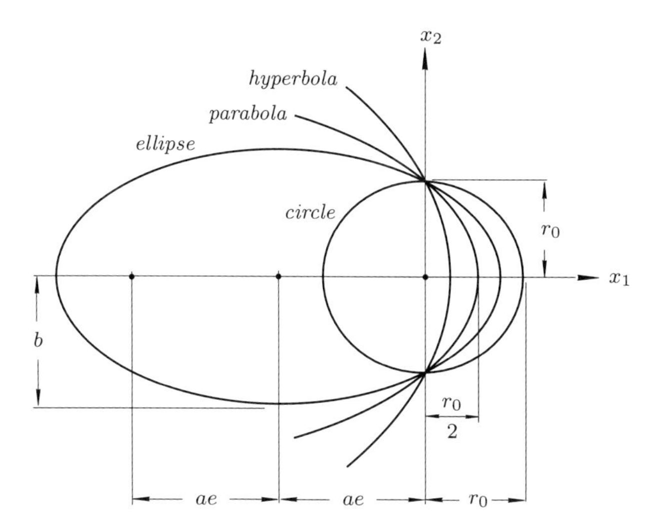

Figure 4-1. The Family of Conic Sections. Conic sections are described by two parameters such as the semi-latus rectum r_0 and the eccentricity e. In a coordinate system whose origin is at the focus, all conics cross the x_2 axis at a distance r_0 from the origin. The circle corresponds to the eccentricity $e = 0$, radius r_0, and both foci lie at the origin. The ellipse ($0 < e < 1$) has one focus at the origin and the other displaced from it by $2ae$ where $a = r_0/(1-e^2)$ is the semi-major axis of the ellipse. The parabola ($e = 1$) separates closed conics from the hyperbolas which are open. The parabola crosses the x_1 axis at a distance $r_0/2$ from the origin. Hyperbolas stream to infinity crossing the x_1 axis at distance $a/(1+e)$ from the origin.

The center points of the bound orbits occur at the points where x_2 reaches a maximum or minimum. Setting

$$\frac{dx_2}{d\phi} = \frac{\cos\phi + e}{(1+e\cos\phi)^2} r_0 = 0,$$

one finds $\cos\phi_{\text{center}} = -e$ and $\cos\phi_{\max} = 1$. The x_2 coordinates of the centers of the bound orbits and the maximum values of x_2 directly follow as

$$(x_2)_{\text{center}} = er_0/(1-e^2), \qquad (x_2)_{\max} = r_0/\sqrt{1-e^2}. \tag{4.5}$$

The results (**4.3**)–(**4.6**) along with the condition that all conics cross the x_2 axis at r_0 may be used to establish the properties of the various cases.

Circle ($e = 0$). The circle crosses the x_1 axis at r_0; it also crosses it symmetrically at $x_1 = -r_0$ and has radius $a = r_0$.

Ellipse ($0 < e < 1$). From the two crossing points of Eqs.(**4.3**) and (**4.4**) one finds the semi-major axis of the ellipse is

$$a = (r_{\text{min}} + r_{\text{max}})/2 = r_0/(1 - e^2).$$

The semi-minor axis is equal to $(x_2)_{\text{max}}$ from the second of Eqs. (**4.5**):

$$b = r_0/\sqrt{1 - e^2}. \tag{4.6}$$

The ratio of the minor to major axis follows as

$$b/a = \sqrt{1 - e^2}.$$

Parabola ($e = 1$). The parabola separates the conics which have a finite extent (circle and ellipse) from those which stream to infinity (hyperbolas). The parabola crosses the x_1 axis at $x_1 = r_{\text{min}} = r_0/2$.

Hyperbola ($e > 1$). The hyperbolas cross the x_1 axis at only one point at $x_1 = r_{\text{min}} = r_0/(1 + e)$.

Note 5
Inversion of the Kepler Equation

The time integral of the orbit results in Kepler's equation relating the eccentric anomaly ψ and the time t:

$$\omega t = \psi - e \sin \psi. \tag{5.1}$$

Newton showed that the solution of Kepler's equation $\psi = \psi(t)$ is transcendental: it is represented as an infinite series in t. This series may be expressed as a Fourier series:

$$\psi(t) = \omega t + \sum_{m=1}^{\infty} a_m(e) \sin m\omega t, \tag{5.2}$$

with the Fourier coefficients a_m given in terms of ψ by

$$a_m = \frac{1}{\pi} \int_0^{2\pi} (\psi - \theta) \sin m\theta \, d\theta,$$

where $\theta \equiv \omega t$. Integration by parts of this expression results in

$$a_m = -\frac{(\psi - \theta)}{m\pi} \cos m\theta \bigg|_0^{2\pi} + \frac{1}{m\pi} \int_0^{2\pi} \cos m\theta \, d\psi - \frac{1}{m\pi} \int_0^{2\pi} \cos m\theta \, d\theta.$$

The first and third terms on the right vanish. Replacing $\theta = \omega t$ by $\psi - e \sin\psi$ from Eq. (**5.1**) one is left with

$$a_m = \frac{1}{m\pi}\int_0^{2\pi} \cos(m\psi - me\sin\psi)\, d\psi. \tag{5.3}$$

This integral is a well known function in mathematical physics known as the Bessel function:

$$J_m(z) \equiv \frac{1}{2\pi}\int_0^{2\pi} \cos(m\psi - z\sin\psi)\, d\psi.$$

The integral (and the Bessel function it represents) can also be represented as the infinite series,

$$J_m(z) = \sum_{k=0}^{\infty} \frac{(-1)^k (z/2)^{m+2k}}{k!(m+k)!}. \tag{5.4}$$

The Fourier coefficients are therefore Bessel functions: $a_m = 2J_m(me)/m$; and the solution of Kepler's equation is expressible as a series in Bessel functions,

$$\psi(t) = \omega t + 2\sum_{m=1}^{\infty} \frac{J_m(me)}{m} \sin m\omega t. \tag{5.5}$$

Note 6
Kepler Motion on the One-, Two-, and Three-Sphere

Kepler motion exists in a four-dimensional phase space $\xi = (\xi_1, \xi_2, \xi_3, \xi_4)$ which, because of its rotational symmetry, lies upon the three-sphere

$${\xi_1}^2 + {\xi_2}^2 + {\xi_3}^2 + {\xi_4}^2 = const.$$

Moreover, because of the exceptional nature of four-dimensional space, the motion may be projected from this three-sphere onto a pair of families of two-spheres. Beyond that, one discovers this phase space is also the phase space of the Hooke oscillator: Kepler and Hooke motion may be transformed into one another! Here is how all these remarkable features of Kepler motion are interwoven.

Kepler motion is most naturally exhibited in the parabolic coordinates unique to the inverse-square force (recall that spherical-polar coordinates are *not* unique to the inverse-square force but are appropriate to all central force laws). The Kepler Hamiltonian (4.38) may be expressed in these coordinates as

$$\mathcal{H} = \frac{4}{\xi + \eta}\left(\xi p_\xi^2/2\mu + \eta p_\eta^2/2\mu\right) + \frac{1}{\xi\eta}{M_3}^2/2\mu - \frac{2k}{\xi + \eta}.$$

The relative motion of the Kepler problem is six-dimensional. Since M_3 is an invariant, the polar mode to which it belongs is immediately integrable and leads to a reduction of two dimensions in the phase space as described in Chap. 3. This integrable mode has the polar action J_ϕ and angle variable $\alpha_\phi \equiv \phi$. The action is the invariant $J_\phi = M_3$ and the phase velocity is $\omega_\phi = \partial\mathcal{H}/\partial J_\phi = M_3/\mu\xi\eta$ so that the separable mode is described by Eqs. (3.32) as

$$\phi = (M_3/\mu\xi\eta)t + const, \qquad J_\phi = M_3.$$

One may remove the polar mode from the Hamiltonian by aligning the coordinates in the natural directions of Kepler motion in parabolic coordinates. These are the three orthogonal directions formed by the vectors $\boldsymbol{M}$, $\mathbf{e}$, and $\boldsymbol{M} \times \mathbf{e}$ of Fig. 4-11. Let the x_3 axis of the coordinate system be aligned with $\mathbf{e}$. Then $M_3 = 0$ and the Hamiltonian becomes

$$\mathcal{H} = \frac{4}{\xi + \eta}\left(\xi p_\xi^2/2\mu + \eta p_\eta^2/2\mu\right) - \frac{2k}{\xi + \eta}. \qquad \mathbf{(6.1)}$$

(Notice that this is in contrast to the development of the orbits in spherical-polar coordinates of Fig. 4-11 for which $\mathbf{M}$ was aligned with the x_3 direction and $e_3 = 0$.)

With removal of the polar mode, the phase space is reduced to four dimensions with coordinates $(\xi, \eta, p_\xi, p_\eta)$. The Hamiltonian (**6.1**) is actually the much-disguised equation of a three-sphere showing that this space is rotationally symmetric. To see that Eq. (**6.1**) is the equation of a three-sphere, introduce the "squared" coordinates,

$$q_1{}^2 \equiv \xi = r + x_3, \qquad q_2{}^2 \equiv \eta = r - x_3,$$

in favor of ξ and η into the Lagrangian corresponding to the Hamiltonian (4.38). The conjugate momenta are

$$p_1 = \partial L/\partial \dot{q}_1 = 2\sqrt{\xi}p_\xi, \qquad p_2 = \partial L/\partial \dot{q}_2 = 2\sqrt{\eta}p_\eta,$$

and the Hamiltonian is

$$\mathcal{H} = \frac{1}{q_1{}^2 + q_2{}^2}\left(p_1{}^2/2\mu + p_2{}^2/2\mu\right) - \frac{2k}{q_1{}^2 + q_2{}^2},$$

which upon multiplication by $(q_1{}^2 + q_2{}^2)$ becomes

$$p_1{}^2/2\mu + p_2{}^2/2\mu + (-2\mathcal{H})(q_1{}^2/2 + q_2{}^2/2) = 2k. \qquad \mathbf{(6.2)}$$

Rescaling the coordinates to

$$\xi_1 = q_1\sqrt{k\mu/2}, \qquad \xi_3 = p_1\sqrt{k/(-4\mathcal{H})},$$
$$\xi_2 = q_2\sqrt{k\mu/2}, \qquad \xi_4 = p_2\sqrt{k/(-4\mathcal{H})},$$

and introducing the Kepler constant $h^2 = k^2\mu/(-2\mathcal{H})$, the equation of a three-sphere of radius $\sqrt{2}h$ embedded in four dimensions results:

$$\xi_1{}^2 + \xi_2{}^2 + \xi_3{}^2 + \xi_4{}^2 = 2h^2. \qquad \textbf{(6.3)}$$

Kepler motion therefore lies upon a three-sphere. But now a striking thing has happened. One finds in passing that Eq. (**6.2**) is also the Hamiltonian of a Hooke oscillator with Hamiltonian $\mathcal{H}_{\text{Hooke}} = 2k$ and force constant $\kappa = -2\mathcal{H}_{\text{Kepler}}$. Hooke and Kepler motions (the only two-body motions with closed orbits) are mutually transformable into one another!

Let us now see how Kepler motion on the three-sphere projects onto two-spheres. This projection is a topological feature of rotations in four-dimensional space and their corresponding three-, two-, and one-spheres. The beautiful family of topological relationships between the three-sphere and its lower dimensional spheres was worked out by the topologist Heinz Hopf in the nineteen thirties and is known as the *Hopf fibration.** Begin by observing that the four coordinates in the equation of the three-sphere (**6.3**) can be written as a pair of complex variables,

$$z_1 = \xi_1 + i\xi_3, \qquad z_2 = \xi_2 + i\xi_4, \qquad \textbf{(6.4)}$$

in terms of which the Kepler Hamiltonian in the form (**6.3**) becomes

$$|z_1|^2 + |z_2|^2 = 2h^2.$$

Let $w = (w_1, w_2, w_3)$ be the coordinates of the three-dimensional space in which a two-sphere is embedded. The projection of the three-sphere onto a pair of two-spheres graphically imaged in Fig. 4-4 is analytically imaged in the mapping

$$\xi = (\xi_1, \xi_2, \xi_3, \xi_4) \begin{array}{l} \nearrow \quad w^+ = (w_1, w_2, w_3) \\ \\ \searrow \quad w^- = (w_1, w_2, w_3) \end{array},$$

where w^+ and w^- are the coordinates of the two three-dimensional spaces into which the four-dimensional space projects. The Hopf map is most simply expressed in the complex representation z_1, z_2:

$$w^+: \begin{array}{rl} w_1 = & z_1^* z_2 + z_1 z_2^*, \\ w_2 = & i(z_1^* z_2 - z_1 z_2^*), \\ w_3 = & z_1^* z_1 - z_2 z_2^*, \end{array} \qquad w^-: \begin{array}{rl} w_1 = & z_1^* z_2^* + z_1 z_2, \\ w_2 = & i(z_1^* z_2^* - z_1 z_2), \\ w_3 = & z_1^* z_1 - z_2^* z_2. \end{array} \qquad \textbf{(6.5)}$$

---

* See, for example, H. Seifert and W. Threlfall, *A Textbook of Topology*, Academic Press (1980) and Chap. III of Raoul Bott and L. W. Tu, *Differential Forms in Algebraic Topology*, Springer-Verlag, New York (1982).

Equations (**6.5**) contain two sets of coordinates. Each set satisfies the equation of the two-sphere,

$$w_1{}^2 + w_2{}^2 + w_3{}^2 = (|z_1|^2 + |z_2|^2)^2 = 4h^4.$$

The sets are the same except that z_2 in w^+ is interchanged with z_2^* in w^-; and it is this interchange that distinguishes one map onto the two-sphere from the other. The switch from z_2 to z_2^* corresponds to a switch in the sign of ξ_4 which means the two sets correspond to the pair of families of two-spheres into which the three-sphere projects. One set covers the three-sphere over the range $\xi_4 > 0$ while the other covers the range $\xi_4 < 0$.

In a beautiful unfolding, the three-sphere of Kepler motion is mapped onto two-spheres, and, accordingly, the four-dimensional rotational symmetry of Kepler motion is made equivalent to a pair of three-dimensional rotational symmetries. In the process Kepler motion is mapped into Hooke motion.

Kepler motion is a bound integrable motion. It therefore consists of a closed orbit lying upon a three-torus. But the motion also lies upon a three-sphere. In general, it is not possible for motion to both lie upon a sphere and a torus; but in the amazing case of the three-sphere it is possible. Simple orthographic projection maps the three-sphere onto a pair of families of two-spheres as shown in Chap. 4. Remarkably, it is also possible to map the three-sphere onto a family of nested two-tori (and two-exceptional circles corresponding to the poles). Each of the two-tori which foliate the three-sphere maps to a circle lying on a two-sphere. Thus, in addition to its foliation by a pair of families of nested two-spheres, the three-sphere may be foliated with a family of nested two-tori. The spherical foliations portray the rotational symmetry of Kepler motion while the toroidal foliations portray its integrability. Since integrability is the result of isolating invariants and the invariants flow from the symmetry, the two features are fully united in the three-sphere.*

Note 7
The Dirac Postulate

P. A. M. Dirac conceived a brilliant postulate which links the commutator of quantum operators and the Poisson bracket of the corresponding dynamical quantities which these operators represent. This postulate is the critical link in the evolution of classical mechanics into quantum mechanics. Here is Dirac's reasoning.

---

* A detailed exposition of the way in which the three-sphere may be foliated with two-tori illustrated with a gallery of computer-generated images may be found in Hüseyin Koçak et. al., "Topology and Mechanics with Computer Graphics," *Adv. Appl. Math.*, **7**, 282-308 (1986) and Hüseyin Koçak and David Laidlaw, "Computer Graphics and the Geometry of S^3," *Mathematical Intelligencer* **9**, 1, 8 (1987).

The Poisson bracket of any two quantities A, B is

$$[A, B] = \left(\frac{\partial A}{\partial q_i} \frac{\partial B}{\partial p_i} - \frac{\partial B}{\partial q_i} \frac{\partial A}{\partial p_i} \right).$$

Poisson brackets behave as

$$[A, B] = -[B, A], \qquad [A, A] = [B, B] = 0.$$

Poisson brackets also possess the properties

$$[A_1 + A_2, B] = [A_1, B] + [A_2, B], \qquad [A, B_1 + B_2] = [A, B_1] + [A, B_2],$$

which flow directly from the behavior of derivatives. Using these properties, the Poisson brackets in which one of the arguments is a product may be shown to be

$$\begin{aligned} [A_1 A_2, B] &= [A_1, B] A_2 + A_1 [A_2, B], \\ [A, B_1 B_2] &= [A, B_1] B_2 + B_1 [A, B_2]. \end{aligned} \tag{7.1}$$

The Poisson brackets in which both arguments are products $A_1 A_2$ and $B_1 B_2$ can be formed in two ways. Using the first of Eqs. (**7.1**), there results

$$\begin{aligned} [A_1 A_2, B_1 B_2] = [A_1, B_1] B_2 A_2 &+ A_1 [A_2, B_1] B_2 \\ &+ B_1 [A_1, B_2] A_2 + A_1 B_1 [A_2, B_2], \end{aligned} \tag{7.2a}$$

and using the second of Eqs. (**7.1**),

$$\begin{aligned} [A_1 A_2, B_1 B_2] = [A_1, B_1] A_2 B_2 &+ A_1 [A_2, B_1] B_2 \\ &+ B_1 [A_1, B_2] A_2 + B_1 A_1 [A_2, B_2]. \end{aligned} \tag{7.2b}$$

Since these two expressions are equal, it follows that

$$\frac{(A_1 B_1 - B_1 A_1)}{[A_1, B_1]} = \frac{(A_2 B_2 - B_2 A_2)}{[A_2, B_2]}. \tag{7.3}$$

Now the pairs A_1, B_1 and A_2, B_2 are independent. Hence, each of the two terms in Eq. (**7.3**) is independent of the other. They must each be the same constant. Thus, for any A, B it is true that

$$(AB - BA) = \gamma [A, B], \tag{7.4}$$

where γ is a universal constant.

The constant in Eq. (**7.4**) can be shown to be pure imaginary with the following argument. Let $C = A + iB$ be a complex quantity where A and B are real but not necessarily commuting quantities. The magnitude of C is then

$$|C|^2 = A^2 + B^2 + i(AB - BA). \tag{7.5}$$

Since $|C|^2$, A^2, and B^2 are real in Eq. (**7.5**), the term $i(AB - BA)$ must also be a real function of A and B, say

$$i(AB - BA) = f(A, B). \qquad (\mathbf{7.6})$$

Comparing Eqs. (**7.4**) and (**7.6**) it follows that $f(A, B) = -i\gamma[A, B]$. Since $[A, B]$ and $f(A, B)$ are real, γ is a pure imaginary constant, say $\gamma = i\hbar$ with $\hbar$ real, and the equality (**7.4**) can be written

$$(AB - BA) = i\hbar[A, B]. \qquad (\mathbf{7.7})$$

Note 8
Canonical Perturbation of Near-Integrable Motion

The canonical perturbation of near-integrable motion takes as starting point an integrable motion in angle-action coordinates (α, J) whose integrable Hamiltonian is $H_0(J)$ and phase velocity is $\omega = \partial H_0/\partial J$. The motion is assumed close to a nonintegrable motion whose Hamiltonian $H(\alpha, J)$ differs from that of the integrable motion by a small quantity $H'(\alpha, J)$:

$$H(\alpha, J) = H_0(J) + H'(\alpha, J). \qquad (\mathbf{8.1})$$

Poincaré proposed a procedure for obtaining an approximate solution to the nonintegrable motion by first neglecting quantities higher order than linear in the perturbation $H'(\alpha, J)$ and then making a canonical transformation to new coordinates,

$$(\alpha, J) \to (\bar{\alpha}, \bar{J}),$$

such that the Hamiltonian in the new coordinates $\bar{H}(\bar{\alpha}, \bar{J})$ is a function only of the new action: $\bar{H}(\bar{\alpha}, \bar{J}) = \bar{H}(\bar{J})$. The motion would then be integrable in the new coordinates.

Poincaré designed the canonical transformation to achieve integrability in the new coordinates by *making* the new Hamiltonian $\bar{H}(\bar{J})$ equal to the old Hamiltonian plus the average of the perturbation Hamiltonian over the angle coordinates,

$$\bar{H}(\bar{J}) = H_0(\bar{J}) + \langle H'(\bar{\alpha}, \bar{J})\rangle. \qquad (\mathbf{8.2})$$

The average over all angle coordinates is defined as

$$\langle H'(\bar{\alpha}, \bar{J})\rangle \equiv (2\pi)^{-s} \int_0^{2\pi} H'(\bar{\alpha}, \bar{J})\, d^s\bar{\alpha}.$$

Poincaré's averaging proposal is the key to the procedure and its validity rests on the following argument. The phase velocity of the motion in the new coordinates is given by

$$\bar{\omega} = \frac{\partial}{\partial \bar{J}}\bar{H}(\bar{J}) = \omega + \frac{\partial}{\partial \bar{J}}\langle H'(\bar{\alpha}, \bar{J})\rangle.$$

The phase velocities in the old and new coordinates differ by a small quantity. The perturbation is a "slow" motion superimposed upon the "fast" unperturbed motion. It therefore changes little over the time scale in which the angle coordinates execute one cycle of their periodic motions. The Hamiltonian perturbation is therefore approximately invariant over one cycle of the angle coordinates; and it is over such a cycle that the average is taken. The perturbation procedure fails if the phase velocity of the perturbation is comparable to the phase velocity of the unperturbed motion.

The generating function of the transformation from $H(\alpha, J)$ to $\bar{H}(\bar{J})$ is the action

$$S(\alpha, \bar{J}) = \alpha_j \bar{J}_j + S'(\alpha, \bar{J}). \tag{8.3}$$

It consists of the identity transformation $\alpha_j \bar{J}_j$ and a perturbation $S'(\alpha, \bar{J})$. The problem is to find $S'(\alpha, \bar{J})$ which transforms the given $H(\alpha, J)$ to $\bar{H}(\bar{J})$.

The canonical transformation (**8.3**) is in the "old position, new momentum" format described by Eqs. (3.29) and (3.30) according to which the remaining coordinates are given by

$$\bar{\alpha} = \partial S/\partial \bar{J}, \qquad J = \partial S/\partial \alpha,$$

or

$$\bar{\alpha} = \alpha + \frac{\partial}{\partial \bar{J}} S'(\alpha, \bar{J}), \qquad J = \bar{J} + \frac{\partial}{\partial \alpha} S'(\alpha, \bar{J}). \tag{8.4}$$

These equations may be inverted to terms linear (or first order) in the perturbations

$$\alpha(\bar{\alpha}, \bar{J}) \approx \bar{\alpha} - \frac{\partial}{\partial \bar{J}} S'(\bar{\alpha}, \bar{J}), \qquad J(\bar{\alpha}, \bar{J}) \approx \bar{J} + \frac{\partial}{\partial \bar{\alpha}} S'(\bar{\alpha}, \bar{J}). \tag{8.5}$$

Note that to this order it is permissible to replace α by $\bar{\alpha}$ in $S'(\bar{\alpha}, \bar{J})$. The Hamiltonian in the new coordinates $\bar{H}(\bar{\alpha}, \bar{J})$ is the same as the Hamiltonian in the old coordinates, $\bar{H}(\bar{\alpha}, \bar{J}) = H(\alpha, J)$. The old coordinate Hamiltonian on the right-hand side of this equation may be expanded into the new coordinates to first order as

$$\bar{H}(\bar{\alpha}, \bar{J}) = H_0(\bar{J}) + \frac{\partial H_0}{\partial \bar{J}}(J - \bar{J}) + H'(\bar{\alpha}, \bar{J}), \tag{8.6}$$

where use has been made of the fact that to first order $H'(\alpha, J) \approx H'(\bar{\alpha}, \bar{J})$. The perturbation $\bar{H}(\bar{\alpha}, \bar{J})$ may be decomposed into its average value and a deviation from the average, $\Delta H'(\bar{\alpha}, \bar{J})$:

$$H'(\bar{\alpha}, \bar{J}) \equiv \langle H'(\bar{\alpha}, \bar{J}) \rangle + \Delta H'(\bar{\alpha}, \bar{J}). \tag{8.7}$$

Substituting from the second of Eqs. (**8.5**) and from Eq. (**8.7**) into Eq. (**8.6**) and using the fact that $\omega = \partial H_0/\partial \bar{J}$, one finds

$$\bar{H}(\bar{\alpha}, \bar{J}) = H_0(\bar{J}) + \omega \frac{\partial}{\partial \bar{\alpha}} S'(\bar{\alpha}, \bar{J}) + \langle H'(\bar{\alpha}, \bar{J}) \rangle + \Delta H'(\bar{\alpha}, \bar{J}). \tag{8.8}$$

The Hamiltonian in new coordinates may now be made purely a function of $\bar{J}$ by selecting it from Eq. (**8.8**) to be $\bar{H}(\bar{\alpha}, \bar{J}) \equiv \bar{H}(\bar{J})$ given by Eq. (**8.2**). This then requires that the action perturbation $S'(\bar{\alpha}, \bar{J})$ is given by Eq. (**8.8**) as

$$\omega \frac{\partial}{\partial \bar{\alpha}} S'(\bar{\alpha}, \bar{J}) = -\Delta H'(\bar{\alpha}, \bar{J}). \tag{8.9}$$

Equation (**8.9**) shows how the action perturbation is related to the Hamiltonian perturbation. The action perturbation may be explicitly expressed in terms of the Hamiltonian perturbation through their Fourier expansions

$$S'(\bar{\alpha}, \bar{J}) = \sum_m S'_m(\bar{J}) e^{i(\bar{\alpha}_1 m_1 + \bar{\alpha}_2 m_2 + \cdots + \bar{\alpha}_s m_s)} \tag{8.10}$$

and

$$\Delta H'(\bar{\alpha}, \bar{J}) = \sum_m \Delta H'_m(\bar{J}) e^{i(\bar{\alpha}_1 m_1 + \bar{\alpha}_2 m_2 + \cdots + \bar{\alpha}_s m_s)}. \tag{8.11}$$

One may now differentiate Eq. (**8.10**) and insert it along with Eq. (**8.11**) into Eq. (**8.9**) to produce the action

$$S(\bar{\alpha}, \bar{J}) = \bar{\alpha}_j \bar{J}_j + \sum_m i \frac{\Delta H'_m(\bar{J}) e^{i(\bar{\alpha}_1 m_1 + \bar{\alpha}_2 m_2 + \cdots + \bar{\alpha}_s m_s)}}{m_1 \omega_1 + m_2 \omega_2 + \cdots + m_s \omega_s} \tag{8.12}$$

from which all the properties of the motion may be found.

Note 9
Hydrogen Atom Spherical-Polar States

It is possible to describe the eigenvalues of the hydrogen atom—the energy levels and the angular momentum and eccentricity values—with minimal reference to a coordinate system. However, a coordinate system must be incorporated into the description to render the eigenstates in detail.

The complete symmetries of the hydrogen atom are manifested in two coordinate systems: spherical-polar and parabolic. Corresponding to these coordinates are the two state specifications, the one vector with Casimir set $(\mathcal{H}, M, M_3)$, the other spinor with $(\mathcal{H}, E_3, M_3)$.

The angular momentum is the bearer of the central force symmetry. The central force symmetry is manifested in spherical-polar coordinates and the angular momentum eigenstates are naturally revealed in these coordinates. The first task in revealing the spherical-polar eigenstates of the hydrogen atom is to establish the angular momentum and energy operators in spherical-polar coordinates. The spherical-polar coordinates r, θ, ϕ of Fig. 4-9 are related to the rectangular Cartesian coordinates by

$$x_1 = r \sin\theta \cos\phi, \qquad x_2 = r \sin\theta \sin\phi, \qquad x_3 = r \cos\theta. \tag{9.1}$$

The operators $\hat{M}^2$ and $\hat{M}_3$ may be calculated from the definition $\hat{M}_i = -i\hbar\epsilon_{ijk}x_j\partial/\partial x_k$ with x_i and $\partial/\partial x_i$ expressed in terms of the spherical-polar variables using Eq. (**9.1**). The angle ϕ describes rotations about the polar axis and the angular momentum represents the rotational aspects of space. The polar angular momentum $M_3 = p_\phi$ therefore has the operator representation

$$\hat{M}_3 = -i\hbar\left(x_1\frac{\partial}{\partial x_2} - x_2\frac{\partial}{\partial x_1}\right) = -i\hbar\frac{\partial}{\partial\phi}. \tag{9.2}$$

The operator $\hat{M}_3$ represents pure rotations about the polar axis. For the total angular momentum operator one obtains in the same manner

$$\hat{M}^2 = -\hbar^2\left[\frac{1}{\sin\theta}\frac{\partial}{\partial\theta}\left(\sin\theta\frac{\partial}{\partial\theta}\right) + \frac{1}{\sin^2\theta}\frac{\partial^2}{\partial\phi^2}\right]. \tag{9.3}$$

The total angular momentum consists of a contribution from the angular variation in θ and a contribution from $\hat{M}_3^2$ representing the rotations in ϕ about the polar axis:

$$\hat{M}^2 = -\hbar^2\frac{1}{\sin\theta}\frac{\partial}{\partial\theta}\left(\sin\theta\frac{\partial}{\partial\theta}\right) + \frac{1}{\sin^2\theta}\hat{M}_3^2. \tag{9.4}$$

Equations (**9.2**) and (**9.4**) may be compared with their classical counterparts, Eqs. (4.27) and (4.28):

$$M_3 = p_\phi, \qquad M^2 = p_\theta^2 + \frac{1}{\sin^2\theta}{M_3}^2,$$

and the operator $\hat{p}_\theta^2$ is found to be

$$\hat{p}_\theta^2 = -\hbar^2\frac{1}{\sin\theta}\frac{\partial}{\partial\theta}\left(\sin\theta\frac{\partial}{\partial\theta}\right).$$

The total angular momentum operator may therefore be expressed in terms of the momentum operator $\hat{p}_\theta^2$ and the polar momentum operator $\hat{M}_3$ in a manner that images the classical connection:

$$\hat{M}^2 = \hat{p}_\theta^2 + \frac{1}{\sin^2\theta}\hat{M}_3^2. \tag{9.5}$$

The operator $\hat{M}^2$ is also directly related to the angular portion of the Laplacian operator in spherical-polar coordinates:

$$\hat{M}^2 = -\hbar^2 r^2\nabla_\perp^2. \tag{9.6}$$

The angular portion of the Laplacian is

$$\nabla_\perp^2 = \frac{1}{r^2 \sin\theta}\frac{\partial}{\partial\theta}\left(\sin\theta\frac{\partial}{\partial\theta}\right) + \frac{1}{r^2\sin^2\theta}\frac{\partial^2}{\partial\phi^2} \qquad \textbf{(9.7)}$$

and is itself composed of angular parts in θ and ϕ which match the total squared angular momentum as a composition of $\hat{p}_\theta^2$ and $\hat{M}_3^2$ given by Eq. (**9.5**).

Having established the angular momentum operators, let us now turn to the energy in spherical-polar coordinates. The kinetic energy operator $\hat{p}^2/2\mu$ is directly related to the Laplacian operator ∇^2 (Chap. 5):

$$\hat{p}^2/2\mu = -(\hbar^2/2\mu)\nabla^2. \qquad \textbf{(9.8)}$$

In spherical-polar coordinates, the Laplacian is

$$\nabla^2 = \frac{1}{r^2}\frac{\partial}{\partial r}\left(r^2\frac{\partial}{\partial r}\right) + \frac{1}{r^2 \sin\theta}\frac{\partial}{\partial\theta}\left(\sin\theta\frac{\partial}{\partial\theta}\right) + \frac{1}{r^2\sin^2\theta}\frac{\partial^2}{\partial\phi^2}. \qquad \textbf{(9.9)}$$

In two-body motion the kinetic energy is decomposable into a radial portion $p_r^2/2\mu$ and an angular portion $M^2/2\mu r^2$ as described in Chap. 3. This decomposition also holds true for the operator representation of these variables. The radial energy operator

$$\hat{p}_r^2/2\mu = -\frac{\hbar^2}{2\mu}\frac{1}{r^2}\frac{\partial}{\partial r}\left(r^2\frac{\partial}{\partial r}\right)$$

corresponds to the radial portion of the Laplacian while the angular energy operator

$$\hat{M}^2/2\mu r^2 = -\frac{\hbar^2}{2\mu}\left[\frac{1}{r^2 \sin\theta}\frac{\partial}{\partial\theta}\left(\sin\theta\frac{\partial}{\partial\theta}\right) + \frac{1}{r^2\sin^2\theta}\frac{\partial^2}{\partial\phi^2}\right]$$

corresponds to the angular portion. The decomposition of the kinetic energy operator

$$\hat{p}^2/2\mu = \hat{p}_r^2/2\mu + \hat{M}^2/2\mu r^2$$

bears a one-to-one correspondence to the decomposition of the Laplacian

$$\nabla^2 = \frac{1}{r^2}\frac{\partial}{\partial r}\left(r^2\frac{\partial}{\partial r}\right) + \nabla_\perp^2.$$

The total energy or Hamiltonian operator $\hat{\mathcal{H}} = \hat{p}^2/2\mu - k/r$ is

$$\hat{\mathcal{H}} = \mathcal{E}_0\left(a_0^2\nabla^2 + 2\frac{a_0}{r}\right), \qquad \textbf{(9.10)}$$

where $\mathcal{E}_0 = -k^2\mu/2\hbar^2$ is the ground state energy and $a_0 = \hbar^2/k\mu$ is the Bohr radius. The Hamiltonian operator may also be expressed in terms of its component parts as

$$\hat{\mathcal{H}} = \mathcal{E}_0 \left[\left(\frac{a_0}{r}\right)^2 \frac{\partial}{\partial r}\left(r^2 \frac{\partial}{\partial r}\right) - \frac{1}{\hbar^2}\left(\frac{a_0}{r}\right)^2 \hat{M}^2 + 2\frac{a_0}{r} \right]. \qquad (\mathbf{9.11})$$

The fact that the Hamiltonian operator splits into a radial and two angular portions is the manifestation of the separation of the Hamilton–Jacobi equation for the two-body problem on the quantum level.

The three operators $\hat{\mathcal{H}}$, $\hat{M}^2$, and $\hat{M}_3$ which describe the spherical-polar states of the hydrogen atom are now established. The eigenvalues of these operators have been determined in Chap. 5. These eigenvalues are characterized by three quantum numbers n, l, m:

$$\mathcal{H} = \mathcal{E}_0/n^2, \qquad M^2 = l(l+1)\hbar^2, \qquad M_3 = m\hbar. \qquad (\mathbf{9.12})$$

The eigenstate corresponding to these eigenvalues is symbolized as Ψ_{nlm}. The operator equations for the eigenstates are therefore

$$\begin{aligned} \hat{M}_3 \Psi_{nlm} &= m\hbar \Psi_{nlm}, \\ \hat{M}^2 \Psi_{nlm} &= l(l+1)\hbar^2 \Psi_{nlm}, \\ \hat{\mathcal{H}} \Psi_{nlm} &= (\mathcal{E}_0/n^2)\Psi_{nlm}. \end{aligned} \qquad (\mathbf{9.13})$$

The integrable nature of two-body motion on the classical level [the fact that one can separate the Hamilton–Jacobi equation into pure r, θ, and ϕ components for which $S(r,\phi,\theta) = -\mathcal{E}t + S_r(r) + S_\phi(\phi) + S_\theta(\theta)$ and the ordinary differential equations are directly integrable] is also manifested on the quantum level. The eigenstates for quantum two-body motion described by the Hamiltonian (**9.11**) are separable. The eigenstate $\Psi = e^{iS/\hbar}$ can be expressed as a product of functions which each separately depend upon r, ϕ, and θ:

$$\Psi_{nlm}(r,\phi,\theta) = R_{nl}(r)\Theta_{lm}(\theta)\Phi_m(\phi),$$

where $R_{nl}(r) = e^{iS_r/\hbar}$, $\Theta_{lm}(\theta) = e^{iS_\theta/\hbar}$, and $\Phi_m(\phi) = e^{iS_\phi/\hbar}$. The corresponding operator equations (**9.13**) for M_3, M^2, and $\mathcal{H}$ also split into ordinary differential equations separately governing $\Phi_m(\phi)$, $\Theta_{lm}(\theta)$, and $R_{nl}(r)$. For Φ_m, one finds

$$-i\hbar \frac{d\Phi_m}{d\phi} = m\hbar\Phi_m. \qquad (\mathbf{9.14})$$

The eigenstates are normalized by the condition $\int \Psi^*_{nlm}\Psi_{nlm} d^3x = 1$ which therefore requires that the solutions of Eq. (**9.14**) be normalized by the condition

$$\int_0^{2\pi} \Phi^*_m \Phi_m \, d\phi = 1.$$

The normalized solutions of Eq. (**9.14**) are

$$\Phi_m(\phi) = \frac{1}{\sqrt{2\pi}} e^{im\phi}. \qquad (\mathbf{9.15})$$

The eigenstates for the motion of the electron about the nucleus must be periodic in azimuthal angle: $\Phi_m(\phi + 2\pi) = \Phi_m(\phi)$. Only those eigenstates with integer values of m are therefore required for the hydrogen atom.

Using Eq. (**9.14**) to eliminate the ϕ dependence in $\hat{M}^2$ given by Eq. (**9.3**), the operator identity for M^2 (**9.3**) becomes an equation for $\Theta_{lm}(\theta)$:

$$\frac{1}{\sin\theta}\frac{d}{d\theta}\left(\sin\theta \frac{d\Theta_{lm}}{d\theta}\right) - \frac{m^2}{\sin^2\theta}\Theta_{lm} + l(l+1)\Theta_{lm} = 0. \qquad (\mathbf{9.16})$$

Although the eigenfunctions which solve (**9.14**) can be found as elementary exponential functions (**9.15**), the solutions of Eq. (**9.16**) are not elementary functions. This differential equation can be thought of as the defining condition for the functions $\Theta_{lm}(\theta)$. The functions $\Theta_{lm}(\theta)$ which are periodic in θ with period π, normalized by the condition

$$\int_0^\pi \Theta_{lm}^* \Theta_{lm} \sin\theta \, d\theta = 1,$$

and satisfying Eq. (**9.16**) are the normalized *associated Legendre polynomials*:

$$\Theta_{lm}(\theta) = (-1)^m i^l \sqrt{\frac{(2l+1)(l-|m|)!}{2(l+|m|)!}} P_l^m(\cos\theta).$$

In the above expression the factor $(-1)^m$ is omitted for $m < 0$. The Legendre polynomials are defined by the recurrence relations

$$P_l^m(\cos\theta) = \frac{1}{2^l l!} \sin^m\theta \frac{d^{l+m}}{d\cos\theta^{l+m}} (\cos^2\theta - 1)^l.$$

The first few angular eigenfunctions in θ are

$$\begin{aligned}
\Theta_{00} &= \frac{1}{\sqrt{2}}, \\
\Theta_{10} &= \frac{\sqrt{6}}{2}\cos\theta, \\
\Theta_{1\pm1} &= \frac{\sqrt{3}}{2}\sin\theta, \\
\Theta_{20} &= \frac{\sqrt{10}}{4}(3\cos^2\theta - 1), \\
\Theta_{2\pm1} &= \frac{\sqrt{15}}{2}\sin\theta\cos\theta, \\
\Theta_{2\pm2} &= \frac{\sqrt{15}}{4}\sin^2\theta.
\end{aligned}$$

Using both Eqs. (**9.14**) and (**9.16**) to eliminate the ϕ and θ dependence from Eq. (**9.11**), the operator equation for the energy, the last of Eqs. (**9.13**) with $\hat{\mathcal{H}}$ given by (**9.11**) becomes an ordinary differential equation in r:

$$\frac{1}{r^2}\frac{d}{dr}\left(r^2\frac{d}{dr}R_{nl}\right) - \frac{l(l+1)}{r^2}R_{nl} - \frac{1}{a_0^2}\left(\frac{1}{n^2} - 2\frac{a_0}{r}\right)R_{nl} = 0. \qquad (\mathbf{9.17})$$

The normalized functions $R_{nl}(r)$ which satisfy Eq. (**9.17**) and remain finite at $r \to 0$ and vanish for $r \to \infty$ then can be found in terms of a pervasive function in mathematical physics: the confluent hypergeometric function $F_\gamma^\alpha(z)$. The confluent hypergeometric function is defined by the series

$$F_\gamma^\alpha(z) = 1 + \frac{\alpha}{\gamma}\frac{z}{1!} + \frac{\alpha(\alpha+1)}{\gamma(\gamma+1)}\frac{z^2}{2!} + \cdots. \qquad (\mathbf{9.18})$$

This series is a terminating polynomial when α is a negative integer which is the case for the hydrogen atom. For the integer case with $\alpha = -n$ and $\gamma = m$, these polynomials may also be defined by the recurrence relation

$$F_m^{-n}(z) = \frac{(-1)^{m-1}}{m(m+1)\cdots(m+n-1)}e^{-z}\frac{d^{m+n-1}}{dz^{m+n-1}}(e^z z^n). \qquad (\mathbf{9.19})$$

The normalization condition for the radial dependence is

$$\int_0^\infty R_{nl}^2 r^2\, dr = 1.$$

With this normalization condition, the radial eigenfunctions are expressible in terms of the confluent hypergeometric function as

$$R_{nl}(r) = \frac{2a_0^{-3/2}}{n^{l+2}(2l+1)!}\sqrt{\frac{(n+l)!}{(n-1-l)!}}(2r/a_0)^l e^{-r/na_0} F_{2l+2}^{-(n-1)+l}(2r/na_0). \qquad (\mathbf{9.20})$$

For this particular combination of exponents, the confluent hypergeometric function is better known as the Laguerre polynomial and the radial eigenfunctions may be expressed alternatively as

$$R_{nl}(r) = -\frac{2a_0^{-3/2}}{n^{l+2}}\sqrt{\frac{(n-l-1)!}{(n+l)!^3}}(2r/a_0)^l e^{-r/na_0} L_{n+1}^{2l+1}(2r/na_0). \qquad (\mathbf{9.21})$$

The functions $L_n^m(z)$ are the *associated Laguerre polynomials* defined by the recurrence relations

$$L_n^m(z) = (-1)^m\frac{n!}{(n-m)!}e^z z^{-m}\frac{d^{n-m}}{dz^{n-m}}(e^{-z}z^n). \qquad (\mathbf{9.22})$$

The first few radial eigenfunctions are

$$
\begin{aligned}
R_{10} &= 2a_0^{-3/2} e^{-r/a_0}, \\
R_{20} &= \frac{1}{\sqrt{2}} a_0^{-3/2} \left[1 - \frac{1}{2}\left(\frac{r}{a_0}\right)\right] e^{-r/2a_0}, \\
R_{21} &= \frac{1}{2\sqrt{6}} a_0^{-3/2} \left(\frac{r}{a_0}\right) e^{-r/2a_0}, \\
R_{30} &= \frac{2}{3\sqrt{3}} a_0^{-3/2} \left[1 - \frac{2}{3}\left(\frac{r}{a_0}\right) + \frac{2}{27}\left(\frac{r}{a_0}\right)^2\right] e^{-r/3a_0}, \\
R_{31} &= \frac{8}{27\sqrt{6}} a_0^{-3/2} \left(\frac{r}{a_0}\right)\left[1 - \frac{1}{6}\left(\frac{r}{a_0}\right)\right] e^{-r/3a_0}, \\
R_{32} &= \frac{4}{81\sqrt{30}} a_0^{-3/2} \left(\frac{r}{a_0}\right)^2 e^{-r/3a_0}.
\end{aligned}
$$

The succession of spherical-polar eigenstates for the hydrogen atom proceeds as

$$
\begin{aligned}
\Psi_{100} &= \frac{1}{\sqrt{\pi}} a_0^{-3/2} e^{-r/a_0}, \\
\Psi_{200} &= \frac{1}{4\sqrt{2\pi}} a_0^{-3/2} \left[2 - \left(\frac{r}{a_0}\right)\right] e^{-r/2a_0}, \\
\Psi_{210} &= \frac{1}{4\sqrt{2\pi}} a_0^{-3/2} \left(\frac{r}{a_0}\right) e^{-r/2a_0} \cos\theta, \\
\Psi_{21\pm1} &= \frac{1}{8\sqrt{\pi}} a_0^{-3/2} \left(\frac{r}{a_0}\right) e^{-r/2a_0} \sin\theta e^{\pm i\phi}, \\
\Psi_{300} &= \frac{1}{3\sqrt{3\pi}} a_0^{-3/2} \left[1 - \frac{2}{3}\left(\frac{r}{a_0}\right) + \frac{2}{27}\left(\frac{r}{a_0}\right)^2\right] e^{-r/3a_0}, \\
\Psi_{310} &= \frac{2\sqrt{2}}{27\sqrt{\pi}} a_0^{-3/2} \left(\frac{r}{a_0}\right)\left[1 - \frac{1}{6}\left(\frac{r}{a_0}\right)\right] e^{-r/3a_0} \cos\theta, \\
\Psi_{31\pm1} &= \frac{2}{27\sqrt{\pi}} a_0^{-3/2} \left(\frac{r}{a_0}\right)\left[1 - \frac{1}{6}\left(\frac{r}{a_0}\right)\right] e^{-r/3a_0} \sin\theta e^{\pm i\phi}, \\
\Psi_{320} &= \frac{1}{81\sqrt{6\pi}} a_0^{-3/2} \left(\frac{r}{a_0}\right)^2 e^{-r/3a_0} \left(3\cos^2\theta - 1\right), \\
\Psi_{32\pm1} &= \frac{1}{81\sqrt{\pi}} a_0^{-3/2} \left(\frac{r}{a_0}\right)^2 e^{-r/3a_0} \sin\theta\cos\theta e^{\pm i\phi}, \\
\Psi_{32\pm2} &= \frac{1}{162\sqrt{\pi}} a_0^{-3/2} \left(\frac{r}{a_0}\right)^2 e^{-r/3a_0} \sin^2\theta e^{\pm i2\phi}.
\end{aligned}
$$

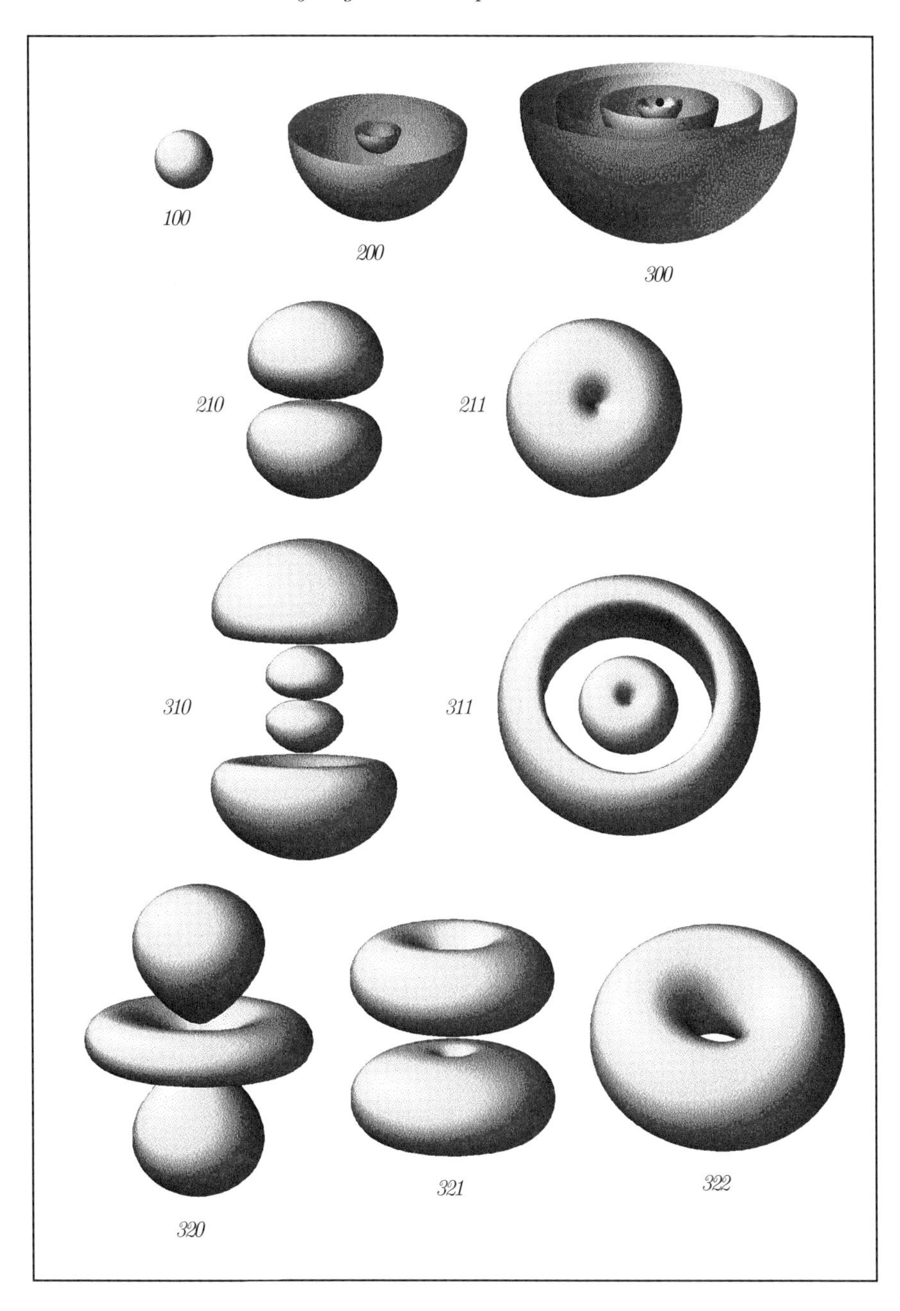

Figure 9-1. Hydrogen Atom Spherical Polar States. States identified by quantum numbers (n, l, m) are exhibited as probability density surfaces. Multiple surfaces of relative maxima in the probability density for the principal levels 200 and 300 are shown which locate nodes between them. Atoms in the 211, 311, and, to a lesser degree, the 322 state have been rotated for better viewing. *(Courtesy George D. Purvis III)*

The parity of spherical-polar states may be established in the following manner. When the coordinates x_i are inverted as $x_i \to -x_i$, spherical-polar coordinates undergo the transformation

$$r \to r, \qquad \phi \to \phi + \pi, \qquad \theta \to \pi - \theta.$$

It can be seen from the forms of $\Phi_m(\phi)$ and $\Theta_{lm}(\theta)$ that these functions undergo the transformation

$$\begin{aligned} P_l^m(\cos\theta) &\to (-1)^{l-m} P_l^m(\cos\theta), \\ e^{im\phi} &\to (-1)^m e^{im\phi}. \end{aligned}$$

As a result, the state function Ψ_{nlm} which is the product of these two eigenfunctions undergoes the transformation

$$e^{im\phi} P_l^m(\cos\theta) \to (-1)^l e^{im\phi} P_l^m(\cos\theta)$$

or $\Psi_{nlm} \to (-1)^l \Psi_{nlm}$. The parity $\mathcal{P}$ of spherical-polar eigenstates therefore turns on the single quantum number l and is given by

$$\mathcal{P} = (-1)^l.$$

States of even l have even parity; states of odd l have odd parity.

Note 10
Hydrogen Atom Parabolic States

The coordinates in which the spinor eigenstates are manifested are the parabolic coordinates (Fig. 4-10) given by

$$x_1 = \rho\cos\phi, \qquad x_2 = \rho\sin\phi, \qquad x_3 = (\xi - \eta)/2, \tag{10.1}$$

where $\rho^2 = {x_1}^2 + {x_2}^2 = \xi\eta$. The magnitude of the position vector in parabolic coordinates is

$$r = \sqrt{{x_1}^2 + {x_2}^2 + {x_3}^2} = (\xi + \eta)/2.$$

The inverse relationships between the two sets of coordinates are

$$\xi = r + x_3, \qquad \eta = r - x_3, \qquad \phi = \tan^{-1}(x_2/x_1). \tag{10.2}$$

The relationships between parabolic and spherical-polar coordinates are also of interest. The polar angle coordinate ϕ is common to both systems. The parabolic coordinates ξ and η are related to the spherical-polar coordinates r and θ by

$$\begin{aligned} \xi &= r(1 + \cos\theta), \\ \eta &= r(1 - \cos\theta). \end{aligned} \tag{10.3}$$

Since the polar angle ϕ is common to both parabolic and spherical-polar coordinates the momentum operator $\hat{p}_\phi = \hat{M}_3$ is similarly common to both and given by Eq. (**9.2**):

$$\hat{M}_3 = -i\hbar\partial/\partial\phi.$$

The total energy operator $\hat{\mathcal{H}}$ and the eccentricity operator $\hat{e}$, however, must be given a parabolic coordinate representation.

The classical Hamilton–Jacobi equation provides a direct transcription for the energy operator. The classical Hamiltonian in parabolic coordinates given by Eq. (4.38) goes over into the Hamiltonian operator

$$\hat{\mathcal{H}} = \frac{2}{\xi+\eta}\left(\xi\hat{p}_\xi^2 + \eta\hat{p}_\eta^2\right)/\mu + \frac{1}{2\xi\eta}\hat{p}_\phi^2/\mu - \frac{2k}{\xi+\eta}. \qquad (\mathbf{10.4})$$

The operators $\hat{p}_\xi^2$ and $\hat{p}_\eta^2$ may be developed from the relationships (**10.1**):

$$\hat{p}_\xi^2 = -\hbar^2\frac{4}{\xi+\eta}\frac{\partial}{\partial\xi}\left(\xi\frac{\partial}{\partial\xi}\right), \qquad \hat{p}_\eta^2 = -\hbar^2\frac{4}{\xi+\eta}\frac{\partial}{\partial\eta}\left(\eta\frac{\partial}{\partial\eta}\right). \qquad (\mathbf{10.5})$$

The eccentricity operator $\hat{e}_3$ in parabolic coordinates may be calculated for $\hat{e}_3$ given by Eq. (5.43):

$$\hat{e}_3 = (\hat{p}_1\hat{M}_2 - \hat{p}_2\hat{M}_1 - i\hbar\hat{p}_3)/k\mu - x_3/r.$$

This operator may be expressed in parabolic coordinates in a manner that images the classical eccentricity, Eq. (4.43), as

$$\hat{e}_3 = \frac{2\xi\eta}{\xi+\eta}\left(\hat{p}_\eta^2 - \hat{p}_\xi^2\right)/k\mu + \frac{\xi-\eta}{\xi\eta}\hat{p}_\phi^2/k\mu - \frac{\xi-\eta}{\xi+\eta}. \qquad (\mathbf{10.6})$$

By combining Eqs. (**10.4**) and (**10.6**) one may separate these equations into pure functions of ξ and η which follow those for the classical separated Hamilton–Jacobi equation (4.45):

$$\begin{aligned} \hat{e}_3 &= 1 + \xi\left(\hat{\mathcal{H}}/k - 2\hat{p}_\xi^2/k\mu - \hat{M}_3^{\,2}/2k\mu\xi^2\right), \\ -\hat{e}_3 &= 1 + \eta\left(\hat{\mathcal{H}}/k - 2\hat{p}_\eta^2/k\mu - \hat{M}_3^{\,2}/2k\mu\eta^2\right). \end{aligned} \qquad (\mathbf{10.7})$$

The operators $\hat{\mathcal{H}}$, $\hat{M}_3$, and $\hat{e}_3$ all commute. Moreover, the eigenvalues of these operators, established in Chap. 5, are

$$\mathcal{H} = \mathcal{E}_0/n^2, \qquad M_3 = m\hbar, \qquad e_3 = q/n,$$

where n takes all the integers greater than zero and q and m take all the positive and negative integers including zero. The wave function has the separable form $\Psi = e^{i[S_\xi(\xi)+S_\eta(\eta)+S_\phi(\phi)]/\hbar}$.

The eigenstates corresponding to these eigenvalues may be separated as

$$\Psi_{nqm} = X(\xi)Y(\eta)\Phi_m(\phi), \tag{10.8}$$

where the polar eigenstates $\Phi_m(\phi)$ are given by Eq. (**9.18**). The separated operator equations for the eccentricity eigenstates $X = e^{iS_\xi/\hbar}$ and $Y = e^{iS_\eta/\hbar}$ which follow from Eq. (**10.7**) are then

$$\frac{1}{\xi}\frac{d}{d\xi}\left(\xi\frac{dX}{d\xi}\right) - \frac{1}{a_0^2}\left(\frac{1}{4n^2} - \frac{(n-q)}{n}\frac{a_0}{2\xi} + \frac{m^2}{4}\frac{a_0^2}{\xi^2}\right)X = 0 \tag{10.9a}$$

and

$$\frac{1}{\eta}\frac{d}{d\eta}\left(\eta\frac{dY}{d\eta}\right) - \frac{1}{a_0^2}\left(\frac{1}{4n^2} - \frac{(n+q)}{n}\frac{a_0}{2\eta} + \frac{m^2}{4}\frac{a_0^2}{\eta^2}\right)Y = 0. \tag{10.9b}$$

Equations (**10.9**) are the governing equations for the spinor eigenstates in parabolic coordinates. These equations are formally identical save for the sign of the quantum number q so the two functions X and Y are the same function with values of the quantum number q of opposite sign. Parabolic quantum numbers q_+, q_- are useful in this representation. They are defined in terms of q and m as

$$q_+ = \frac{(n-1-|m|)+q}{2}, \qquad q_- = \frac{(n-1-|m|)-q}{2}$$

and are fixed by the triplet (n, q, m). The state specified by (q_+, q_-, m) is equivalent to that specified by (n, q, m) since the eccentrum quantum number q is given in terms of (q_+, q_-, m) by

$$q = q_+ - q_-,$$

and the energy level quantum number n by

$$n = q_+ + q_- + |m| + 1.$$

The quantum number q may take positive and negative integers lying between $-(n-1)$ and $(n-1)$ but the quantum numbers $q_\pm$ are always positive and take integer values in the range

$$0 \leqslant q_\pm \leqslant (n-1).$$

One sees that $X(\xi) = X_{nq_-m}(\xi)$ and $Y(\eta) = X_{nq_+m}(\eta)$. One may therefore express the eigenstate in terms of the single function $X_{nq_\pm m}(w)$ as

$$\Psi_{nqm} = X_{nq_-m}(\xi)X_{nq_+m}(\eta)\Phi_m(\phi).$$

Note that it is q which appears in Ψ_{nqm} while $q_\pm$ appears in $X_{nq_\pm m}(w)$. The angular momentum quantum number m appears only as m^2 in Eqs. (**10.9**) so the state functions $X_{nq_\pm m}$ depend only upon the absolute value of m. The parabolic eigenstate equations have a structure similar to that of the Laguerre differential equation (**9.17**) governing the radial eigenstates $R_{nl}(r)$ in spherical-polar coordinates. The solutions of these differential equations are all expressible in terms of the confluent hypergeometric function F_γ^α of Eq. (**9.18**).

The volume in parabolic coordinates is $d^3x = (\xi + \eta)\, d\xi\, d\eta\, d\phi/4$ and the parabolic eigenstates are normalized as

$$\int_0^\infty \int_0^\infty X_{nq_-m}(\xi) X_{nq_+m}(\eta) X^*_{nq_-m}(\xi) X^*_{nq_+m}(\eta) \frac{(\xi+\eta)}{4}\, d\xi\, d\eta = 1.$$

Both the ξ and η eigenstates are represented in terms of the same function $F_{|m|+1}^{-q_\pm}$ but with different arguments and quantum number indices $q_\pm$. The solutions to Eq. (**10.9**) for the states $X_{nq_\pm|m|}(w)$ are

$$\begin{aligned} X_{nq_\pm|m|}(w) =& \frac{2^{1/4} a_0^{-3/4}}{n|m|!} \sqrt{\frac{(q_\pm + |m|)!}{q_\pm!}} (w/na_0)^{|m|/2} \\ & \times e^{-w/2na_0} F_{|m|+1}^{-q_\pm}(w/na_0). \end{aligned} \tag{10.10}$$

The first few eigenstates $X_{nq_\pm|m|}(w)$ are

$$\begin{aligned}
X_{100}(w) =& 2^{1/4} a_0^{-3/4} e^{-w/2a_0}, \\
X_{200}(w) =& \frac{2^{1/4}}{2} a_0^{-3/4} e^{-w/4a_0}, \\
X_{210}(w) =& \frac{2^{1/4}}{2} a_0^{-3/4} \left[1 - \frac{1}{2}\left(\frac{w}{a_0}\right)\right] e^{-w/4a_0}, \\
X_{201}(w) =& \frac{2^{1/4}}{2\sqrt{2}} a_0^{-3/4} \left(\frac{w}{a_0}\right)^{1/2} e^{-w/4a_0}, \\
X_{300}(w) =& \frac{2^{1/4}}{3} a_0^{-3/4} e^{-w/6a_0}, \\
X_{310}(w) =& \frac{2^{1/4}}{3} a_0^{-3/4} \left[1 - \frac{1}{3}\left(\frac{w}{a_0}\right)\right] e^{-w/6a_0}, \\
X_{320}(w) =& \frac{2^{1/4}}{3} a_0^{-3/4} \left[1 - \frac{2}{3}\left(\frac{w}{a_0}\right) + \frac{1}{18}\left(\frac{w}{a_0}\right)^2\right] e^{-w/6a_0}, \\
X_{301}(w) =& \frac{2^{1/4}}{3\sqrt{3}} a_0^{-3/4} \left(\frac{w}{a_0}\right)^{1/2} e^{-w/6a_0}, \\
X_{311}(w) =& \frac{2^{1/4}\sqrt{2}}{3\sqrt{3}} a_0^{-3/4} \left(\frac{w}{a_0}\right)^{1/2} \left[1 - \frac{1}{6}\left(\frac{w}{a_0}\right)\right] e^{-w/6a_0}, \\
X_{302}(w) =& \frac{2^{1/4}\sqrt{2}}{9} a_0^{-3/4} \left(\frac{w}{a_0}\right) e^{-w/6a_0}.
\end{aligned}$$

The first few parabolic eigenstates $\Psi_{nqm}(\xi, \eta, \phi)$ [with the complete state function $\Psi_{nqm}(\xi, \eta, \phi)$ expressed in terms of the quantum number q rather than the quantum numbers $q_\pm$] are

$$\Psi_{100} = \frac{1}{\sqrt{\pi}} a_0^{-3/2} e^{-(\xi+\eta)/2a_0},$$

$$\Psi_{20\pm1} = \frac{1}{8\sqrt{\pi}} a_0^{-3/2} \left(\frac{\xi}{a_0}\right)^{1/2} \left(\frac{\eta}{a_0}\right)^{1/2} e^{-(\xi+\eta)/4a_0} e^{\pm i\phi},$$

$$\Psi_{210} = \frac{1}{4\sqrt{\pi}} a_0^{-3/2} \left[1 - \frac{1}{2}\left(\frac{\xi}{a_0}\right)\right] e^{-(\xi+\eta)/4a_0},$$

$$\Psi_{2-10} = \frac{1}{4\sqrt{\pi}} a_0^{-3/2} \left[1 - \frac{1}{2}\left(\frac{\eta}{a_0}\right)\right] e^{-(\xi+\eta)/4a_0},$$

$$\Psi_{300} = \frac{1}{9\sqrt{\pi}} a_0^{-3/2} \left[1 - \frac{1}{3}\left(\frac{\xi}{a_0}\right)\right] \left[1 - \frac{1}{3}\left(\frac{\eta}{a_0}\right)\right] e^{-(\xi+\eta)/6a_0},$$

$$\Psi_{31\pm1} = \frac{\sqrt{2}}{27\sqrt{\pi}} a_0^{-3/2} \left(\frac{\xi}{a_0}\right)^{1/2} \left(\frac{\eta}{a_0}\right)^{1/2} \left[1 - \frac{1}{6}\left(\frac{\xi}{a_0}\right)\right] e^{-(\xi+\eta)/6a_0} e^{\pm i\phi},$$

$$\Psi_{3-1\pm1} = \frac{\sqrt{2}}{27\sqrt{\pi}} a_0^{-3/2} \left(\frac{\xi}{a_0}\right)^{1/2} \left(\frac{\eta}{a_0}\right)^{1/2} \left[1 - \frac{1}{6}\left(\frac{\eta}{a_0}\right)\right] e^{-(\xi+\eta)/6a_0} e^{\pm i\phi},$$

$$\Psi_{30\pm2} = \frac{1}{162\sqrt{\pi}} a_0^{-3/2} \left(\frac{\xi}{a_0}\right) \left(\frac{\eta}{a_0}\right) e^{-(\xi+\eta)/6a_0} e^{\pm i2\phi},$$

$$\Psi_{320} = \frac{1}{9\sqrt{\pi}} a_0^{-3/2} \left[1 - \frac{2}{3}\left(\frac{\xi}{a_0}\right) + \frac{1}{18}\left(\frac{\xi}{a_0}\right)^2\right] e^{-(\xi+\eta)/6a_0},$$

$$\Psi_{3-20} = \frac{1}{9\sqrt{\pi}} a_0^{-3/2} \left[1 - \frac{2}{3}\left(\frac{\eta}{a_0}\right) + \frac{1}{18}\left(\frac{\eta}{a_0}\right)^2\right] e^{-(\xi+\eta)/6a_0}.$$

The common factor $(\xi+\eta)/2na_0$ appears in all the parabolic eigenstates; but in parabolic coordinates the factor $(\xi + \eta)/2$ is the magnitude of the position vector r as indicated by Eq. (**10.1**). The exponential decay terms are therefore of the form

$$e^{-(\xi+\eta)/2na_0} = e^{-r/na_0}$$

and are identical to the exponential decay factors of the state functions Ψ_{nlm} in spherical-polar coordinates. Since these factors depend only upon the energy level quantum number n and this quantum number is common to both state specifications, the two factors are identical. Moreover, since the ground state is completely symmetrical in both state specifications, the ground state functions Ψ_{100} are identical for both spherical-polar and parabolic coordinates.

There are no parabolic states of the form Ψ_{n00} when n is even. It is not possible for the angular momentum and eccentrum to vanish simultaneously in such states.

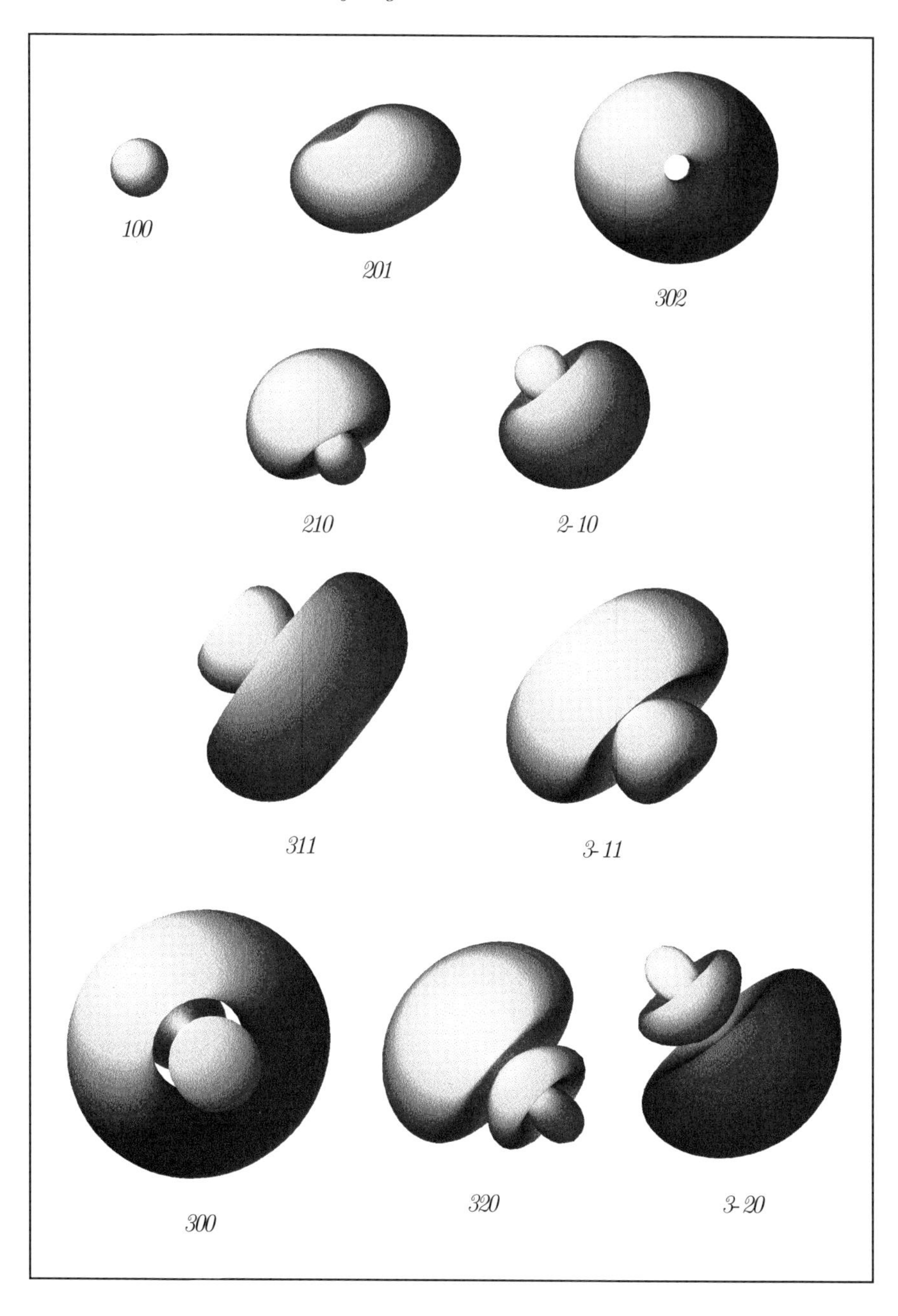

Figure 10-1. Hydrogen Atom Parabolic States. Parabolic states are identified by the quantum numbers (n, q, m). The parabolic states 100, 201, and 302 are identical to the (n, l, m) spherical-polar states 100, 211, and 322 of **Fig. 9-1** but are shown here from a different angle. The polar axis x_3 is the symmetry axis of the structures. The atoms have been rotated for the best viewing angle. *(Courtesy George D. Purvis III)*

Consider now the parity of the parabolic states. The polar eigenfunction is common to both the spherical-polar and parabolic states. Upon coordinate inversion, the polar angular eigenfunction undergoes

$$e^{im\phi} \to (-1)^m e^{im\phi}.$$

The parity of the parabolic eigenstates may be established by noting that on coordinate inversion $x_i \to -x_i$, the radial coordinate r is unchanged and the parabolic coordinates undergo

$$\begin{aligned} \xi = r - x_3 &\to r + x_3 = \eta, \\ \eta = r + x_3 &\to r - x_3 = \xi, \\ \phi &\to \phi + \pi. \end{aligned}$$

Coordinate inversion therefore interchanges ξ and η.

The eccentricity is a function of odd parity (it changes sign when the coordinates are inverted). When its quantum number q is inverted, the quantum numbers undergo the change

$$\begin{aligned} q &\to -q, \\ q_+ &\to q_-, \\ q_- &\to q_+. \end{aligned}$$

Since $|m|$ does not change sign, inversion of the quantum number q results in an interchange of the parabolic eigenfunctions

$$\begin{aligned} X_{nq_-|m|}(\xi) &\to X_{nq_+|m|}(\eta), \\ X_{nq_+|m|}(\eta) &\to X_{nq_-|m|}(\xi). \end{aligned}$$

Thus, when coordinate inversion switches the arguments of $X_{nq_-|m|}(\xi)$ and $X_{nq_+|m|}(\eta)$, inversion of the quantum number q returns these functions and arguments to their original state. The eigenstates Ψ_{nqm} involve the product of the two parabolic eigenfunctions $X_{nq_-|m|}(\xi)$ and $X_{nq_+|m|}(\eta)$ and the polar eigenfunction. They are therefore transformed under parity as

$$\Psi_{nqm} \to (-1)^m \Psi_{n-qm}.$$

The parity $\mathcal{P}$ of parabolic eigenstates therefore turns on the quantum numbers m and q.

Index

References in boldface are to the Notes